LASERS AND THEIR APPLICATIONS
LAZERY I IKH PRIMENENIE
ЛАЗЕРЫ И ИХ ПРИМЕНЕНИЕ

The Lebedev Physics Institute Series

Editors: Academicians D. V. Skobel'tsyn and N. G. Basov

P. N. Lebedev Physics Institute, Academy of Sciences of the USSR

Recent Volumes in this Series

Volume 35	Electronic and Vibrational Spectra of Molecules
Volume 36	Photodisintegration of Nuclei in the Giant Resonance Region
Volume 37	Electrical and Optical Properties of Semiconductors
Volume 38	Wideband Cruciform Radio Telescope Research
Volume 39	Optical Studies in Liquids and Solids
Volume 40	Experimental Physics: Methods and Apparatus
Volume 41	The Nucleon Compton Effect at Low and Medium Energies
Volume 42	Electronics in Experimental Physics
Volume 43	Nonlinear Optics
Volume 44	Nuclear Physics and Interaction of Particles with Matter
Volume 45	Programming and Computer Techniques in Experimental Physics
Volume 46	Cosmic Rays and Nuclear Interactions at High Energies
Volume 47	Radio Astronomy: Instruments and Observations
Volume 48	Surface Properties of Semiconductors and Dynamics of Ionic Crystals
Volume 49	Quantum Electronics and Paramagnetic Resonance
Volume 50	Electroluminescence
Volume 51	Physics of Atomic Collisions
Volume 52	Quantum Electronics in Lasers and Masers, Part 2
Volume 53	Studies in Nuclear Physics
Volume 54	Photomesic and Photonuclear Reactions and Investigation Method with Synchrotrons
Volume 55	Optical Properties of Metals and Intermolecular Interactions
Volume 56	Physical Processes in Lasers
Volume 57	Theory of Interaction of Elementary Particles at High Energies
Volume 58	Investigations in Nonlinear Optics and Hyperacoustics
Volume 59	Luminescence and Nonlinear Optics
Volume 60	Spectroscopy of Laser Crystals with Ionic Structure
Volume 61	Theory of Plasmas
Volume 62	Methods in Stellar Atmosphere and Interplanetary Plasma Research
Volume 63	Nuclear Reactions and Interaction of Neutrons and Matter
Volume 64	Primary Cosmic Radiation
Volume 65	Stellarators
Volume 66	Theory of Collective Particle Acceleration and Relativistic Electron Beam Emission
Volume 67	Physical Investigations in Strong Magnetic Fields
Volume 68	Radiative Recombination in Semiconducting Crystals
Volume 69	Nuclear Reactions and Charged-Particle Accelerators
Volume 70	Group-Theoretical Methods in Physics
Volume 71	Photonuclear and Photomesic Processes
Volume 72	Physical Acoustics and Optics: Molecular Scattering of Light; Propagation of Hypersound; Metal Optics
Volume 73	Microwave–Plasma Interactions
Volume 74	Neutral Current Sheets in Plasmas
Volume 75	Optical Properties of Semiconductors
Volume 76	Lasers and Their Applications
Volume 78	Research in Molecular Laser Plasmas

In preparation

Volume 77	Radio, Submillimeter, and X-Ray Telescopes
Volume 79	Luminescence Centers in Crystals
Volume 80	Synchrotron Radiation
Volume 81	Pulse Gas-Discharge Atomic and Molecular Lasers

Proceedings (Trudy) of the P. N. Lebedev Physics Institute

Volume 76

LASERS AND THEIR APPLICATIONS

Edited by
N. G. Basov
P.N. Lebedev Physics Institute
Academy of Sciences of the USSR
Moscow, USSR

Translated from Russian by
Albin Tybulewicz
Editor: *Journal of Quantum Electronics*

CONSULTANTS BUREAU
NEW YORK AND LONDON

Library of Congress Cataloging in Publication Data

Main entry under title:

Lasers and their applications.

(Proceedings (Trudy) of the P. N. Lebedev Physics Institute; v. 76)
Translation of Lazery i ikh primenenie.
Includes bibliographical references and index.
1. Lasers–Addresses, essays, lectures. 2. Quantum electronics–Addresses, essays, lectures. I. Basov, Nikolaĭ Gennadievich, 1922- II. Series: Akademiía nauk SSSR. Fizicheskiĭ institut. Proceedings; v. 76. QC1.A4114 vol. 76 [QC689] 530'.08s [535.5'8]
ISBN 978-1-4684-1622-0 ISBN 978-1-4684-1620-6 (eBook) 76-26590
DOI 10.1007/978-1-4684-1620-6

The original Russian text was published by Nauka Press in Moscow in 1974 for the Academy of Sciences of the USSR as Volume 76 of the Proceedings of the P. N. Lebedev Physics Institute. This translation is published under an agreement with the Copyright Agency of the USSR (VAAP).

A Division of Plenum Publishing Corporation
227 West 17th Street, New York, N. Y. 10011

CONTENTS

EXPERIMENTAL AND THEORETICAL INVESTIGATIONS OF THE DYNAMICS OF HIGH-POWER RADIATION-EMITTING ELECTRIC DISCHARGES IN GASES

B. L. Borovich, P. G. Grigor'ev, V. S. Zuev, V. B. Rozanov, A. V. Startsev, and A. P. Shirokikh

An experimental investigation was made of the characteristics of high-current (up to 400 kA) discharges in gases at pressures up to 1 atm supplied with energies up to ~40 kJ. Measurements were made of the current and voltage, rate of expansion of the discharge channel, and absolute intensities of the radiation emitted at various frequencies. This made it possible to compute the energy balance in the discharge. An investigation was made of the stability of such discharges. A general physical description of the dynamics of high-current discharges in gases and a self-similar theory of these discharges were developed. The theory was based on the solution of gasdynamic equations with a nonlinear thermal conductivity and of the equation for an electric circuit in the case when a shock wave traveled ahead of a thermal wave. The self-similar theory was compared with the results of exact numerical calculations and a satisfactory agreement was obtained. The relationships obtained explained the principal experimental observations and made it possible to predict the discharge parameters on the basis of the characteristics of the power supply.

The present paper reports the results of investigations of high-current discharges in gases at pressures up to 1 atm, which were carried out in the Quantum Electronics Laboratory at the Lebedev Physics Institute 1967-1971. These discharges are of interest because they can be used as extremely powerful light sources for the optical pumping of lasers, particularly in the ultraviolet part of the spectrum, since the brightness temperature of the discharge plasma may reach 40,000-50,000°K if the current is sufficiently high [1, 2].

The main task of our investigations [3-7] was the development of a general physical picture of the dynamics of high-current discharges, analysis of the energy balance, and determination of the optical and gasdynamic characteristics. We investigated discharges up to 1.5 m long which were supplied from a sufficiently large capacitor bank (the energy supplied was up to 40 kJ). Therefore, the transverse dimensions of the discharge plasma were greater than the Rosseland photon range and the radiative heat transfer played an important role in the energy balance. This distinguishes such discharges from electric sparks [8-10] in which heat is transported mainly by electronic heat conduction.

§ 1, Methods for Initiating Discharges with Large Radiating Surfaces

The simplest method for initiating discharges of considerable length and regular cylindrical shape is the technique employing explosions of thin metal wires. This method is well known and it is described in detail in the literature [11, 12]. Explosions of long wires are known to occur in two stages. In the first stage the wire material is sublimated and it is converted into a high-density ($\sim 10^{22}$ cm^{-3}) gas column. This gas has a low conductivity so that the current falls nearly to zero and a "current pause" is observed. The wire continues to expand for some time, its electric strength falls, and this is finally followed by a secondary breakdown and a discharge in the gas column. If the energy stored in capacitors is sufficiently high, it is dissipated mainly during the second stage. This stage produces a plasma column with a brightness temperature of several tens of thousands of degrees which expands at a velocity of 1-2 km/sec and generates a shock wave.

We usually employed a bank of IM-3-50 capacitors with a total capacitance of 30 μF, which could be charged to 50 kV. The current was switched by an air discharger in which use was made of the breakdown on the surface of an insulator. The combined self-inductance of the capacitance bank and the discharger was 180 nH.

Short discharges (below 20 cm) were initiated between two plane busbars [3, 4]. Long discharges were investigated in a special cylindrical discharge chamber with an internal diameter of 112 mm, made of stainless steel. A schematic diagram of the discharge chamber is given in Fig. 1. The ends of the chamber were closed with glass windows and this made it possible to investigate discharges in different gases. A discharge was initiated along the axis of the chamber. The chamber itself provided the return path for the electric current. A high voltage was applied to a copper rod of 5 mm diameter which was protected by a polyethylene

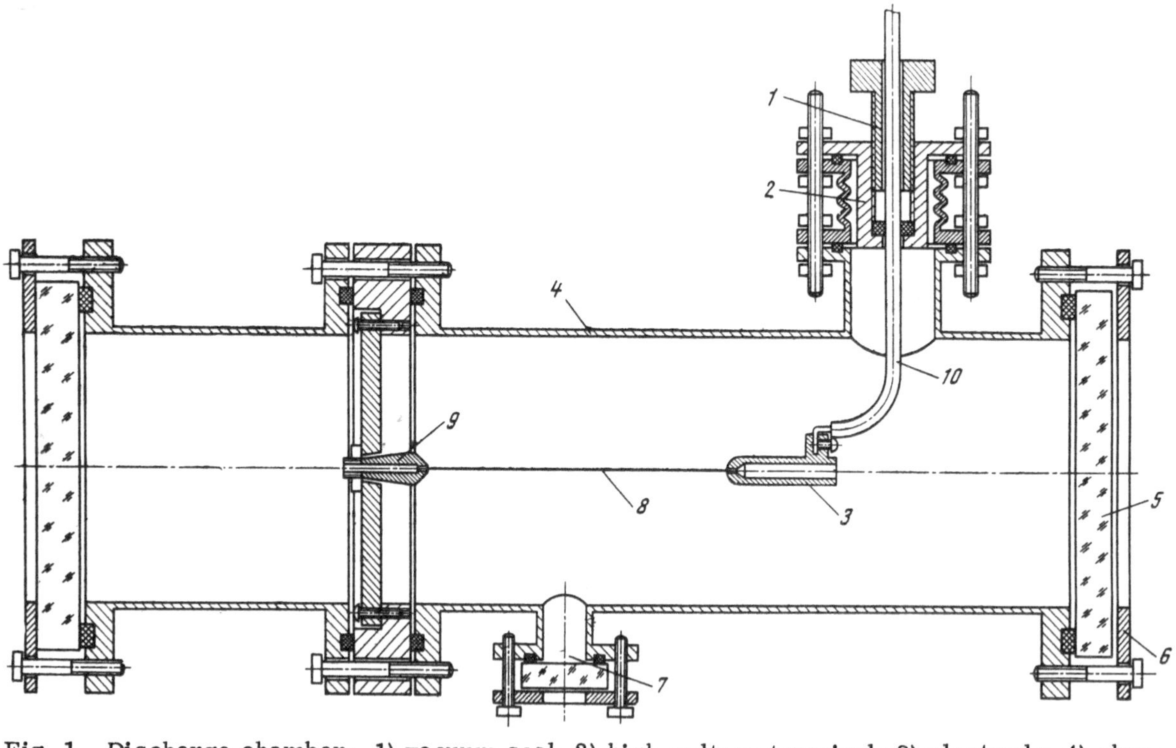

Fig. 1. Discharge chamber: 1) vacuum seal; 2) high-voltage terminal; 3) electrode; 4) chamber casing; 5) end window; 6) flange; 7) side window; 8) exploding wire; 9) electrode; 10) current conductor.

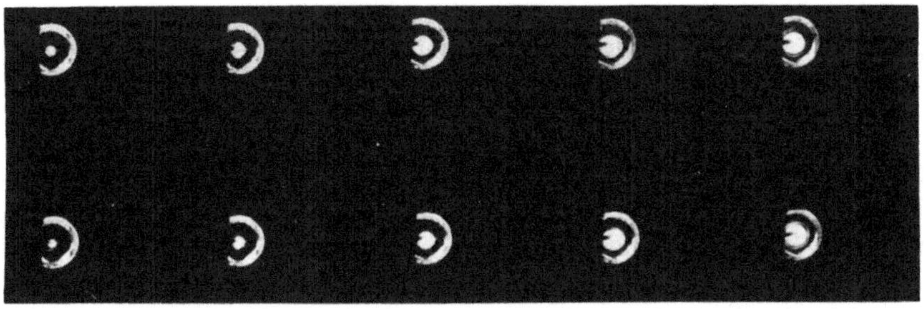

Fig. 2. Streak patterns of a discharge 1-m long in air. The frames were obtained at intervals of 2 μsec; C = 30 μF, U_0 = 50 kV.

insulator of 12 mm diameter. A rubber washer, surrounding the polyethylene insulation, provided the vacuum seal of this terminal. An electrode was attached to the end of the high-voltage terminal. The second electrode was placed at the center of a copper ring which was clamped between the two sections of the chamber. The end windows of the chamber were located at a distance of 40-50 cm from the discharge and this prevented scorching of the windows.

A streak pattern of a 1-m-long discharge in air, recorded through the end window of the chamber with an SFR-2M streak camera (Fig. 2), revealed clearly the region occupied by the plasma and the shock-wave region. The discharge region was clearly of regular cylindrical shape.

When the wire was too long or the voltage across the capacitor band was too low, the secondary breakdown stage did not occur and the discharge did not develop. For a given voltage across the capacitors the maximum length of the wire which would give rise to a high-power discharge depended on the wire material, diameter, and the ambient gas. Our investigations demonstrated that the most suitable wire material was molybdenum or tungsten. Such wires enabled us to initiate discharges several meters long even in electronegative gases of the SF_6 type when the voltage across the capacitor band was 50 kV. Moreover, when refractory wire materials were used, the first and second stages of the explosion merged and the current pause disappeared.

Strong magnetic fields were generated during high-power discharges involving the flow of currents amounting to hundreds or thousands of kiloamperes. These magnetic fields could be undesirable in some applications. They could be eliminated by placing an insulated conductor inside a cylindrical discharge and passing through it the return discharge current. Then, the complete symmetry of the geometric configuration should compensate the magnetic field outside the discharge.

The main problem in the initiation of discharges of this type was the difficulty encountered in the generation of a homogeneous plasma tube around the return current terminal. We in-

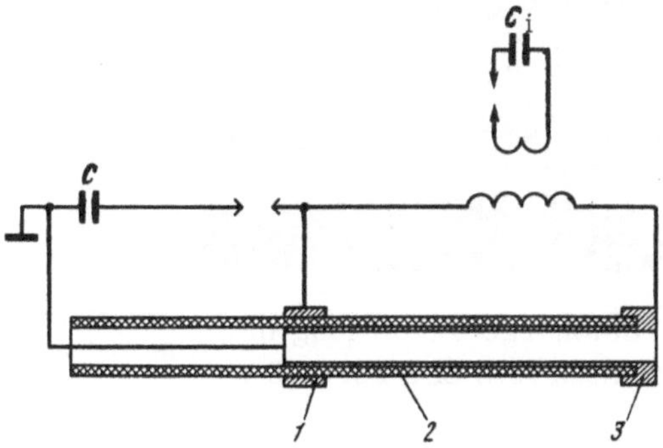

Fig. 3. Construction of a unit for investigating surface discharges: 1, 2) electrodes; 3) polyethylene insulation.

vestigated two possible initiation methods, one of which involved discharges on the surface of an insulator [13, 14] and the other in which a large number of thin wires was exploded simultaneously.

The construction of a unit used in investigations of surface discharges is shown schematically in Fig. 3. Experiments demonstrated that the formation of a more or less uniform plasma tube could be achieved if at the moment of initiation a fairly large number of equivalent streamers developed simultaneously and filled the gap. When one or two of these streamers were more intense than the rest, the main capacitor band was discharged along channels formed by these streamers and a homogeneous plasma tube did not appear.

An analysis of the nature of the influence of various parameters on the number and length of streamers formed in this way led to the following conclusions:

1. The nature of the streamer breakdown (number of streamers, degree of their branching, etc.) was governed mainly by the insulator employed; the initiation parameters and the specific (per unit surface area) capacitance of the initiation electrode governed primarily the streamer length.

2. A large number of equivalent streamers was formed on insulators with a high permittivity ε.

3. When the value of ε was high (~100-1000), the streamer length decreased considerably.

The nature of the development of streamers depended largely on the gas in which the discharge took place. The densest streamer networks were observed in inert gases. Figure 4 shows the streak patterns of a surface discharge with a return-current conductor in xenon; it is clear from this figure that a fully satisfactory plasma tube was formed under these conditions.

A dense network of streamers was not normally obtained in air, SF_6, and other gases whose electric strength was considerably higher than that of xenon. We studied the influence of an inhomogeneous electric field on the streamer formation process. The return-current conductor was surrounded by a polyethylene tube and we stretched a series of thin wires along this tube; these wires were connected to the return-current conductor. A thin polyethylene terephthalate film was then wound on this structure and ring electrodes were placed around it. The electrode not connected to the return-current conductor had sharp-point terminals located above the wires. When this system was used, up to 10 more or less identical streamers about

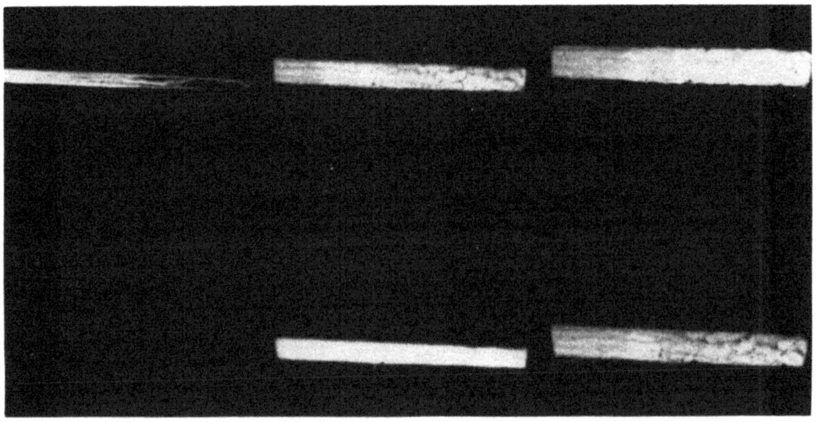

Fig. 4. Streak patterns of a surface discharge 0.5 m long in xenon. The frames were recorded at intervals of 2 μsec; C = 33 μF, U_0 = 45 kV.

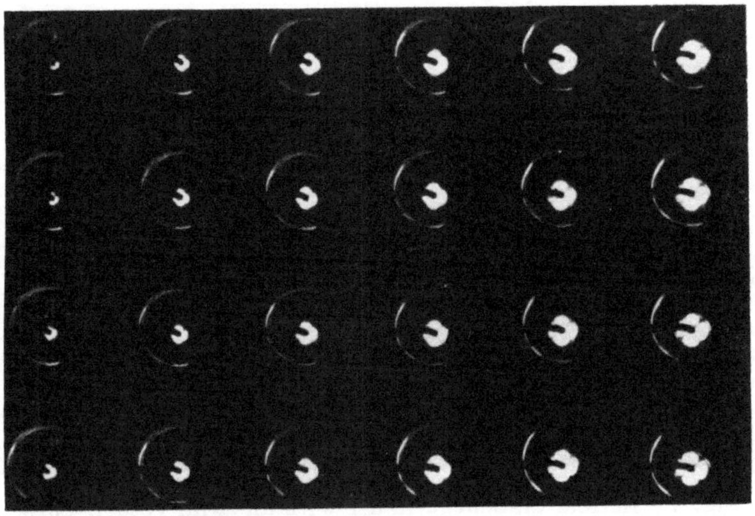

Fig. 5. Four-channel discharge 60 cm long in air with an axial return-current conductor. The frames were recorded at intervals of 0.5 μsec; C = 30 μF, U_0 = 50 kV.

0.5 m long appeared on the surface of a polyethylene tube of 30 mm diameter surrounded by air. The main disadvantage of this structure was its unreliability because the polyethylene terephthalate film frequently broke down and when the number of layers of this film was increased, the surface was no longer covered uniformly by streamers.

A more reliable method for initiating cylindrical discharges with compensated magnetic fields was a parallel explosion of many thin wires stretched symmetrically along the return-current conductor. It was established that practically any number of wires could be exploded simultaneously and this produced a symmetric distribution of the plasma around the return-current conductor (Figs. 5 and 6), which guaranteed a reliable compensation of the magnetic field outside the discharge. One should note that the separate "lobes" of the discharge did not merge even after a fairly long time and the shock wave formed in this way was not circular.

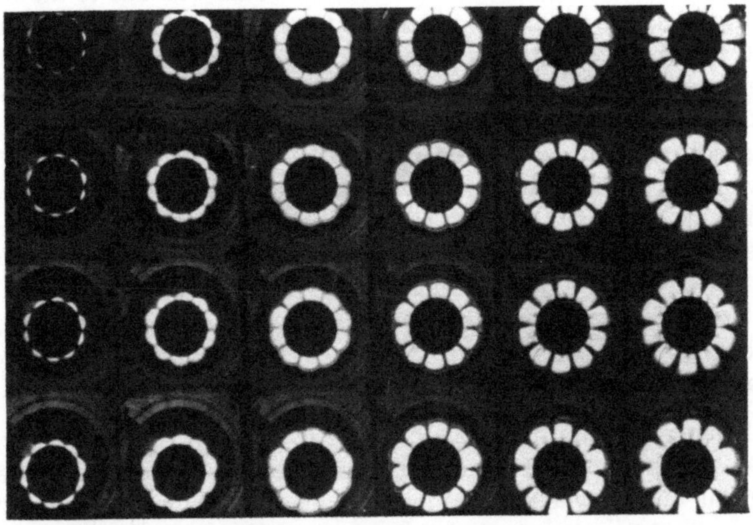

Fig. 6. Ten-channel discharge in air with an axial return-current conductor. The frames were recorded at intervals of 0.5 μsec; C = 30 μF, U_0 = 50 kV.

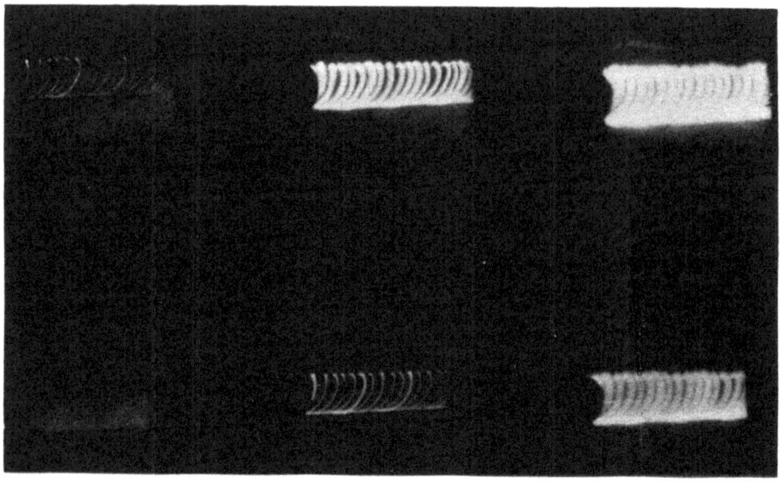

Fig. 7. Helical discharge in air with an axial return-current conductor. The frames were recorded at intervals of 2 μsec; C = 30 μF, U_0 = 50 kV.

We also tried a method in which a plasma tube was formed by exploding a wire wound helically on the return-current conductor (Fig. 7). Although the discharge symmetry was fully satisfactory, the high inductance reduced the discharge current and the brightness of the emitted radiation was low. Therefore, we did not investigate any further this type of discharge.

§ 2. Optical, Gasdynamic, and Energy Characteristics of High-Power Electric Discharges

The results of our experimental investigations of high-power electric discharges up to 20 cm long in air were reported in [3, 4]. These discharges were initiated by exploding thin wires. We established that the discharge characteristics were practically independent of the wire material and thickness (up to a certain limit). This was due to the fact that when a wire was sufficiently thin, the number of electrons formed as a result of ionization of air heated by radiative heat transfer exceeded considerably, even during the first few microseconds, the number of electrons emitted directly from the wire material. In these experiments the discharge current exhibited damped oscillations with a period of 15-20 μsec, and the amplitude of the current during the first half-period reached 370 kA when the voltage was 50 kV. Under these conditions we observed a plasma column in air and this column expanded at a velocity of 2-2.5 km/sec during the first 10-15 μsec. In the range of wavelengths exceeding 2000 Å the plasma emitted as if it were an absolute black body and its brightness temperature reached 40,000-50,000°K. A shock wave traveled ahead of the luminous region and the velocity of this wave was slightly greater than the velocity of expansion of the plasma column. Streak patterns of one of such discharges are shown in Fig. 8, whereas the time dependences of the brightness temperature and of the plasma column radius are plotted in Fig. 9.

The energy balance of a discharge 20-cm long (C = 30 μF, U_0 = 50 kV) was as follows: during the first 20 μsec about 60% of the energy stored in the capacitors was evolved in the plasma and 20% of this energy was emitted in the transparency band of air, 70% was converted into the thermal energy of the plasma, and 10% was used to form a shock wave.

We investigated experimentally discharges ~1 m long. The experiments were carried out using a bank of capacitors whose parameters were C = 30 μF, U_0 = 50 kV. The self-induc-

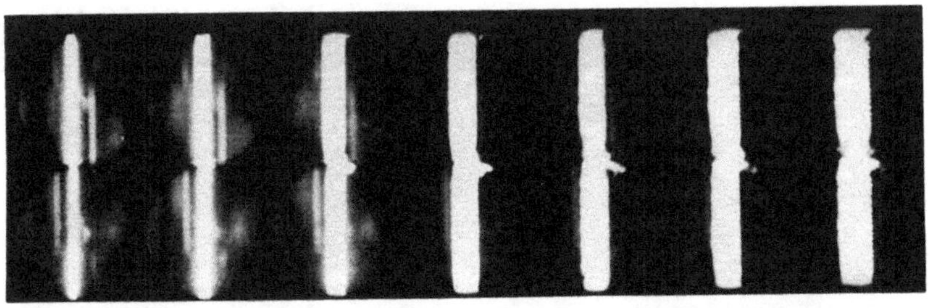

Fig. 8. Streak patterns of a discharge 24 cm long in air. The frames were recorded at intervals of 2 μsec; C = 30 μF, U_0 = 45 kV.

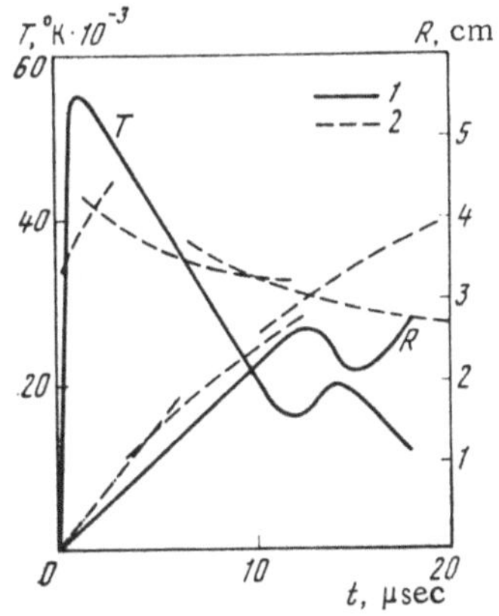

Fig. 9. Time dependences of the plasma radius and brightness temperature near λ = 2400 Å for a discharge 20 cm long in air (C = 30 μF, U_0 = 50 kV); 1) experimental results; 2) self-similar theory.

tance of this bank was ~190 nH. The current was also measured with a Rogowski loop. The voltage was measured employing two dividers from which signals were applied directly to the plates of an OK-17 oscillograph. The plates were insulated from the ground in order to avoid picking up electric strays. The first divider measured the voltage across the capacitors and the second across the plasma column. Differentiation of the signal from the first divider enabled us to determine the current independently of the Rogowski loop. Oscillograms obtained from the Rogowski loop and the two dividers are shown in Figs. 10 and 11. When the discharge in air was 67 cm long, the duration of the first half-period of the current oscillations was ~18 μsec and the duration of the second half-period was 15 μsec; the amplitude of the current during the first half-period was ~185 kA and during the second it was 150 kA.

The ohmic resistance of the plasma and the electrical energy supplied to the discharge were determined bearing in mind that the voltage was measured with an RL circuit, where L

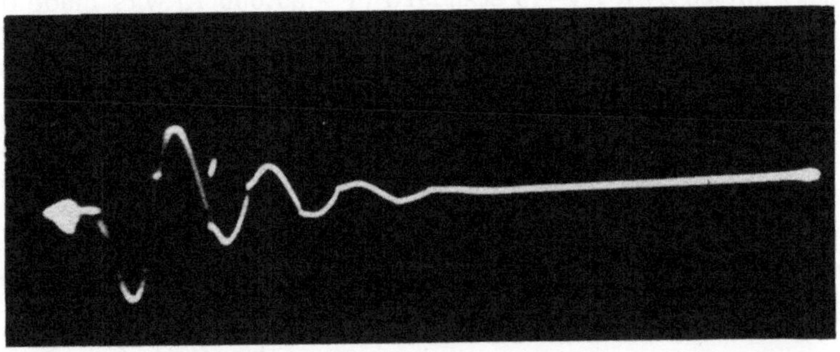

Fig. 10. Oscillogram of a signal picked up by a Rogowski loop (current) for a discharge 67 cm long in air; C = 30 μF, U_0 = 50 kV.

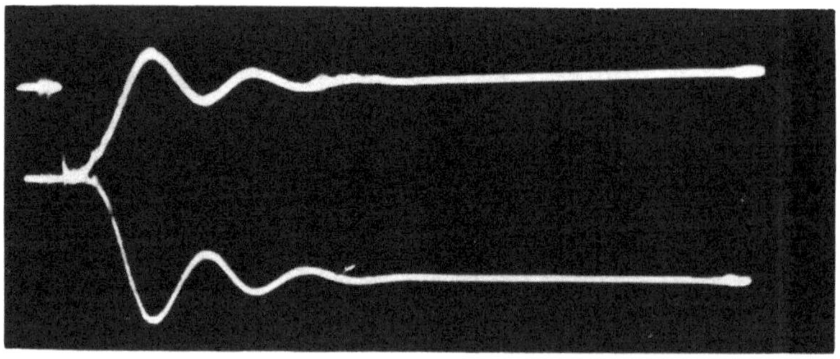

Fig. 11. Oscillograms of signals picked up from voltage dividers for a discharge 67 cm long in air; C = 30 μF, U_0 = 50 kV. The upper trace represents the voltage across the plasma and the lower trace the voltage across the capacitor bank.

was the variable discharge capacitance. We used the current oscillation period and the self-inductance of the capacitor bank to demonstrate that the inductance of the RL circuit had a constant component (~460 nH) and a variable one [$\sim 2\mathscr{L}\ln(R_c/R)$nH, where R_c is the radius of the discharge chamber, $\mathscr{L}$ is the length of the discharge, and R is the radius of the discharge]. We took this point into account and calculated the ohmic resistance of the plasma r_{ohm} and the electrical energy E_{el} supplied to the discharge (Fig. 12). The value of r_{ohm} was determined with a precision not exceeding 20-30% and the error increased with decreasing value of r_{ohm}.

In the calculation of the electrical energy we took into account the fact that the integral $\int_0^t UIdt$, calculated at the moments when the current I passed through zero, was independent of the RL circuit and the values of this integral at such moments were used as the reference points in the determination of the electrical energy E_{el}. During intermediate moments the value of E_{el} was calculated from $\int_0^t I^2 r_{ohm} dt$.

An independent method for the determination of the energy input was provided by a comparison of the energy remaining in the capacitors at the moment when the current passed through zero with the initial stored energy. However, since it was not possible to separate the ohmic losses in the plasma, supply busbars, and discharger, such measurements could only by used to determine the upper limit of the energy input.

The brightness temperature of a discharge was determined through a side window in the discharge chamber employing a photoelectric method and a IF-88 unit whose basic optical system is shown schematically in Fig. 13. A comparison of the radiation emitted by the investigated source with that emitted by a ÉV-45 standard (T = 39,000°K) was made during successive

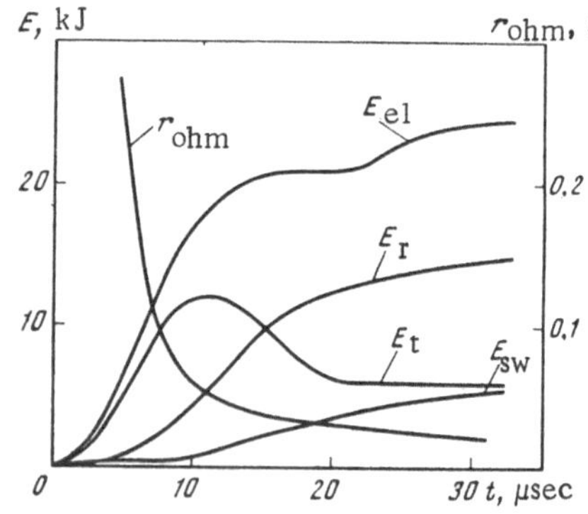

Fig. 12. Ohmic resistance and energy balance in a discharge 67 cm long in air; C = 30 μF, U_0 = 50 kV.

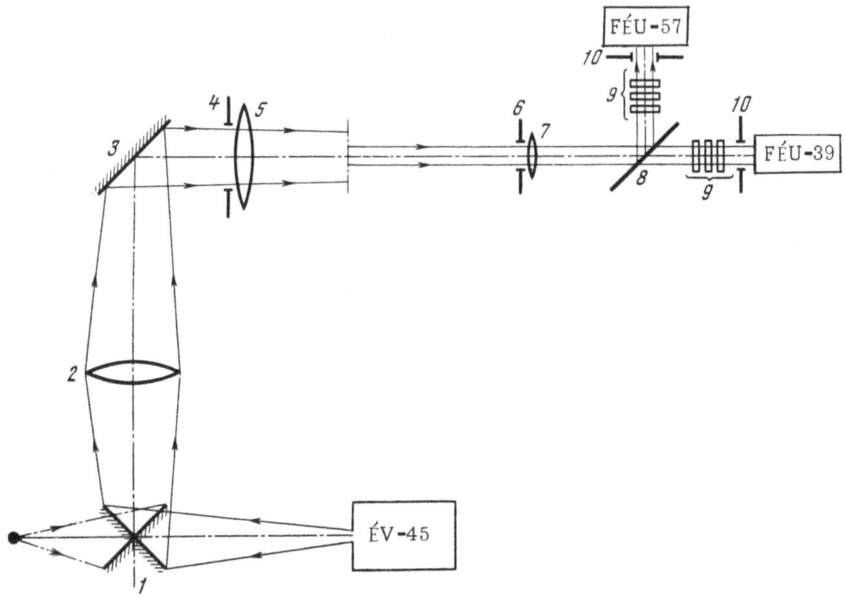

Fig. 13. Optical system of a IF-88 unit: 1, 3) deflecting mirrors; 2, 5) achromatic objectives; 4, 6, 10) screens with apertures; 7) converging lens; 8) interference beam splitter; 9) filters.

time intervals. In the recording of the radiation emitted by our source the optical system consisted of plane deflecting mirrors 1 and 3, which were aluminized externally, and achromatic objectives 2 and 5, which produced a doubly magnified image of the light source on the aperture in a screen 6 placed in front of a converging lens 7, which projected a tenfold-reduced image of the objective 5 onto the cathodes of a photoelectric multiplier.

An interference beam splitter 8 separated the light beam into two channels so that measurements could be carried out simultaneously at two wavelengths. A particular wavelength was selected by a set of interference filters. The intensity of the recorded signal could be varied by mutual filters placed in front of the photomultiplier and also by changing screens 4 and 6.

The results of measurements of the brightness temperature of the plasma carried out at two wavelengths (2640 and 4000 Å) are plotted in Fig. 14. It should be pointed out that, with the exception of the first few microseconds, the brightness temperature was the same at both wavelengths, i.e., the discharge emitted basically as a black body. The difference between the

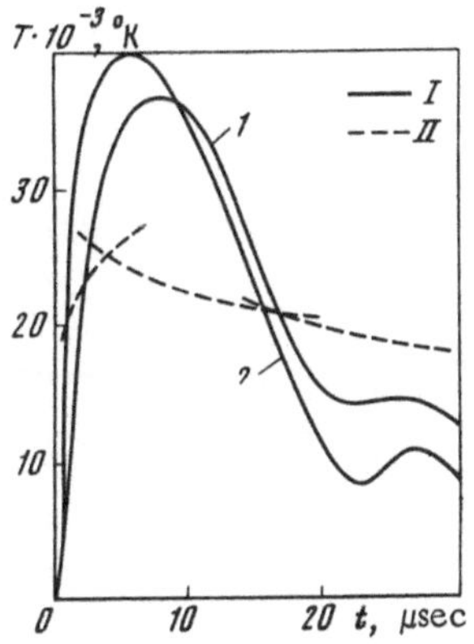

Fig. 14. Time dependences of the brightness temperature of a discharge 67 cm long in air ($C = 30\ \mu F$, $U_0 = 50$ kV) for $\lambda = 2640$ Å (1) and $\lambda = 4000$ Å (2): I) experimental results; II) self-similar theory.

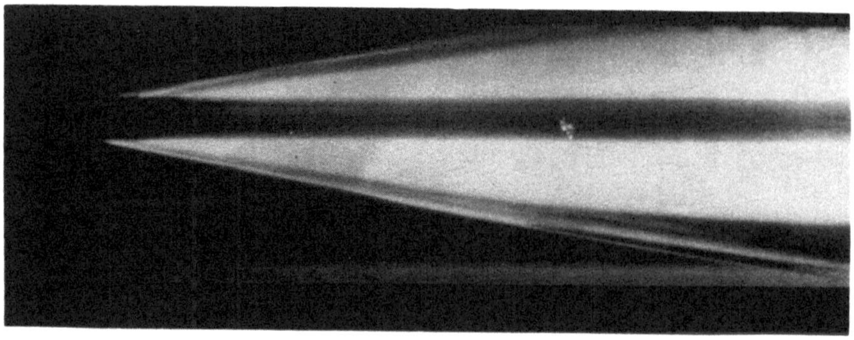

Fig. 15. Slit scan of a discharge 67 cm long in air (view through end window); $C = 30\ \mu F$, $U_0 = 50$ kV.

brightness temperatures during the first few microseconds was due to the fact that the absorption coefficient of the plasma at 2640 Å was less than at 4000 Å and, therefore, the plasma became black in this part of the spectrum only when the thickness increased sufficiently.

Figure 15 shows a slit scan of a discharge obtained through an end window using an SFR-2M streak camera. This photograph shows clearly the position of the shock wave and of the edge of the luminous region. In general, we found that several shock waves traveled ahead of the luminous region. They were probably due to waves causing thermal dissociation of the matter in a compressed layer. Beginning from a certain moment the intensity of the radiation emitted by the central part of the discharge became considerably lower than the radiation emitted by the peripheral regions and this was probably due to the skin effect which affected the current and reduced the plasma density in the axial part of the discharge.

Figure 16 shows the time dependences of the radii of the first shock wave R_{sw} and of the thermal (luminous) region R_t.

Slit scans of such a discharge indicated that after 22 μsec the velocity of expansion of the plasma column decreased considerably and the shock wave traveled well ahead of the luminous region. This moment corresponded approximately to the second maximum of the current and was clearly due to the compression of the plasma by the magnetic field.

The electrical energy E_{el} supplied to the discharge consisted of the energy of the thermal (luminous) region E_t, energy of the shock wave E_{sw}, and the energy emitted in the transparency band of cold air ($\lambda > 1860$ Å) E_r. Since the absorption coefficient of air at wavelengths shorter than 1860 Å was very high, the far ultraviolet radiation did not carry away the energy from the region surrounded by the shock wave.

A detailed energy balance in the discharge could be determined only when the radial distributions of the temperature, density, and velocity of particles were known. We assumed that

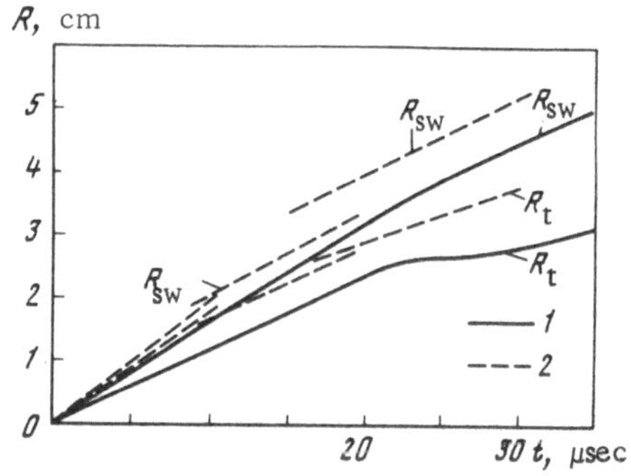

Fig. 16. Time dependences of the radii of shock and thermal waves in a discharge 67 cm long in air ($C = 30\ \mu F$, $U_0 = 50$ kV); 1) experimental results; 2) self-similar theory.

almost all the particles (> 90%) were expelled from the thermal region into the shock wave (this was in agreement with the results of numerical calculations and with the self-similar theory given below). Therefore, we could ignore the energy of the directional motion of the particles in the thermal region compared with the shock-wave energy. The shock-wave energy was estimated from the formula

$$E_{sw} \approx 0.9Mv^2,$$

where $v = \sqrt{\frac{2}{\gamma_0+1}\frac{P}{\rho_0}} = \frac{2}{\gamma_0+1}D$ is the velocity of matter accelerated by the shock wave; M is the total mass of the gas set in a motion. The pressure P and the adiabatic exponent γ_0 in the shock-wave front were deduced from the results reported in [15]. The results of these calculations of E_{sw} are plotted in Fig. 12.

We calculated the radiation losses on the assumption that the plasma emitted as a black body with a temperature measured in the region of λ = 2640 Å. Clearly, the values of the energy lost by radiatiion (Fig. 12) were somewhat overestimated because a comparison of the brightness temperatures in the region of 2640 and 4000 Å indicated that at shorter wavelengths the plasma became more transparent and probably did not emit as a black body.

Since no information was avaliable on the pressure profile in the discharge, the energy in the thermal region could not be determined sufficiently reliably at all moments. However, the thermal energy could be estimated at the moment of the onset of magnetic compression of the plasma when the magnetic pressure $P_m = H^2/4\pi$ became comparable with the gasdynamic pressure $[P \approx (\gamma - 1) E_t/\mathscr{L}\pi R^2]$. This estimate gave the thermal energy $E_t \approx 7$ kJ after 22 μsec. A comparison of E_t, E_{sw}, and E_r at 22 μsec with E_{el} indicated that the energy balance was correct to within 10%, so that we could calculate E_t simply from $E_t = E_{el} - E_r - E_{sw}$. In such calculations the value of E_r should be reduced by 10% (Fig. 12).

Thus, the energy balance of a 67-cm-long discharge was as follows: during the first 33 μsec the discharge absorbed 70% of the energy stored in the capacitor bank and about 50% of this energy was emitted as radiation in the transparency band of air, 20% was converted into the shock-wave energy, and about 30% remained in the form of the thermal energy of the plasma.

The time dependence of E_t was of particular interest because the thermal region of the discharge acted initially as a reservoir which received the electrical energy from the capacitor bank. During the early stages of the discharge the radiation emitted in the transparency band of air carried away a negligible proportion of the energy and the radiation outside this transparency band did not carry away any energy from the discharge region and simply increased the mass of the heated gas which subsequently reemitted the absorbed energy in the form of radiation in the transparency band. Therefore, the radiative efficiency E_r/E_{el} of such discharges exceeded considerably the spectral efficiency which could be defined as

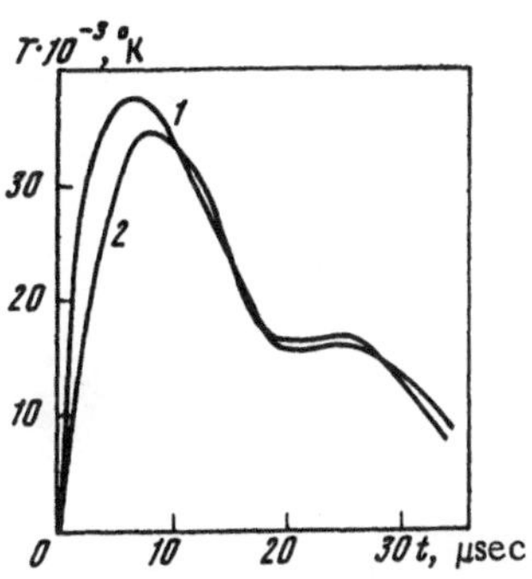

Fig. 17. Time dependences of the brightness temperature of a discharge 67 cm long in SF_6 (C = 30 μF, U_0 = 50 kV): 1) λ = 4000 Å; 2) λ = 2640 Å.

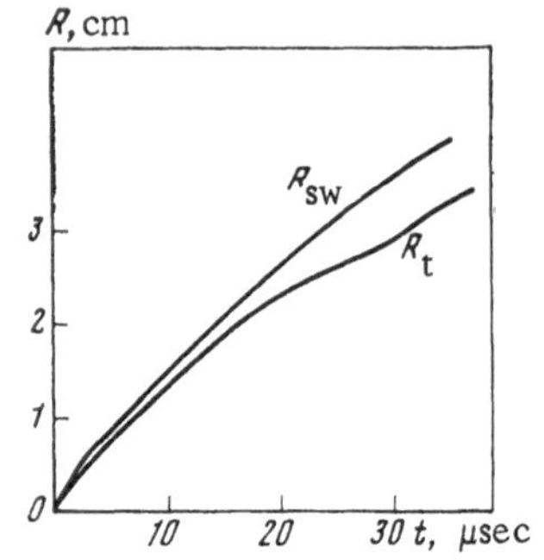

Fig. 18. Time dependences of the radii of shock and thermal waves in a discharge 67 cm long in SF_6; C = 30 μF, U_0 = 50 kV.

the ratio of the emitted energy to the energy which would be emitted by an absolute black body at the same temperature and with the same surface area as the discharge under consideration.

A similar cycle of measurements was carried out also in discharges in sulfur hexafluoride (SF_6) at pressures of 0.4-1 atm. By way of example, we plotted in Figs. 17 and 18 the time dependences of the brightness temperature of a discharge in SF_6 and of the radii of the thermal and shock waves. The greatest difference between the discharge in SF_6 and the discharge in air was manifested by the fact that the separation between the shock and thermal waves in SF_6 was considerably less than in air. This difference was a natural consequence of the fact that the adiabatic exponent of SF_6 was less than that of air and, consequently, the gas could be compressed more strongly in a shock-wave front in SF_6 than in air. Moreover, the brightness temperature of SF_6 and the rate of expansion of the plasma were somewhat lower than the corresponding parameters of air.

The main characteristics of a four-channel discharge 60 cm long in air with an axial return-current conductor are plotted in Figs. 19 and 20. A characteristic feature of the discharge with a return-current conductor was a higher initial rate of increase of the length of

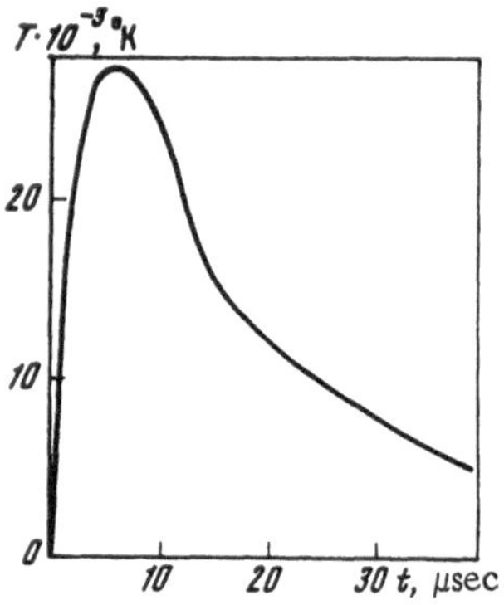

Fig. 19. Time dependence of the brightness temperature of a four-channel discharge 60 cm long in air; C = 30 μF, U_0 = 50 kV.

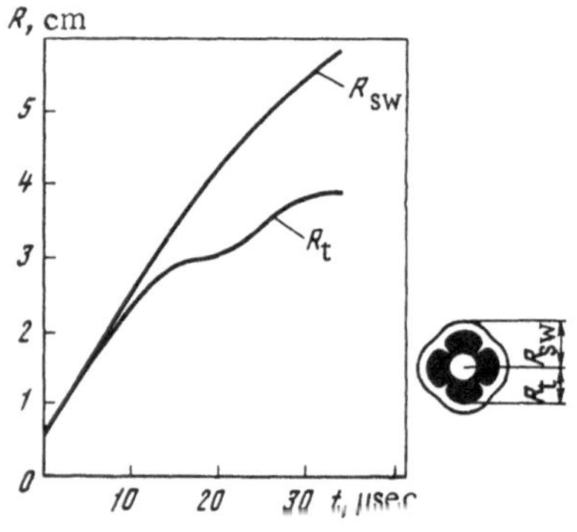

Fig. 20. Time dependences of the radii of shock and thermal waves in a four-channel discharge 60 cm long in air; C = 30 μF, U_0 = 50 kV.

the discharge "lobes," which was due to the magnetic interaction between the forward and reverse currents.

The brightness of the radiation emitted by a plasma produced in a discharge with a return-current conductor was considerably less than the brightness of the emission from an ordinary discharge. This was primarily due to the fact that in the former case the stored energy was dissipated in four basically independent discharges. Moreover, an increase in the transverse cross section of the discharge reduced its ohmic resistance and, consequently, the rate of supply of energy to the discharge. Moreover, the contribution to the kinetic energy of the gas motion increased and the radiation losses became greater because of the larger surface area of the discharge.

Experimental investigations of high-power discharges enabled us to develop a clear physical picture of these phenomena and provided the basis for the calculation of the main characteristics of the discharges. A high-power discharge in a gas could be described as follows.

A discharge produces a hot-plasma region where the temperature is several tens of thousands of degrees. This region expands at a velocity of 1-2 km/sec and it generates a shock wave. Immediately behind the shock-wave front the gas is strongly compressed and its temperature, governed by the specific heat of the gas and by the shock-wave velocity, may reach several thousands of degrees. Since the plasma conductivity depends strongly on temperature, the electrical energy is evolved mainly in the inner hot region. The mass of the hot region increases with time because of the heating of the gas layer at the boundary between the plasma and shock wave. When the rate of supply of the Joule heat slows down (or stops completely), the energy of the hot region is partly emitted as radiation and is partly converted into the shock-wave energy giving rise to the well-known Sedov strong explosion conditions. Spatial distributions of the plasma temperature, density, and velocity existing at some particular moment are shown qualitatively in Fig. 21. This type of motion is known as a thermal wave of the second kind, which differs from an ordinary nonlinear thermal wave traveling in matter at rest.

Up to this moment the optical thickness of the hot plasma is small and the dominant energy transfer process is the electronic heat conduction. The plasma is optically transparent when the energy input to the discharge is relatively small or during early stages of the discharge and in this case the discharge is known as a spark. This type of discharge has been investigated experimentally and theoretically in [8-10] and the detailed references can be found in [1]. When the optical thickness of the plasma is large, the energy transfer is dominated by radiation and this type of discharge is of greatest interest to us.

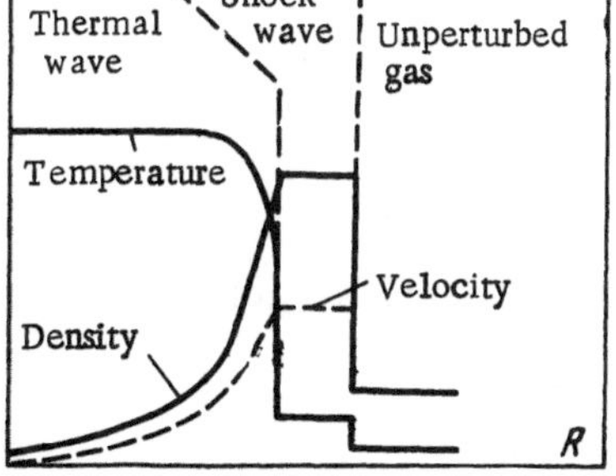

Fig. 21. Qualitative spatial distributions of gasdynamic quantities in a thermal wave of the second kind.

§ 3. Self-similar Theory of a Discharge in an Unbounded Medium

The physical description of high-power electric discharges in gases given in the preceding section can also be replaced satisfactorily by gasdynamic equations with radiative (nonlinear) heat transfer ignoring the magnetic forces and the skin effect experienced by the current (limits of validity of this approach will be analyzed in detail later). The equations for planar, cylindrical, and spherical (ν = 0, 1, 2) cases and for the nonlinear thermal diffusivity coefficient of the general kind $\varkappa_0(T^{m+3}/\rho^n)$ (in the case of the electronic thermal conductivity we have m = −1/2 and n = 0, whereas for radiative heat transfer we have m = 1.5–3, n = 1.5–2) are as follows:

$$\left.\begin{aligned} &\frac{d}{dt}\ln\rho+\frac{\partial v}{\partial r}+\frac{\nu v}{r}=0,\\ &\frac{dv}{dt}=-\frac{1}{\rho}\frac{\partial P}{\partial r},\\ &\varepsilon\rho\frac{d}{dt}\ln\frac{T}{\rho^{\gamma-1}}=Q(r,\ t)+\frac{1}{r^{\nu}}\frac{\partial}{\partial r}r^{\nu}\varkappa_0\frac{T^{m+3}}{\rho^{n}}\frac{\partial T}{\partial r}.\end{aligned}\right\} \tag{1}$$

Here, ρ is the density; v is the velocity; $P = (\gamma - 1)AT\rho$ is the pressure; ε is the average energy (ergs/g); A is the specific heat per unit mass; T is the plasma temperature; Q is the energy evolved per unit volume. In the case of radiative heat transfer we have $\varkappa_0 = {}^4/_3\, c\sigma l_0$, where $c\sigma/4$ is constant for a flux emitted from a unit surface area of an absolute black body, $\sigma = (8\pi k^4/h^3c^3)/(\pi^4/15)$, $k = 1.6 \cdot 10^{-12}$ erg/eV; l_0, m, and n govern the dependence of the Rosseland photon range l_R on the temperature and density: $l_R = l_0(T^m/\rho^n)$.

Well-known dimensional considerations [16] demonstrate that self-similar solutions are generally not obtained for gasdynamic problems with a nonlinear thermal conductivity. In fact, in such problems there are usually three dimensional parameters, which are the density of the unperturbed matter ρ_0, nonlinear thermal conductivity $\varkappa_0$, and a parameter associated with the energy input, so that the variables r and t and the parameters of the problem yield not one but two dimensionless quantities. Under these conditions we can obtain self-similar solutions either for special values of the power exponents of the quantities occurring in this problem or as a result of an approximate analysis which is justified by conditions in the problem and allows us to eliminate one of the parameters mentioned above. An example of the first type is the solution obtained by Marshak [17] for $\varkappa_0 \propto T^6$ and $T \propto t^{1/5}$; approximate solutions were obtained by Sakharov [18].

In the problems considered here the self-similar approach is possible if the energy input into the system is so small that the front of a thermal wave does not overtake the front of a shock wave. The total energy supplied to the system is governed either by the boundary condition (temperature or energy flux at a boundary is described by a power-law dependence on time) or by an internal energy source. The corresponding term in the energy equation may vary in a power-law manner with the coordinate, temperature, and time.

We shall seek a solution of the system (1) in the form

$$T=\Pi(t)\tau(\xi),\qquad \rho=M(t)g(\xi),\qquad v=V(t)u(\xi), \tag{2a}$$

where $\xi = r/R(t)$; R(t) is the radius of the discharge channel (thermal region).

The time-dependent parts of the functions in Eq. (2a) are

$$R=Bt^{k_1},\qquad \Pi=Ct^{k_2},\qquad M=Dt^{k_3}. \tag{2b}$$

Following the self-similar approach, an energy source Q(r, t) can also be represented in the form $Q(r, t) = \tau p(\xi)q(t)$. For example, we may assume that a current of density j(r, t) maintained by a homogeneous electric field $\mathcal{E}$ flows through a discharge channel. Then, if $\sigma = \sigma_0 T^p$ is the plasma conductivity, we find that

$$j = \sigma\mathcal{E}(t) = \sigma_0 \Pi^p \mathcal{E} \tau^p$$

and

$$Q(r, t) = j\mathcal{E} = \sigma_0 \Pi^p \mathcal{E}^2 \tau^p = q(t)\,\tau^p(\xi).$$

The rate of energy evolution per unit length of the discharge channel is

$$W(t) = \int_0^R Q(r, t)\, 2\pi r dr = \pi R^2 q(t)\,\theta,$$

where

$$\theta = 2\int_0^1 \xi \tau^p d\xi.$$

If $E = \mathcal{L} E_0 t^\alpha$ is the energy supplied to a discharge of length $\mathcal{L}$ in a time t, we find that

$$W = \alpha \mathcal{L} E_0 t^{\alpha-1}.$$

After the formal substitution of Eq. (2) into Eq. (1), we obtain the following equations (as usual, a prime denotes differentiation with respect to ξ and a dot represents differentiation with respect to t):

$$\left.\begin{aligned} &-\xi\frac{g'}{g} + \frac{V}{\dot R}\left(u\frac{g'}{g} + u' + \frac{\nu u}{\xi}\right) + \frac{\dot M R}{M \dot R} = 0, \\ &\frac{d}{d\xi}(g\tau) = \frac{\dot R^2}{c^2} g\left[\frac{R\dot V}{\dot R^2}u + \frac{V}{\dot R}u'\left(\frac{V}{\dot R}u - \xi\right)\right], \qquad c^2 = (\gamma-1)A\Pi, \\ &\frac{AM^{n+1}R\dot R}{\varkappa_0 \Pi^{m+3}} g\tau\left\{\frac{\dot\Pi R}{\Pi \dot R} - (\gamma-1)\frac{\dot M R}{M\dot R} + \left(\frac{V}{\dot R}u - \xi\right)\left[\frac{\tau'}{\tau} - (\gamma-1)\frac{g'}{g}\right]\right\} = \frac{q(t)R^2M^n}{\varkappa_0\Pi^{m+4}}\tau^p(\xi) + \frac{1}{\xi^\nu}\frac{d}{d\xi}\xi^\nu\frac{\tau^{m+3}}{g^n}\frac{d\tau}{d\xi}. \end{aligned}\right\} \quad (3)$$

We shall now formulate the boundary conditions. If $\xi = 0$, we have

$$g(0) = 1, \quad \tau(0) = 1, \quad \tau'(0) = 0, \quad u(0) = 0. \qquad (4)$$

In the front of a thermal wave at the point $\xi = 1$ the density, pressure, and velocity are continuous but the values of these quantities in a shock wave in our problem are not known. It is shown in [19] that the problem of an expanding piston with a power law of the piston motion ($R = Bt^{k_1}$) is self-similar, i.e., the radius of a shock wave R_{sw} is related to the piston radius R by $R_{sw} = \lambda R$, where λ is a constant. The thermal region is the expanding piston in high-power discharges.

Since the pressure, density, and velocity behind the shock-wave front vary only slowly, a shock wave can be represented by the average values of the characteristics. Therefore, the conditions corresponding to $\xi = 1$ can be expressed in the following form:

$$\left.\begin{aligned} &\xi = 1, \quad Mg(1) = \delta\rho_0, \quad Vu(1) = \frac{1}{\lambda}\sqrt{\frac{\gamma_0+1}{2}\frac{P}{\rho_0}}, \\ &P = (\gamma-1)AM\Pi g(1)\,\tau(1). \end{aligned}\right\} \quad (5)$$

The constants λ and δ (the latter denotes the average compression) can be found from the law of conservation of mass and energy in a shock wave. We shall distinguish the values of the adiabatic exponent in the thermal region (γ) and in the shock-compression region (γ_0). It follows from Eq. (5) that

$$\tau(1) = \frac{(\gamma_0+1)(\gamma-1)}{2\lambda^2\delta} u^2(1) \frac{V^2}{\dot{R}^2} \frac{\dot{R}^2}{c^2}. \tag{6}$$

The system (3) and the relationship (6) contain the parameter $\dot{R}^2/c^2$, which is the ratio of the squares of the velocities of the thermal-wave front and of sound in the thermal wave region. Experimental results indicate that for a wide range of initial parameters a high-current discharge obeys the inequality $\dot{R}^2/c^2 \ll 1$; we shall now consider the consequences of this inequality. The first three equations in the system (3) contain the following integrals obtained in the $\dot{R}^2/c^2 = 0$ approximation subject to an additional condition $V/\dot{R} = w = \text{const}$:

$$g\tau = 1, \tag{7}$$

$$wu(\xi) = \xi - \left(\nu + 1 + \frac{k_3}{k_1}\right) \xi^{-\nu} \tau(\xi) \int_0^\xi \frac{\xi^\nu}{\tau} d\xi. \tag{8}$$

The relationship $u(1) = 1$ provides a scale of the velocity V; it then follows from Eq. (3) that

$$w = \frac{V}{\dot{R}} = 1 - \left(\nu + 1 + \frac{k_3}{k_1}\right) \tau(1) \int_0^1 \frac{\xi^\nu}{\tau} d\xi \tag{9}$$

and in the approximation employed we find that $w = 1$. The condition $\dot{R} = V$ means that the thermal wave velocity is basically equal to the gasdynamic velocity and the diffusion of photons alters it only slightly.

Thus, if we just assume that $\dot{R}^2/c^2 \ll 1$, it follows from Eq. (3) that the pressure is independent of the coordinate, the velocity of the thermal-wave front is equal to the velocity of matter accelerated by the shock wave and, finally, the temperature (density) behind the shock wave is low (high) compared with the temperature (density) in the heated region. The condition $\tau(1) \ll 1$ is the criterion of validity of the self-similar relationships given below. In the $\dot{R}^2/c^2 = 0$ approximation the above condition is independent of time in the thermal-wave front: $\tau(1) = 0$. This equality eliminates the additional dimensionless parameter and makes the problem self-similar.

The last equation in the system (3), taken together with Eqs. (7), (8), and (9), includes all the profile dependences and it can be reduced to the form

$$\alpha\left[\frac{k_2}{k_1} - (\gamma-1)\frac{k_3}{k_1} - \gamma\left(\nu + 1 + \frac{k_3}{k_1}\right)\frac{\tau'(\xi)}{\xi^\nu} \int_0^\xi \frac{\xi^\nu}{\tau} d\xi\right] = \beta\tau^p + \frac{1}{\xi^\nu}\frac{d}{d\xi}\xi^\nu \tau^{m+n+3}\frac{d\tau}{d\xi}, \tag{10}$$

where

$$\alpha = \frac{A\dot{R}RM^{n+1}}{\varkappa_0 \Pi^{m+3}}, \quad \beta = \frac{qR^2M^n}{\varkappa_0 \Pi^{m+4}}.$$

The variables t and ξ can be separated if α and β are constants. We shall consider the physical factors which make the combinations of α and β constant. We shall integrate Eq. (10) over the volume bearing in mind that in the wave front subject to $\xi = 1$ the radiation flux should vanish because of considerations of continuity:

$$\tau^{m+n+3}\tau'|_{\xi=1} = 0.$$

As a result of such integration, we find [subject to the condition $\tau(1) = 0$] that α and β

proportional to one another:

$$\alpha\left[\gamma+\frac{k_2+k_3}{k_1(\nu+1)}\right]=\beta\int_0^1 \xi^\nu\tau^p d\xi=\frac{\beta\theta}{\nu+1}, \quad \theta=(\nu+1)\int_0^1 \xi^\nu\tau^p d\xi. \tag{11}$$

The above relationship represents the law of conservation of energy. The quantity β is the ratio of the energy evolved in a volume $r \leq R$ per unit time to the energy which is carried away from this region by the emitted radiation, multiplied by the optical thickness of the heated region:

$$\beta \propto \frac{qR^\nu}{\frac{c\sigma}{4}T^4R^{\nu-1}}\frac{R}{l_0\frac{T^m}{\rho^n}},$$

We may expect the process to be self-regulating so that at any moment we have β = const ~ 1. In fact, if for a given source the temperature is high and the density is low ($\beta \ll 1$), the radiation from the whole of the heated volume reaches the cold gas enclosed by the shock wave, increases the mass of the heated gas, lowers the temperature, and raises the density. If the temperature is low and the density is high (corresponding to a high optical thickness of the gas so that $\beta \gg 1$), the radiation flux from the heated region is small and, therefore, the amount of gas entering the thermal-wave region is small which tends to reduce the optical thickness during gasdynamic expansion. A numerical value of the constant β can be found from the condition that the integral curve $\tau(\xi, \beta)$ of Eq. (10) passes through the point $\tau(1, \beta) = 0$.

Combining Eqs. (5), (10), and (11), we obtain the following relationships from which we can deduce the coefficients B, C, and D as well as the power exponents k_1, k_2, and k_3:

$$\left.\begin{array}{c}\dot{R}^2=\frac{1}{2}(\gamma_0+1)(\gamma-1)\frac{AM\Pi}{\lambda^2\rho_0},\\ \frac{AR\dot{R}M^{n+1}}{\varkappa_0\Pi^{m+3}}=\frac{\beta\theta}{(\nu+1)\left[\gamma+\frac{k_2+k_3}{k_1(\nu+1)}\right]}, \quad \frac{q(t)R^2M^n}{\varkappa_0\Pi^{m+4}}=\beta.\end{array}\right\} \tag{12}$$

We shall consider once again the physical meaning of the system (12) for a cylindrical discharge. The first equation represents the statement that the velocity of the thermal wave is equal to the velocity of matter accelerated by the shock wave. The second and third equations represent the law of conservation of energy: the energy entering the discharge E(t) is equal to the thermal energy of the heated region $E_t(t)$ added to the shock-wave energy $E_{sw}(t)$:

$$E(t)=E_t(t)+E_{sw}(t),$$

and the energy carried away from the heated region by radiation is equal to the energy brought in by the gas flowing into the discharge:

$$S2\pi R=\varepsilon\frac{d\mathcal{M}}{dt}.$$

Here, S is the light flux calculated at the boundary between the heated and cooled parts of the thermal wave; $\mathcal{M}$ is the mass of the gas in the thermal region; ε is the specific energy of the thermal region; $2\pi R$ is the surface of the cylindrical discharge.

The system (12) yields formulas which give the time dependences of the temperature, density, and radius of the thermal-wave front:

$$k_1=\frac{1}{4}(\alpha+2), \quad k_2=\frac{\alpha(n+2)}{2(m+n+4)}-\frac{n+1}{m+n+4}, \quad k_3=2k_1-k_2-2; \tag{13}$$

$$B = a_1^{1/4}\rho_0^{-1/4}E_0^{1/4}, \quad C = \left[a_1^{n/2}a_2\frac{E_0^{1+\frac{n}{2}}\rho_0^{\frac{n}{2}}}{\varkappa_0 A^n}\right]^{\frac{1}{m+n+4}}, \quad D = \frac{2\rho_0\lambda^2 k_1^2}{(\gamma_0+1)(\gamma-1)A}\frac{B^2}{C}. \tag{14}$$

Here,

$$a_1 = \frac{(\gamma_0+1)(\gamma-1)a}{4\pi\lambda^2 k_1^2[k_1(\gamma+1)-1]},$$
$$a_2 = \frac{2a\lambda^2 k_1^2}{\pi(\gamma_0+1)(\gamma-1)\theta\beta}.$$

We shall now describe a simple approximate method for the determination of the self-similarity constant β. We shall do this by rewriting Eq. (10) with the aid of Eq. (11) for the case $\nu = 1$:

$$\left.\begin{aligned} &\frac{1}{\xi}\frac{d}{d\xi}\xi\tau^{m+n+3}\tau' = \beta\varphi(\xi),\\ &\varphi(\xi) = \frac{\theta}{2\left[\gamma+\frac{k_2+k_3}{2k_1}\right]}\left[\frac{k_2}{k_1}-(\gamma-1)\frac{k_3}{k_1}-\gamma\left(2+\frac{k_3}{k_1}\right)\xi^{-1}\tau'\int_0^{\xi}\xi\tau^{-1}d\xi\right]-\tau^p(\xi). \end{aligned}\right\} \tag{15}$$

Using the conditions of Eq. (4), we find that integration of the system (15) gives

$$\tau^{m+n+4}(\xi) = 1+\beta(m+n+4)\int_0^{\xi}\frac{dy}{y}\int_0^{y}x\varphi(x)\,dx, \tag{16}$$

and since $\tau(1) = 0$, we find that

$$\beta = -\frac{1}{(m+n+4)\int_0^1\frac{dy}{y}\int_0^y x\varphi(x)\,dx}.$$

The integrals in the denominator of the above expression can be transformed using an explicit expression for $\varphi(\xi)$; consequently, we obtain the following relationship for β (on the assumption that $p = 0$, $\tau^p = 1$):

$$\beta = \frac{2\left[\gamma+\frac{k_2+k_3}{2k_1}\right]}{(m+n+4)\gamma\left(2+\frac{k_3}{k_1}\right)\int_0^1\frac{\tau dy}{y}\int_0^y\frac{x}{\tau(x)}dx}. \tag{17}$$

In Eq. (17) the quantity $\tau(\xi)$ depends on β. The method under consideration involves successive approximations. Starting with $\tau_0 = 1$, we use Eq. (17) to find β_1, and then employing Eq. (16) we refine the temperature profile $\tau_1(\xi)$, and so on. In the first approximation, we find that

$$\tau_0 = 1, \quad \beta_1 = \frac{8\left[\gamma+\frac{k_2+k_3}{2k_1}\right]}{(m+n+4)\gamma\left(2+\frac{k_3}{k_1}\right)}. \tag{18}$$

In the next approximation (for $m + n + 4 = 7$), we obtain

$$\tau_1 = (1-\xi^2)^{\frac{1}{m+n+4}}, \quad \beta_2 = \frac{7{,}35\left[\gamma+\frac{k_2+k_3}{2k_1}\right]}{(m+n+4)\gamma\left(2+\frac{k_3}{k_1}\right)}. \tag{19}$$

In this approximation we have $\theta_2 = 2\int_0^1 \xi\tau^p d\xi = \frac{m+n+4}{m+n+4+p}$. A numerical calculation carried out in one of the variants shows a rapid convergence of the approximation for β_n. If m + n + 4 = 8, γ = 1.17, and α = 1, the successive approximations give

$$\beta_1 = 0,41804, \quad \beta_2 = 0.453090, \quad \beta_3 = 0.45727,$$
$$\beta_4 = 0.45767, \quad \beta_5 = 0,45776, \quad \beta_6 = 0.45776.$$

The constants λ and δ are found from the laws of conservation of mass and energy. If we assume that all matter is expelled into the shock wave, we find that

$$\rho_0 \pi (\lambda R)^2 = \delta \rho_0 \pi R^2 (\lambda^2 - 1). \tag{20}$$

The shock-wave energy is equal to the work done by an expanding piston (thermal wave)

$$\int_R^{\lambda R} \left(\frac{\rho v^2}{2} + \frac{P}{\gamma_0 - 1}\right) 2\pi r dr = \int_0^R P 2\pi R dR. \tag{21}$$

Using Eqs. (20) and (21), assuming that $\rho = \delta\rho_0$, $v = \sqrt{\frac{2}{\gamma_0 + 1}\frac{P}{\rho_0}}$, and P = ($\gamma$ − 1), and employing Eqs. (2b) and (12), we obtain the relationships

$$\lambda^2 = \frac{h_0 (2k_1 - 1) + k_1 (\gamma_0 + 1)}{(h_0 + 1)(2k_1 - 1)}, \qquad h_0 = \frac{\gamma_0 + 1}{\gamma_0 - 1}, \tag{22}$$

$$\delta = \frac{h_0 (2k_1 - 1) + k_1 (\gamma_0 + 1)}{(\gamma_0 - 1) k_1 + 1}. \tag{23}$$

The values of λ^2 and δ for different values of k_1 are as follows:

	$k_1 = 1/2$	$k_1 = 1$	$k_1 = \infty$
λ^2	∞	$\frac{\gamma_0 + 1}{2}$	$\frac{(\gamma_0 + 1)^2}{4\gamma_0}$
δ	1	h_0	h_0^2

The case k_1 = 1/2 corresponds to a Sedov strong explosion.

Using the solutions (22) and (23), we can readily show that the energy E supplied to a discharge is converted into the shock-wave energy E_{sw} and the thermal energy E_t in the following ratio:

$$E_t = \frac{2k_1 - 1}{k_1(\gamma + 1) - 1} E, \qquad E_{sw} = \frac{k_1 (\gamma - 1)}{k_1 (\gamma + 1) - 1}. \tag{24}$$

Self-similar solutions of the system (1) can be found also in the case when instead of the energy E we specify the discharge current $I = I_0 t^r$. This approach is used to solve the problem in [5]. The same paper gives formulations of the cases when the temperature on the discharge axis (r = 0) varies in accordance with the power law $\Pi = Ct^{k_2}$ or when the energy flux is

$$r^{\nu} \varkappa_0 \frac{T^{m+3}}{\rho^n} \frac{\partial T}{\partial r}\bigg|_{r=0} = -St^k.$$

All these cases are self-similar problems and in all cases we can obtain relationships describing the temperature, density, and other characteristics of the heated region, which are similar to Eqs. (13) and (14).

In the case of a current growing in accordance with a power law ($I = I_0 t^r$), we can readily obtain expressions relating E_0 and α with I_0 and r:

$$\alpha = \frac{4}{3} r \left[1 + \frac{p(n+1)}{2r(m+n+4)}\right] : \left[1 + \frac{p(n+2)}{3(m+n+4)}\right],$$

$$E(t) = E_0 t^{\alpha} = \frac{I_0^2 t^{2r+1}}{\alpha \sigma_0 \pi R^2(t)\, \Pi^p(t)\, \theta \mathscr{L}},$$

which can then be used to transform Eqs. (13) and (14) and to express all the characteristics of a discharge in terms of the parameters of the current [5]. It follows from such an analysis that the temperature of a discharge in air remains constant ($k_2 = 0$) if r = 1.1, i.e., the current should rise linearly with time. In particular, in order to maintain a temperature of 30,000°K in a thermal wave in air the discharge current should rise in accordance with the law $I = 6 \cdot 10^4 (t/10^{-6})^{1.1}$ A. It is interesting to note also that the development of a spark channel in the case of a current growing with time in accordance with the law $t^{3/4}$, considered theoretically by Braginskii [10], is obtained from the formulas of [5] by substituting m = −1/2, n = 0, which correspond to the electronic thermal conductivity that governs the transport of heat in a spark discharge.

§ 4. Allowance for the Discharge Circuit Equation in Self-Similar Description of a Discharge. Comparison with the Experimental Results

In a complete description of a discharge it is necessary to supplement the three gas-dynamic equations (1) with an equation describing the electric circuit. It the power supply is a capacitor bank, the circuit equation is of the form

$$U_0 - \frac{1}{C} \int_0^t I dt - L \frac{dI}{dt} - r_{\text{ohm}} I = 0, \tag{25}$$

where U_0 is the initial voltage across the capacitor bank; I is the current; L is the constant inductance; C is the capacitance of the bank; r_{ohm} is the ohmic resistance of the discharge plasma.

Equation (25) ignores the change in the inductance during expansion of the plasma and the intrinsic ohmic losses in the circuit.

Equations (1) and (25) contain six dimensional constants and, strictly speaking, the self-similar approach is inapplicable. However, approximate self-similar solutions can still be obtained. Such solutions are founded on the observation that even in the case of an oscillatory current the supply of energy to a discharge is monotonic. Therefore, the time dependence of the energy input can be divided into separate stages and in each stage this dependence can be approximated by a power law, and then the formulas of the self-similar theory can be applied. In this approach one meets the problem of "matching" the solutions at the boundaries of the stages into which the dependence is divided.

A characteristic feature of the self-similar solutions (13) and (14) is that the radius, temperature, and other parameters of a discharge depend strongly on the energy input E and not on E_0 or α ($E = \mathscr{L} E_0 t^{\alpha}$). In fact it follows from Eqs. (13) and (14) that

$$R = a_1^{1/4} \rho_0^{-1/4} E^{1/4} t^{1/2} \mathscr{L}^{-1/4},$$

$$T = a_1^{n} a_2 \rho_0^{n} (\varkappa_0 A^{n})^{-1} E^{-1+\frac{n}{2}} t^{-(n+1)} \mathscr{L}^{\frac{n}{2}+1}.$$

The dependence on α occurs only in the coefficients a_1 and a_2. The coefficients a_1 and a_2 essentially determine the proportion of the input energy E converted into the thermal energy of the plasma and into the shock-wave energy. It follows from Eq. (24) that

$$\frac{E_t}{E}=\frac{2\alpha}{(\alpha+2)(1+\gamma)-4}. \tag{26}$$

The dependence of E_t/E on α is plotted in Fig. 22. If α is sufficiently large (for example, $\alpha > 0.4$), the proportion of the thermal energy in the overall energy balance of a discharge is practically independent of α. In the case of smaller values of α, a considerable proportion of the energy is represented by the gasdynamic motion. Thus, if $\alpha > 0.4$, the discharge characteristics are governed mainly by the energy E and not by the rate of its supply to the discharge. Therefore, in "matching" the solutions at the boundaries between the stages we must ensure that the power functions approximating the energy have the same values at the boundaries and that the power exponents are sufficiently large ($\alpha > 0.4$).

In the self-similar approach the ohmic resistance of the plasma is also represented in the form

$$r_{\text{ohm}}=Ft^{-k_4},$$

where F and k_4 depend in a certain way on the approximation parameters E_0 and α. Solving Eq. (25), we can find the current and energy supplied to the discharge:

$$E=\int_0^t I^2 r_{\text{ohm}}\,dt. \tag{27}$$

Equations (25) and (27) can be solved numerically and then Eq. (27) is represented in the form $E = \mathscr{L} E_0' t^{\alpha'}$, where the matching conditions $E_0 = E_0'$ and $\alpha = \alpha'$ are used to find the values of E_0 and α.

An analysis of the numerical solutions of Eq. (25) shows that there is a universal dependence $\eta = \eta(q)$ which governs the parameters of a discharge channel. Here, η is the proportion of the energy stored in the capacitor bank, which reaches the discharge in a time $\pi\sqrt{LC}$, and the parameter q is given by the formula

$$q=\frac{\mathscr{L}(\varkappa_0 A^n)^{\frac{p}{m+n+4}}\left(\frac{3\mathscr{L}\sqrt{LC}}{CU_0^2/2}\right)^{\frac{1}{2}+\frac{p(n+2)}{2(m+n+4)}}}{\pi\sigma_0 a_1^{\frac{1}{2}+\frac{np}{2(m+n+4)}} a_2^{\frac{p}{m+n+4}} \rho_0^{-\frac{1}{2}+\frac{np}{2(m+n+4)}} \sqrt{\frac{L}{C}}\,(\sqrt{LC})^{2k_1+pk_2}}, \tag{28}$$

where a_1, a_2, k_1, and k_2 are calculated for $\alpha = 1$. The parameter q represents the ratio of the ohmic resistance of the plasma to the wave resistance and is essentially a parameter describing the matching of a capacitor bank to a discharge supplied by this bank. Figure 23 shows

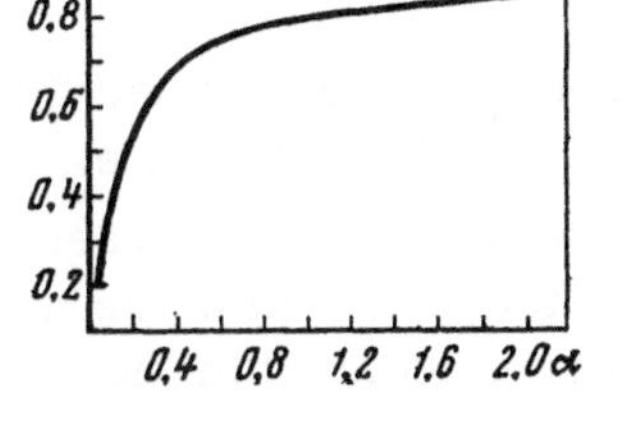

Fig. 22. Dependence of E_t/E on α.

Fig. 23. Dependence η (q).

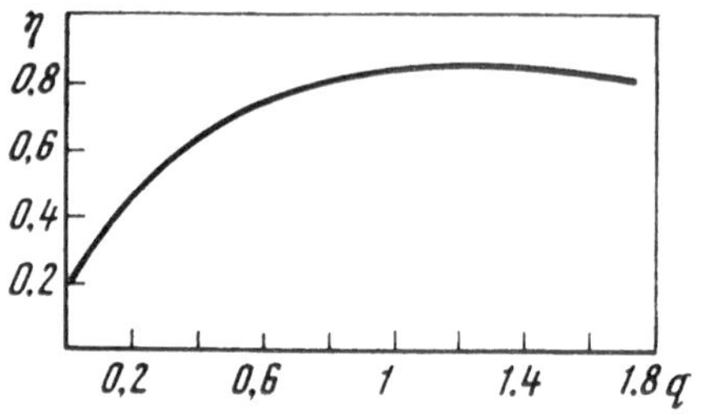

the dependence η (q) for air. The existence of this dependence can be demonstrated as follows. The substitutions

$$I = U_0\sqrt{\frac{C}{L}}\,i, \quad t = \sqrt{LC}\,\tau, \quad r_{ohm} = \frac{\mathscr{L}}{\pi\sigma_0 a_1^{1/2}\rho_0^{-1/2}E_0^{1/2}(LC)^{\frac{\alpha+2}{4}}\tau^{\frac{\alpha+2}{2}}}$$

reduce Eq. (25) to the dimensionless form (for simplicity it is assumed that p = 0):

$$1 - \int_0^{\tau} i\,d\tau - \frac{di}{d\tau} - \frac{C_1}{\tau^{\frac{\alpha+2}{2}}}\,i = 0,$$

where

$$C_1 = \frac{\mathscr{L}\sqrt{\frac{C}{L}}}{\pi\sigma_0 a_1^{1/2}\rho_0^{-1/2}E_0^{1/2}(LC)^{\frac{\alpha+2}{4}}};$$

by definition,

$$\eta = \frac{E\,(t = \pi\sqrt{LC})}{CU_0^2/2} = \frac{2}{CU_0^2}\int_0^{\pi\sqrt{LC}} r_{ohm}\,I^2 dt = 2C_1\int_0^{\pi} \frac{i^2}{\tau^{\frac{\alpha+2}{2}}}\,d\tau.$$

In the case of an almost sinusoidal discharge characterized by a weak damping ($C_1 \to 0$, $i = \sin\tau$), we find that $\eta = 2.8C_1$ for $\alpha = 1$. In the other limiting case when we can ignore the derivative $di/d\tau$ compared with $\left(C_1/\tau^{\frac{\alpha+2}{2}}\right)i$ ($C_1 \to \infty$) we find that for $\alpha = 1$ the corresponding relationship is $\eta = [1 - \exp(-4\pi^{5/2}/5C_1)]$. Bearing in mind that $C_1 = q/\sqrt{\pi}$ for η (q), we obtain

$$\eta = 2q^{2/3} \quad \text{for} \quad q \ll \frac{1}{2.8},$$

$$\eta \sim 1 \text{ in the intermediate range,}$$

$$\eta \sim \frac{200}{q^2}\left(1 - \frac{100}{q^2}\right) \quad \text{for} \quad q \gg 10.$$

In numerical estimates and in the calculations reported below we use the following properties of air: the Rosseland photon range $l_R = 6.8 \cdot 10^{-8}\,T_{eV}^{4/3}\,\rho^{-7/4}$ cm (i.e., n = 7/4, m = 4/3); the radiative thermal conductivity is $\varkappa_0 = 3.7 \cdot 10^5$ $g^{2.75} \cdot eV^{-5.33} \cdot cm^{-4.25} \cdot sec^{-3}$. These radiative characteristics of air were very important because they determine the mass and, consequently, the temperature of the heated region. They were calculated in the hydrogen-like ion approximation [15] ignoring the contribution of spectral lines to radiative energy transfer. Allowance for the lines should reduce the Rosseland range.

The interpolation formula for the plasma conductivity $\sigma = 6.8 \cdot 10^{13} T^{0.7}\ \mathrm{sec}^{-1}$ (T in eV) was obtained on the basis of tabulated data given in [20]. This formula could be used satisfactorily in the density range $(1-10) \cdot 10^{18}\ \mathrm{cm}^{-3}$ at temperatures of 1.5-8 eV. Lower values of the plasma conductivity were used in [5-7].

The specific heat of the plasma was also interpolated using the data given in [20] and assumed to be $A = 0.5 \cdot 10^{12}\ \mathrm{ergs \cdot g^{-1} \cdot eV^{-1}}$. The range of validity of this value of the specific heat was $\Delta T = 1.5-3.5$ eV and $\Delta N = (1-10) \cdot 10^{18}\ \mathrm{cm}^{-3}$.

The effective adiabatic exponent of the hot region was assumed to be $\gamma = 1.17$ [15]. It should be noted that our values of γ and A were calculated allowing for the ionization and dissociation processes. At temperatures of 3-4 eV there should be practically no neutral atoms in the thermal region and a considerable contribution should be made by singly and doubly ionized atoms.

The formulas were simplified by assuming the maximum compression in the shock-wave front although at velocities of 1-2 km/sec the compression should be somewhat less. Therefore, it was assumed that $a_0 = 5$, corresponding to $\gamma_0 = 1.5$, in order to obtain a better agreement with the experimental results.

It follows from the results of the numerical calculations that the discharge process and the time dependence of the energy can be divided into three stages.

1. During the first stage ($t < \sqrt{LC}$), corresponding to the temperature rise front, we may assume that $E(t) = (\eta/5)/(CU_0^2/2)/(t/\sqrt{LC})^2$. During this time a discharge receives approximately $\eta/5$ of the energy stored in a capacitor bank. During this stage the radius temperature on the axis of a discharge in air are given by (here and later we shall use $\beta = 1/2$ and $\theta = 1$):

$$\left.\begin{aligned} T &= 0.48\left(\frac{\eta C U_0^2}{\mathscr{L} LC}\right)^{0.265} \rho_0^{0.12} (\varkappa_0 A^{7/4})^{-0.14} t^{0.14}\ \ \mathrm{eV}, \\ R &= 0.26\left(\frac{\eta C U_0^2}{\mathscr{L} LC \rho_0}\right)^{1/4} t\ \ \mathrm{cm}. \end{aligned}\right\} \quad (29)$$

2. During the second stage, $\sqrt{LC} \le t \le \pi\sqrt{LC}$, lasting up to the moment $t = \pi\sqrt{LC}$ the discharge receives the major part of the energy, which is η. We may assume that during this stage the energy rises linearly with time:

$$E(t) = \eta \frac{CU_0^2}{2} \frac{t}{\pi\sqrt{LC}}. \quad (30)$$

During the second stage the temperature on the discharge axis and the radius of the thermal wave are

$$\left.\begin{aligned} T &= 0.48\left(\frac{\eta C U_0^2}{\mathscr{L} \sqrt{LC}}\right)^{0.265} \rho_0^{0.12} (\varkappa_0 A^{7/4})^{-0.14} t^{-0,12}\,\mathrm{eV}, \\ R &= 0.3\left(\frac{\eta C U_0^2}{\mathscr{L} \sqrt{LC}\, \rho_0}\right)^{1/4} t^{0.75}\,\mathrm{cm}. \end{aligned}\right\} \quad (31)$$

3. During the third stage, $\pi\sqrt{LC} \le t \le 2\pi\sqrt{LC}$, the ohmic resistance of the plasma is low, little energy enters the discharge, and the thermal energy of the plasma is transformed mainly into the energy of gasdynamic motion. During this stage it is desirable to select the power exponent α in such a way that the distribution of energy between the thermal region and the shock wave is the same as at the moment $2\pi\sqrt{LC}$ if the plasma had been expanding adiabatically during the time interval $\pi\sqrt{LC} \le t \le 2\pi\sqrt{LC}$. Thus, α_3 is found, subject to the assumption that

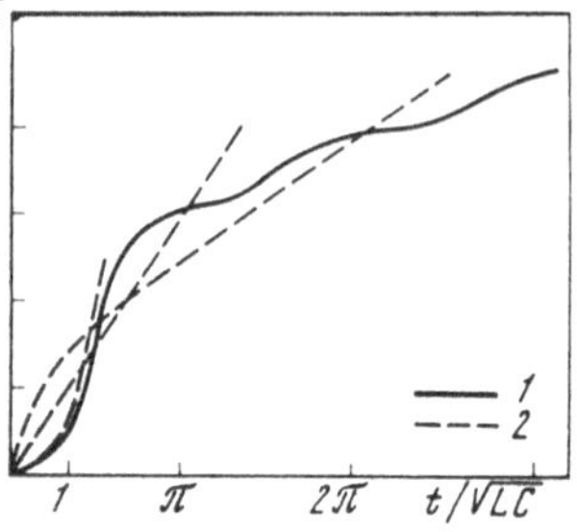

Fig. 24. Method for approximation of the time dependence of the electrical energy supplied to a discharge: 1) real time dependence; 2) calculated dependences.

$\alpha_2 = 1$, from the equation

$$\frac{\alpha_3}{(\alpha_3+2)(\gamma+1)-4}=\frac{1}{[3(\gamma+1)-4]\,2^{\frac{(\alpha_3+2)(\gamma-1)}{4}}}, \tag{32}$$

which can easily be solved graphically.

In the case of air, we have $\alpha_3 \approx 0.50$.

The temperature on the axis of the discharge and the wave radius during the third stage are

$$\left.\begin{aligned} T &= 0.48\left[\frac{\eta C U_0^2}{\mathscr{L}(LC)^{1/4}}\right]^{0.265}\rho_0^{0.12}\,(\varkappa_0 A^{7/4})^{-0.14} t^{-0.265}\ \mathrm{eV},\\ R &= 0.35\left[\frac{\eta C U_0^2}{\mathscr{L}(LC)^{1/4}\rho_0}\right]^{1/4} t^{0.62}\ \mathrm{cm}. \end{aligned}\right\} \tag{33}$$

Figure 24 shows schematically how the real rate of supply of the energy to a discharge can be approximated. Figure 25 shows the results of a calculation of the current during the first half-period, carried out for different values of q on the assumption that the energy supplied to a discharge rises linearly with time.

Let us compare the calculated characteristics of a discharge 67-cm long with the experimental results. About 60% of the stored energy reaches a discharge during the first half-period (Fig. 12) and this corresponds to a matching parameter $q \approx 0.4$. On the other hand, a comparison of the discharge current with the curves in Fig. 25 gives $q \approx 1.1$ ($\eta \approx 0.8$). The latter value is close to reality because losses in the supply busbars and in the capacitor bank itself are ignored in the measurements. A calculation based on Eq. (28) gives $q \approx 1.6$ ($\eta \approx 0.8$).

If we use the value of the energy input, we find that in the case of a shorter discharge ($\mathscr{L}$ = 20 cm) we obtain $\eta \approx 0.3$ and $q \approx 0.1$. Comparing the experimental and calculated values of the current, we obtain $q \approx 1$ ($\eta \approx 0.8$). Finally, a calculation based on Eq. (28) gives $q \approx 0.6$ ($\eta \approx 0.75$). In this variant of a discharge the losses in the supply busbars and in the capacitor bank are even more important.

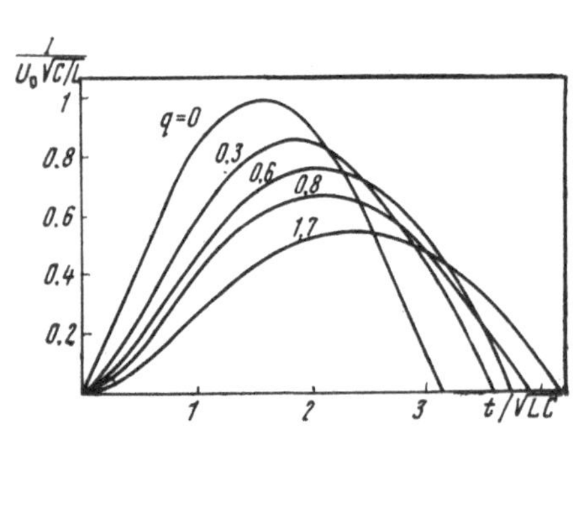

Fig. 25. Time dependences of the discharge current calculated for different values of the matching parameter q.

Thus, we can see that a calculation based on Eq. (28) overestimates the value of the matching parameter and, consequently, the amount of energy supplied to a discharge. The discrepancy is particularly large in the case of short discharges. A possible cause of this discrepancy is the inaccuracy of the value of radiative thermal conductivity. If this quantity is too high, a calculation overestimates the number of particles in the thermal region and also the temperature of this region so that the ohmic resistance of the discharge is too large. The calculated value of q can be made to agree with the experimental value for long discharges by reducing $\varkappa_0$ by a factor of approximately 30.

It should be pointed out that it follows from Eqs. (29), (31), and (33) that the difference between the real dependence E(t) and that adopted in the schematic approximation of the energy supplied to a discharge affects the discharge characteristics to a very slight degree. For example, an error in the energy by a factor of 1.5 alters the temperature and the radius of a discharge only by 10%.

A comparison of the calculated and experimental values of the radii of the thermal and shock waves is made in Figs. 9 and 16. The calculated values are based on the experimentally determined η. As pointed out above, the use of a calculated value of η increases slightly the calculated radii ($R \propto \eta^{1/4}$). The agreement between the theoretical and experimental curves in these figures is fully satisfactory. The greatest discrepancy is observed during the second maximum of the current when a slight compression occurs in the thermal region and the shock wave separates from the thermal wave. We should mention here the investigations reported in [21, 22] in which a study has been made of discharges during the compression (pinch) stage.

An important feature of Eqs. (29), (31), and (33) describing the radius of a thermal region in a discharge is the absence of any dependence on the radiative characteristics of air (if η is ignored). These characteristics do occur in the formulas for the temperature of a discharge because they are contained both in η and appear directly; they determine the power exponents in the functional dependences. Therefore, we cannot expect a good agreement between the experimental and calculated values of the temperature (Figs. 9 and 14). In a comparison of this type it is found that the calculated temperature is lower than the experimental value. We must bear in mind that the calculations give the plasma temperature on the axis of a discharge and the measurements give the brightness temperature which can only be lower than the plasmas temperature. We can also use the ohmic resistance of a discharge in estimating the plasma temperature. In measurements of this kind it is necessary to know the radial distributions of the temperature and pressure. If we assume that the temperature and pressure are constant along the radius, we find that the temperature deduced from the conductivity [20] is in good agreement with the brightness temperature. Allowance for a smooth fall of the temperature toward the edge of the thermal region increases the plasma temperature. This makes the discrepancy between the theory and experiment even greater. As pointed out in the course of our discussions of the matching parameter η, this situation is possible if the radiative thermal conductivity is far too high and, therefore, the mass in the heated region is overestimated. The calculated and experimental values of the temperature can be made to agree by reducing the radiative thermal conductivity by a factor of 20-30.

The discrepancy may also be due to an underestimate of the mass of the heated region because the Rosseland range is far too large and the plasma is transparent to the emitted radiation. In this case the radiant energy flux is weak and, consequently, the amount of matter arriving in the thermal region is small. Such a situation is unlikely because the radiative transfer is dominated by photons of (4-5)T energy [15] and if the temperature is 3-4 eV, these energies lie in the region of the first and second ionization potentials of atoms in molecules which have sufficiently large absorption cross sections.

We shall now estimate the optical thickness of a discharge channel if – in order to match the calculated and experimental results – we reduce the Rosseland range by a factor of 30:

$$\frac{2R}{l_R}=\frac{2P^{7/4}}{6.8\cdot 10^{-8}A^{7/4}T^{3.08}(\gamma-1)^{7/4}}=15,$$

where R = 1 cm, $P = 2.5\cdot 10^7$ ergs/cm^3, $\dot{R}_{sw} = 1.6\cdot 10^5$ cm/sec, and T = 3.4 eV (40,000°K). This is an underestimate because we have used the maximum value of the temperature and we have ignored its spatial distribution.

The plasma temperature may be affected somewhat by ponderomotive forces. Magnetic compression slows down strongly the plasma motion during the second current maximum and raises the temperature somewhat. The magnetic field of the current also slows down slightly the plasma expansion during the later stages of a discharge. However, this effect cannot alter significantly the plasma temperature because in the adiabatic compression case we have $T \propto V^{1-\gamma}$ and the adiabatic exponent for an ionized gas is $\gamma \approx 1$. This magnetic compression during the first half-period is unimportant subject to a condition which can be obtained easily by comparing the magnetic pressure [$P_m = H^2/4\pi = 4I^2/(4\pi\cdot 100R^2)$ ergs/cm^3, I is measured in amperes] and the gasdynamic pressure {$P = [(\eta/2)(1/2)(CU_0/2)(\gamma-1)]/\pi R^2\mathscr{L}$; it is assumed here that up to the first current maximum a discharge receives $\eta/2$ of the stored energy and 50% of the energy input is converted into a shock wave and radiation}:

$$\frac{P}{P_m}=\frac{\eta(\gamma-1)L\cdot 10^{-9}}{8\mathscr{L}(I/I_{max})^2}>1. \tag{34}$$

Here, $I_{max} = U_0(C/L)^{1/2}$; the inductance L is measured in Henrys, C in Farads, U_0 in volts, I in amperes, and $\mathscr{L}$ in centimeters.

Thus, the magnetic field is unimportant if a discharge is sufficiently slow. In particular, if $\mathscr{L} = 65$ cm, $\eta = 0.8$, and $I/I_{max} = 0.6$ (q = 1), we find that L > 1500 nH. If a discharge is initiated by exploding a wire, there may be no compression during the first current maximum in spite of the fact that the condition (34) is not satisfied, because in this case the number of ions and electrons formed directly from the wire material may be quite large compared with the total number of particles in the thermal region at that particular moment. This circumstance may increase the gasdynamic pressure and prevent magnetic compression. However, by the second maximum of the current the mass of the wire material becomes small compared with the mass of air captured by the discharge and this material no longer affects the compression process.

An important shortcoming of the self-similar theory is the fact that it ignores the losses of the discharge energy as a result of emission of wavelengths corresponding to the transparency "window" of the cold gas. These losses may amount to 30-50% and they are particularly large during the later stages of a discharge when the radiating surface is large and the brightness temperature has not yet fallen significantly. These radiation losses may explain the rapid fall of the discharge temperature during the third stage, which occurs much more rapidly than predicted by the theory.

During the first stage of a discharge the calculations predicted a more rapid fall of the temperature than that found experimentally and this may be due to the transparency of the discharge plasma when the diameter of the discharge column is still small. Moreover, during this stage the discharge parameters may be affected considerably by the initiation method since the mass of a wire is not small compared with the mass of air inside a discharge column.

Thus, the self-similar theory describes satisfactorily the radius of a discharge and the rate of its expansion as long as the magnetic field is unimportant. When the necessary corrections are made to the radiative thermal conductivity, this theory gives the correct scale of the plasma temperature. It cannot give the exact variation of the temperature because the losses due to radiation are ignored.

§ 5. Numerical Calculations of Characteristics of High-Current Discharges in Air

We shall now compare the results obtained by considering a high-current discharge on the basis of the self-similar theory and the experimental results discussed above with the results of numerical calculations reported in [23, 24]. In these numerical calculations a solution was obtained of a system of gasdynamic equations with electronic, ionic, and radiative thermal conductivity allowing for the electromagnetic field, ponderomotive forces (particularly the pinch effect), skin effect of the current, intrinsic losses in the circuit, etc. This provided a very full description, and the mathematical aspects as well as some physical considerations were discussed in [25, 26]. The calculations were carried out for external conditions (capacitor bank, discharge length, etc.) actually used in experiments. Figures 26 and 27 give the radial distributions of the temperature and density, as well as of the pressure, velocity, electric field, and a current density at one particular moment. The dependences T(t) are shown in Fig. 28 and the trajectories of the thermal and shock waves are given in Fig. 29. The numerical calculations confirmed the assumptions made in the self-similar theoretical description: the pressure is approximately constant in the perturbed region; the skin effect of the current is of little significance; the temperature profile is characteristic of nonlinear heat conduction and the temperature varies with time as predicted. It is interesting to compare the results in Figs 26 and 27 with the qualitative picture of Fig. 21.

In these numerical calculations the initial state of the plasma represented a zone of ~0.5 cm radius on the axis of the system and the temperature in this zone is assumed to be 3 eV. It is clear from Fig. 28 that the temperature first decreases and then begins to rise because of the arrival of the energy from the capacitor bank. During this initial state the en-

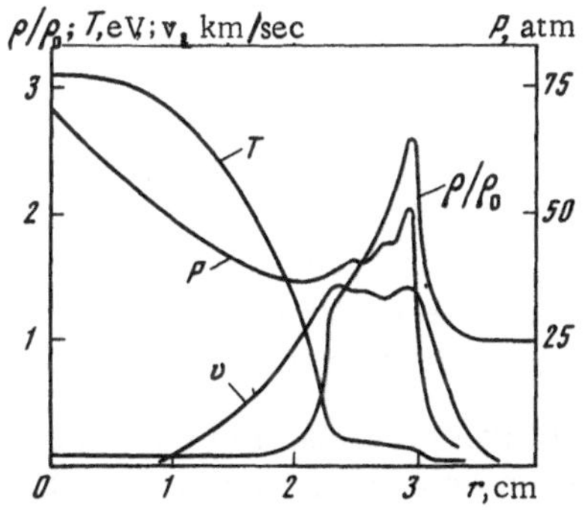

Fig. 26. Results of numerical calculations of spatial distributions of gasdynamic characteristics of a discharge 50 cm long in air, obtained for t = 8 μsec, C = 30 μF, U_0 = 50 kV, r_0 = $1.3 \cdot 10^{-2}$ Ω, L_0 = $500 \cdot 10^{-9}$ H, $\varkappa_0 = 3.7 \cdot 10^4$ $g^{2.75} \cdot eV^{-5.33} \cdot cm^{-4.25} \cdot sec^{-3}$.

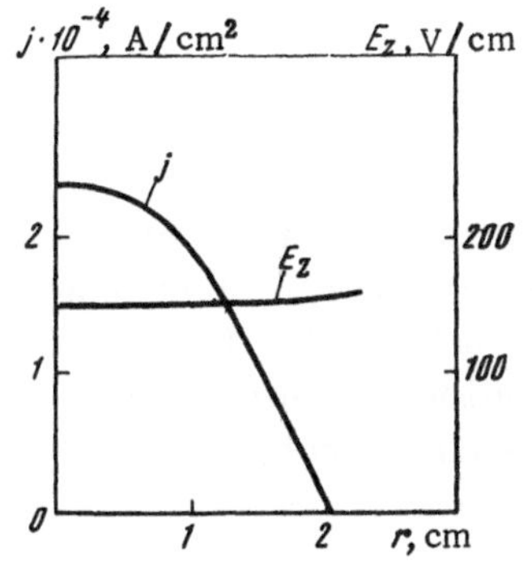

Fig. 27. Results of numerical calculations of the current density and electric field for a discharge 50 cm long in air under the same conditions as in Fig. 26.

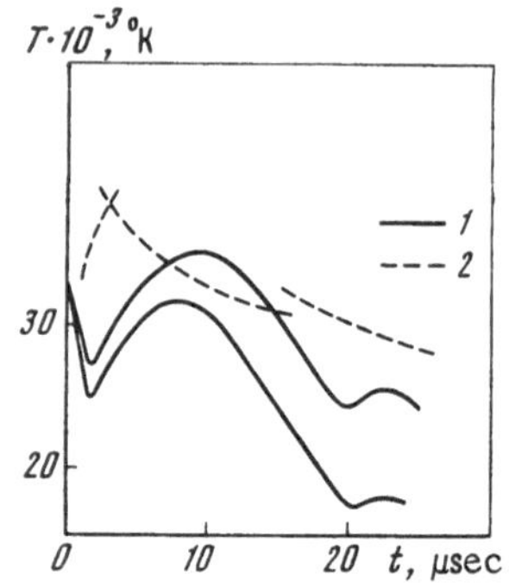

Fig. 28. Time dependences of the plasma temperature on the axis of a discharge in air (under the same conditions as in Fig. 26): 1) numerical calculations (upper curve is obtained without allowance for the emission of radiation at wavelengths corresponding to the transparency "window" of air and the lower curve allows for the losses due to such emission); 2) self-similar theory.

ergy supplied to the system is about 10 kJ and this gives rise to a somewhat more intense shock wave and a slightly overestimated plasma temperature. During the first 10-20 μsec the initial energy input is about 40% of the total energy in the system.

The influence of emission of radiation on the discharge characteristics was determined by carrying out these calculations in two variants. In the first variant the absorption coefficient of the radiation averaged in accordance with the Rosseland theory was assumed to be $k_R = 1.47 \cdot 10^8 \rho^{1.75} T^{-1.33}$ cm^{-1} (T is in electron-volts so that the radiation was absorbed completely in the gas and did not escape from the system. (It should be pointed out that in the preceeding section we used a value of k_R which was 10 times higher.) In the second variant the absorption coefficient was taken from tables in [27]; the cold gas was then partly transparent to the radiation. In the second variant the escape of the radiation from the system was found to be considerable. During the first 10-20 μsec the radiation corresponding to the transparency "window" of air should carry away about 40% of the energy input. The radiation losses reduce somewhat the dimensions of the hot zone and the temperature in this zone falls (Fig. 28). The confinement of the central part of the hot zone by the magnetic forces becomes more significant. However, during the first 5-10 μsec the discharge characteristics are not greatly affected by the radiation losses.

The experimental measurements yielded the brightness temperature which depended on the optical thickness of the plasma and the corresponding plasma temperature could only be higher (the calculated results gave the plasma temperature on the discharge axis). An increase in the optical thickness of the plasma could explain the slower rise of the measured temperature with time than that predicted by the calculations. At the moment when the temperature had its maximum value the numerical calculations and the self-similar theory predicted values of the temperature lower than those found experimentally.

The experimental and calculated results agree if the value of $\varkappa_0$ calculated in the hydrogen-like ion approximation [15] is reduced approximately by a factor of 20-30. This procedure can be used in determining the Rosseland range in gases. The proposed method of finding the

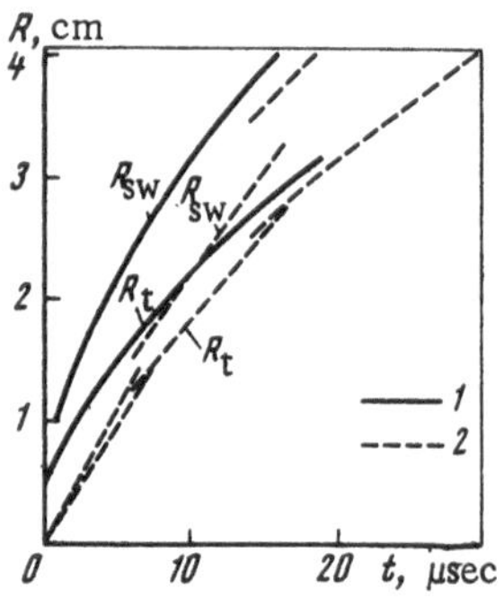

Fig. 29. Time dependences of the radii of thermal and shock waves in a discharge in air (under the conditions given in Fig. 26): 1) numerical calculation (without allowance for the radiation losses); 2) self-similar theory.

Rosseland range is based on the experimental results and on relatively simple calculations of the thermophysical properties. It should be particularly useful in the case of gases of complex composition because it is then very difficult to calculate the optical characteristics. We should bear in mind that the precision of such determination of $\varkappa_0$ is low because the temperature is not greatly affected by the variation of $\varkappa_0$.

On the whole the description of electric discharges in dense gases based on the self-similar theory and on the numerical calculations mentioned above is close to the experimental results. The relationships obtained explain the main experimental observations and can be used to predict the discharge parameters on the basis of the characteristics of the power supply source. In those cases when the magnetic forces are unimportant and the radiation losses can be ignored the self-similar theory gives results of satisfactory accuracy. The numerical calculations are not restricted in this way and the refinement of the thermophysical characteristics of a plasma, mainly by introduction of a spectral dependence of the absorption coefficient of the radiation, should make it possible to achieve not only a quantitative agreement between the calculations and experiments but also to predict reliably those plasma parameters which are difficult to measure directly.

We shall conclude by making some comments on the work of Aleksandrov et al. [22] concerned with the characteristics of high-current discharges in air. It is not clear why the self-similar calculations of the radius of the thermal region and the experimental results differed so strongly. The energy supplied in these experiments was fairly high and the rate $\dot{R} \sim 0.5$ km/sec seems far too low. If we assume that the energy introduced into a discharge is concentrated in a volume $\pi R^2 \mathscr{L}$, the resultant pressure should give rise to a rate of expansion approximately twice as high as that given in [22]. Moreover, it seems incorrect to calculate the energy balance on the assumption that all the particles are captured by the discharge. Clearly, the density in the discharge zone is less than the density of the unperturbed gas. Finally, it is incorrect to say that the energy losses due to ionization are ignored in our investigation reported in [5]. We used an effective adiabatic exponent $\gamma = 1.17$ for the hot region and the specific heat $A = 0.5 \cdot 10^{12}$ erg$\cdot$g$^{-1}\cdot$eV^{-1}, which allowed for the ionization and dissociation processes.

§ 6. Stability of High-Current Discharges in Gases

Numerous investigations of the high-temperature pinch effect have shown that self-pinched discharges are characterized by hydrodynamic instabilities in the form of axially symmetric compression of the plasma (constrictions) or twisting of a plasma column in accordance with some particular geometry (flexural and helical instability modes). These instabilities interrupt the current or force the plasma to the walls of the containing chamber in a time of the order of several microseconds. The generally accepted cause of these instabilities is the absence of the compensation of fluctuations of the magnetic pressure by fluctuations of the kinetic pressure of the plasma. In addition to these most dangerous (at high temperatures) force instabilities, an optically transparent plasma may also exhibit overheating instabilities of the type described in [31]. Such instabilities may also appear under the conditions of interest to us. A theory of the instability of self-pinched discharges under conditions of considerable radiative heat transfer is developed in [28-30]. An analysis is given there of the nature of the development of the various force and overheating stabilities, and the growth increments of these instabilities are estimated. The instability range of a plasma is given by the condition $v_a > v_s$, where v_a is the Alfven velocity ($v_a = (B^2/4\pi\rho)^{1/2}$, B is the magnetic induction); v_s is the velocity of sound in the plasma. In the discharges of interest to us the gas-dynamic pressure is greater than the magnetic pressure (i.e., $v_s > v_a$), at least during the initial stages of the expansion of the thermal wave, and the plasma should not be affected by the force instabilities. Moreover, the overheating instabilities should not develop either because the radiative heat transfer is suppressed by the temperature oscillations and, moreover, the conductivity under these conditions depends weakly on temperature.

However, hydrodynamic instabilities may develop in high-current discharges. As shown in [31], an interface between two media of different densities is unstable in the presence of turbulent mixing if the vectors ∇P (P is the pressure) are opposed. This instability is analogous to the instability exhibited in the gravity field by an interface between two liquids when the heavier liquid is poured from above (gravitational instability). In our case there is a boundary between the thermal shock-wave regions and these regions differ greatly in respect of the density. This boundary collapses if $\ddot{R} > 0$, i.e., if the expansion of the plasma is accelerated. According to the results reported in [31], the size of the turbulent mixing region is

$$y = 0.1 \ln \frac{\rho_2}{\rho_1} \left(\int_0^t \sqrt{\ddot{R}}\, dt \right)^2 .$$

An experimental investigation of the discharge instabilities was carried out by us using a capacitor bank with C = 33 μF charged to U_0 = 45 kV. A 24-cm-long discharge was initiated by exploding a copper or tungsten wire of 0.015 mm diameter. A streak pattern of the discharge in air is shown in Fig. 8. We can easily see the expanding region of the plasma which is pushing a shock wave ahead of it. This shock wave is not luminous and it is visible because of the plasma radiation scattered in the wave. The shock wave moves at an almost constant velocity which is initially 2.15 km/sec. For the first ~12 μsec the plasma also moves at an almost constant velocity of 1.7 km/sec. After 15 μsec the plasma edge stops, but subsequently begins to expand again. In these experiments the period of the oscillations of the current was 22 μsec and the amplitude of the current was 370 kA at the first maximum and 230 kA at the second maximum.

A comparison of the streak pattern with the time dependence of the current shows that the plasma stops after 15 μsec due to the magnetic compression during the second maximum of the current. Up to this moment the plasma column is a regular cylinder. A special investigation showed that when a discharge was initiated by a very homogeneous and well stretched wire, the plasma cylinder was even to within 0.1 mm (the limit of resolution of the apparatus) during the first 12-15 μsec. A significant unevenness appeared only after 15 μsec.

Etching or deposition of a lacquer on some parts of a wire enabled us to induce artificially unevenness of the plasma column. Once again the growth of such unevenness increased strongly at a time of 15 μsec.

The stopping of the plasma at 15 μsec is clearly due to the fact that at this moment the magnetic pressure exceeds the gasdynamic pressure. Consequently, the plasma is then unstable and this is why the unevenness grows.

We measured directly the velocity of sound in the plasma. This was done by placing an obstacle in the form of a thin filament in the path of the plasma column. When the plasma came in contact with the filament, a perturbation appeared (slightly later) on the opposite side of the plasma column. Allowance for the motion of the plasma made it possible to determine the average (over the diameter) velocity of sound which was found to be $v_s = 6.5 \pm 1.5$ km/sec during the period between 3 and 10 μsec.

Knowing the velocity of sound we were able to estimate the plasma density. In air at atmospheric pressure a shock-wave velocity of 2.15 km/sec corresponds to a pressure of 50 atm [15]. If the velocities of the thermal and shock waves are constant, the pressure is independent of the radius and, consequently, the pressure in the plasma is also 50 atm. It then follows from $P = v_s^2 \rho$ that $\rho = 1.1 \cdot 10^{-4}$ g/cm^3.

If we assume that up to the second maximum of the current the plasma density remains practically constant, we find that the Alfven velocity is $\sim 2 \cdot 10^6$ cm/sec, i.e., the plasma during this time is indeed unstable and we can apply the theory developed in [28-30].

The subsequent experiments were intended to obtain information on the development of instabilities during the first stage of the discharge when the magnetic forces were weaker than the gasdynamic pressure and the plasma expanded at an almost constant velocity. Since a plasma was hardly accelerated, the gravitational instability was unimportant.

We carried out experiments in which we excited surface waves in a plasma. This was done by placing an insulator wedge or a metal ring in the path of the discharge plasma. The velocity of propagation of the surface perturbations was found to be 6 ± 1.5 km/sec. These perturbations decayed in about 7-9 μsec.

As pointed out earlier, the force instabilities during the first stage of the discharge had to be induced deliberately. Periodic perturbations in the plasma were generated by etching parts of the wire used for initiation. A check of the axial symmetry of the resultant constrictions was made by synchronous photography of the discharge from three sides.

An analysis of the resultant plasma oscillations was carried out as follows. Perturbations of the plasma surface at any given moment can be represented by a function

$$r = r_0 + r(t, \varphi, z),$$

where φ is the azimuthal angle; z is the coordinate along the discharge axis. The function $r(t, \varphi, z)$ can be expanded as a complex Fourier series for two variables, considering t as a parameter:

$$r(t, \varphi, z) = \sum_{m, n=-\infty}^{\infty} C_{mn}(t) e^{i(m\varphi + nz)}.$$

If we restrict the treatment to short time intervals and to perturbations which are small compared with the discharge radius, we can introduce an explicit time dependence in the form

$$r(t, \varphi, z) = \sum_{m, n=-\infty}^{\infty} C'_{mn} e^{i(\omega_{mn} t + m\varphi + nz)}, \quad (35)$$

where C'_{mn} no longer depends on time and ω_{mn} may be complex.

The experimental results give, at some moment t_k and for a fixed value of $\varphi = \varphi_i$, the function $r = r(t_k, \varphi_i, z)$. Expanding this function as a series in terms of z and comparing it with Eq. (35), we obtain a system of equations with 6MN unknowns:

$$\sum_{m=1}^{M} e^{-\beta_{mn} t_k} [R_{mn} \cos m\varphi_i \sin(\alpha_{mn} t_k + \delta_{mn}) + H_{mn} \sin m\varphi_i \cos(\alpha_{mn} t_k + \delta_{mn})] = a_{ikn}, \quad (36)$$

where a_{ikn} are the coefficients of the expansion of $r(t_k, \varphi_i, z)$ as a series in z, $\omega_{mn} = \alpha_{mn} + i\beta_{mn}$, and n varies from 1 to N.

We have to determine the frequency ω_{mn} as a function of the mode number. The system (36) is fully determinate and, in principle, the solution of the system should give the frequencies and growth increments of the modes.

In a description of the surface of a discharge 10 cm long with minimum perturbations of about 3 mm we had to use at least 10 harmonic functions of the angle and 40-50 harmonics of z. The resultant high-order system of equations was soluble but the precision of the results was low. It is clear from [28-30] that the flexural oscillations developed more slowly than the purely axial modes (constrictions), so that we separated the pure constriction modes in order to increase the precision.

The characteristic velocities of the growth of perturbations in these experiments were approximately constant during the first 10-12 μsec and their values were 0.3 km/sec when the spatial period of the constrictions was 2.3 mm; 0.58 km/sec for the constriction period 26 mm; 0.82 km/sec for the constriction period 40 mm.

The frequencies and growth increments were found by determining the function $r(t_k, \varphi_i, z)$ for a region of the plasma 120 mm long; a deliberate continuation of this region was made and the whole period was divided into 240 parts. In this way the wavelength corresponding to the highest harmonic in the expansion was 2 mm and the longest wavelength was 240 mm. The expansion as a Fourier series was made on a computer (when the harmonics were added, the function was reproduced to within 20%).

The growth increments of the modes with $n \leq 30$ were determined from a series of experiments and they were found to vary in a random manner from one experiment to another; no correlation was found between the neighboring modes in the same expansion. For frequencies with $n > 30$, corresponding to wavelengths shorter than 8 mm, the dependence of the growth increment was close to $\beta_n \propto \sqrt{n}$, in agreement with the results obtained in [28-30].

The following conclusions can be drawn from the above results. A thermal-wave type of a high-current discharge in air is fairly stable. As long as the magnetic pressure is less than the gasdynamic pressure, the surface perturbations of the plasma due to the inhomogeneity of the initial conditions develop fairly slowly. During this stage of the discharge the gravitational instability plays the dominant role but since the acceleration is not significant, such instabilities do not distort very greatly the plasma profile. If the discharge conditions are such that at some moment the magnetic pressure exceeds the gasdynamic value, the usual force instabilities (constrictions or bending) develop in accordance with the theoretical representations given in [28-30]. However, under the conditions in our experiments the current is oscillatory and the time during which such force instabilities can develop is relatively short. When the maximum of the current is passed, the magnetic pressure again becomes less than the gasdynamic value, plasma begins to expand once more, and again the conditions are stable. The development of the force instabilities then slows down and the discharge does not collapse but its surface is no longer even.

Literature Cited

1. M. P. Vanyukov and A. A. Mak, Usp. Fiz. Nauk., 64:301 (1958).
2. H. Bartels and J. Bortfeldt, in: Exploding Wires (Proc. Third Conf., 1964), Vol. 3, Plenum Press, New York (1964), p. 9.
3. N. G. Basov, B. L. Borovich, V. S. Zuev, and Yu. Yu. Stoilov, Zh. Tekh. Fiz., 38:2079 (1968).
4. N. G. Basov, B. L. Borovich, V. S. Zuev, V. B. Rozanov, and Yu. Yu. Stoilov, Zh. Tekh. Fiz., 40:516 (1970).
5. N. G. Basov, B. L. Borovich, V. S. Zuev, V. B. Rozanov, and Yu. Yu. Stoilov, Zh. Tekh. Fiz., 40:805 (1970).
6. N. G. Basov, B. L. Borovich, V. S. Zuev, et al., Paper presented at Second All-Union Conf. on Physics of Low-Temperature Plasma [in Russian], Moscow (1969).
7. B. L. Borovich and V. B. Rozanov, Preprint No. 147 [in Russian], Lebedev Physics Institute, Academy of Sciences of the USSR, Moscow (1970); Kratk. Soobshch. Fiz., No. 12, 3 (1970).
8. S. I. Drabkina, Zh. Eksp. Teor. Fiz., 21:473 (1951).
9. S. L. Mandel'shtam and N. K. Sukhodrev, Zh. Eksp. Teor. Fiz., 24:701 (1953).

10. S. I. Braginskii, Zh. Eksp. Teor. Fiz., 34:1548 (1958).
11. A. A. Rukhadze (ed.) Exploding Wires [Russian translation], IL, Moscow (1963).
12. A. A. Rukhadze and I. S. Shpigel' (eds.), Electrical Explosions of Conductors [Russian translation], Mir, Moscow (1965).
13. E. Fünfer, Z. Angew. Phys., 1:295 (1949).
14. E. V. Daniél', Thesis for Candidate's Degree [in Russian], Minsk (1968).
15. Ya. B. Zel'dovich and Yu. P. Raizer, Physics of Shock Waves and High-Temperature Hydrodynamic Phenomena, Vols. 1-2, Academic Press, New York (1966, 1967).
16. L. I. Sedov, Similarity and Dimensional Methods in Mechanics [in Russian], Nauka, Moscow (1967).
17. R. E. Marshak, Phys. Fluids, 1:24 (1958).
18. A. D. Sakharov, Usp. Fiz. Nauk, 88:725 (1966).
19 N. L. Krasheninnikova, Izv. Akad. Nauk SSSR Otd. Tekh. Nauk, No. 8, 22 (1955).
20. N. N. Kalitkin, L. V. Kuz'mina, and V. S. Rogova, Tables of Thermodynamic Functions and Transport Coefficients of Plasma (Preprint) [in Russian], Institute of Applied Mathematics, Academy of Sciences of the USSR, Moscow (1972).
21. A. F. Aleksandrov, V. V. Zosimov, A. A. Rukhadze, et al., Kratk. Soobshch. Fiz., No. 8, 72 (1970).
22. A. F. Aleksandrov, V. V. Zosimov, S. P. Kurdyumov, Yu. P. Popov, A. A. Rukhadze, and I. B. Timofeev, Zh. Eksp. Teor. Fiz., 61:1841 (1971).
23. V. Ya. Gol'din, D. A. Gol'dina, G. V. Danilova, et al., Preprint No. 36 [in Russian], Institute of Applied Mathematics, Academy of Sciences of the USSR, Moscow (1971).
24. P. P. Volosevich, V. Ya. Gol'din, and N. N. Kalitkin, Preprint No. 40 [in Russian], Institute of Applied Mathematics, Academy of Sciences of the USSR, Moscow (1971).
25. B. N. Chetverushkin, Thesis for Candidate's Degree [in Russian], Institute of Applied Mathematics, Academy of Sciences of the USSR, Moscow (1971).
26. Yu. P. Popov, Thesis for Candidate's Degree [in Russian], Institute of Applied Mathematics, Academy of Sciences of the USSR, Moscow (1971).
27. N. N. Kuznetsov, Thermodyanmic Functions and Shock Adiabats of Air at High Temperatures [in Russian], Mashinostroenie, Moscow (1965).
28. V. B. Rozanov and A. A. Rukhadze, Preprint No. 132 [in Russian], Lebedev Physics Institute, Academy of Sciences of the USSR, Moscow (1969).
29. A. A. Rukhadze and S. A. Triger, Zh. Prikl. Mekh. Tekh. Fiz., No. 3, 11 (1968); Zh. Eksp. Teor. Fiz., 56:1029 (1969); Preprints Nos. 168 and 26 [in Russian], Lebedev Physics Institute, Academy of Sciences of the USSR, Moscow (1968).
30. V. B. Rozanov, A. A. Rukhadze, and S. A. Triger, Zh. Prikl. Mekh. Tekh. Fiz., No. 5, 18 (1968).
31. S. Z. Belen'kii and E. S. Fradkin, Tr. Fiz. Inst. Akad. Nauk SSSR, 29:207 (1965).

RADIATION, DYNAMICS, AND STABILITY OF A HIGH-CURRENT LITHIUM DISCHARGE PLASMA

F. A. Nikolaev, V. B. Rozanov, and Yu. P. Sviridenko

A theoretical and experimental investigation was made of a cylindrical discharge in a lithium plasma excited by pulses of the order of 200-300 kA and 10^{-4} sec duration. A quasisteady state was observed near the current maximum, the plasma was confined by a magnetic field, and all the energy supplied to the discharge was converted into radiation. The spectrum of this radiation was selective: the plasma emitted as a black body in the visible range but became optically transparent at shorter wavelengths and the emission of radiation was a three-dimensional process. A plasma of this kind should be an effective laser-pumping source. Measurements were made of the distributions of the plasma parameters (density, temperature, optical thickness, radiation flux and spectrum), the evolution of the discharge was investigated, and the characteristics of plasma inhomogeneities due to the competition of the overheating and force instabilities were measured.

The present paper reports the results of theoretical and experimental investigations of a high-current discharge in a lithium plasma (Z-pinch geometry) and discusses the results of measurements of all the important parameters of this plasma. A detailed comparison is made with the theoretical calculations and conclusions are drawn as to the radiative capabilities of this low-temperature density plasma, its stability, and suitable calculation methods.

We shall begin with several general comments on the essence of the problem under discussion and on some parameters of plasmas of this type.

INTRODUCTION

The physics of low-temperature dense plasmas, which have numerous practical applications, is a rapidly developing subject. The term "low-temperature plasma" applies to a very wide range of temperatures from 10^3 to 10^6 °K and it is used to distinguish this plasma from the plasma of thermonuclear interest. Similarly, the division between the "dense" and "rare" plasmas is arbitrary and depends on the actual application; a "dense" plasma is usually one with a density ranging from 10^{16}-10^{17} cm^{-3} to densities enoountered in solids.

The physical description of the processes and the results of investigations given in the present paper are directed mainly to the problem of high-intensity light sources. Therefore, the range of plasma temperatures and densities, linear dimensions, and the nature of the gas to be considered are restricted by the condition that the radiation flux emitted by the plasma in the range of wavelengths of interest should have the required intensity and duration. During a radiation pulse emitted by a plasma, the plasma and discharge characteristics should remain constant or vary slowly so that a general feature of the phenomena described below is the slowness of variation (or even constancy) of the external conditions. We shall be interested in the range of temperatures 1-10 eV, range or densities 10^{18}-10^{20} cm^{-3}, and range of plasma lifetimes 10^{-4}-10^{-3} sec.

Except at the lowest temperatures that we shall consider, under the conditions stated above a plasma consists of monatomic gases (or their mixtures) ionized to a greater or lesser extent. Consequently, we shall not consider the complex process of emission of radiation from molecular gases. Under the conditions of interest to us, the material properties of a plasma can be regarded as well known and the methods suitable for calculations of these properties are reviewed quite fully in [1] and in monographs [2-4].

Collisions between atoms, ions, and electrons in a plasma rapidly establish thermodynamic equilibrium distributions of the velocity of the particles, degree of ionization, and excited states of atoms and ions. Estimates indicate that, under such conditions, even the slowest equilibria are established in 10^{-8}-10^{-9} sec. Thus, in problems concerned with high-current radiation-emitting discharges, when the total pulse duration is 10^{-4}-10^{-3} sec and changes in the external conditions take at least 10^{-5}-10^{-6} sec, there is every reason to expect that a plasma is not only characterized by a Maxwellian velocity distribution with a single temperature for the heavy and light components, but is also in ionization equilibrium governed by the Saha formula and has a Boltzmann distribution over the excited states.

The presence of these equilibria allows us to calculate quite readily the thermodynamic characteristics of a plasma such as the specific energy per unit mass ε and pressure p. For a plasma of known composition, density, and temperature, these quantities can be calculated quite accurately on a computer, allowing for the ionization and excitation processes, on the assumption that the plasma departs only slightly from the ideal state.

In detailed calculations, it is reasonable to distinguish two cases: 1) in numerical calculations on a computer, it is desirable to use numerical data which allow for the specific structure of the ionization potentials of the atoms, statistical weights of the levels, etc.; 2) in analytic solutions and estimates, it is convenient to use the equation of state of an ideal gas $\varepsilon = \rho/(\gamma - 1)\rho$, $p = (1 + \alpha)nT$, $\rho = Mn$ (α is the average degree of ionization and M is the mass of an ion) with an adiabatic exponent that allows for the energy lost due to ionization (Q_{ion}, erg/cm^3): $\gamma = 1 + (1 + \alpha)kT/[\frac{3}{2}(1 + \alpha)kT + Q_{ion}]$. The interpolation parameters α and γ can be found both from exact tables and using an approximate procedure [2]. Practice shows that, in the problems under consideration, the results are not very sensitive to the precision with which these parameters are determined.

In experimental studies, a plasma is frequently formed as a result of electrical explosion of metal conductors. The explosion process has to be described by an equation of state which allows for the phase transition involving the evaporation of the metal. However, the evaporation and formation of a plasma occupy only a small fraction of the total duration of the process and the energy used in the evaporation is negligible compared with that supplied to the plasma and emitted as radiation. For these reasons, the evaporation is ignored in the energy calculations.

In the problems under discussion, the viscosity and electronic thermal conductivity are unimportant. We can use the elementary formula for the viscosity $\mu = \frac{1}{3}\rho v^2 \tau_{ii}$ (τ_{ii} is the ion-ion collision time) and readily show that the viscosity forces are practically always weak. The only exception are the problems in which the cumulation (collapse) of the Z pinch [5] or the thickness of a shock-wave front [2] is considered. However, these problems are outside the scope of the present paper. The electronic thermal conductivity should be compared with radiative energy transfer. If we consider the diffusion of photons and electrons in an optically opaque plasma, we can easily show that the higher velocity and the longer range of photons ensures that they carry 10^2-10^3 times greater energy than electrons.

A magnetic field (induction) B is always present in a high-current discharge plasma. Under steady-state conditions, the gas kinetic pressure is balanced out by the magnetic pressure and, therefore, $B \sim (8\pi p)^{1/2}$, so that B is of the order of $5 \cdot 10^4$ G for typical plasma temperatures and densities. The role of the magnetic field is described by the parameter $\omega_B \tau$, where ω_B is the Larmor frequency ($1.7 \cdot 10^7 B$ sec^{-1}) and $1/\tau$ is the Coulomb frequency. We can readily see that plasmas of interest to us obey the inequality $\omega_B \tau < 1$ [$n_0 > (4 \cdot 10^{12} T^{3/2} B)/\alpha \ln \Lambda$]. In this case, the transfer coefficients are isotropic and independent of the magnetic field. Bearing in mind that the electron and ion temperatures are equal and that there is no spatial polarization (because the Debye radius is small), we can provide a satisfactory description of the processes in such a plasma with the aid of one-liquid magnetohydrodynamic equations [6].

In spite of the basic simplicity of the situation, calculations of thermodynamic functions and transport coefficients of a low-temperature dense plasma require considerable work on a computer. There are relatively few systematic data and, among these, one should mention the detailed tables of Kuznetsov for air [7] as well as the results reported for several elements and compunds (inert gases, copper, lithium, etc.) in [8]. Even less published information is available on the experimentally determined thermodynamic functions and transport coefficients and few comparisons have been made with calculations. This is due to the fact that, in the range of temperatures and densities of interest to us, the experimental studies are fairly difficult and one has to know independently determined fairly accurate temperatures, densities, conductivities, etc. Plasma in such experiments is usually unstable and spatially inhomogeneous so that it is difficult to interpret the results.

We shall consider in some detail the radiative characteristics of a plasma. The radiation flux S (erg · cm^{-2} · sec^{-1} · eV^{-1}) from a unit surface area of a planar plasma layer with a temperature constant across the layer is given by

$$S = S_e(\varepsilon, T)\,[1 - 2E_3(k(\varepsilon))], \tag{1}$$

where $k(\varepsilon) = \varkappa'(\varepsilon)L$; $\varkappa'(\varepsilon) = \varkappa(\varepsilon)[1 - \exp(-\varepsilon/T)]$ is the absorption coefficient which allows for the stimulated emission; L is the thickness of the layer; $S_e(\varepsilon, T) = (2\pi k^4/\hbar^3 e^2)[\varepsilon^3/(e^{\varepsilon/T} - 1)]$ is the radiation flux from the surface of a black body; $E_n(Z) = \int_1^\infty \frac{e^{-Zt}}{t^n}\,dt$ is the exponential integral (Z is the integration parameter). If $k(\varepsilon)$ is greater than unity at all photon energies ε in the range (6-10)T, it follows that $S \approx S_e(\varepsilon, T)$ and the plasma layer emits as a black body. In this case, the plasma is regarded as optically opaque. Radiative transfer in an optically opaque plasma is governed by the average Rosseland photon range

$$l_R = \frac{15}{\pi^4} \int \frac{du\, u^4 e^4}{\varkappa(u)\,(e^4 - 1)^2} = \frac{7.5 \cdot 10^{36} T^{7/2}}{n^2 \alpha\,(\alpha + 1)}, \tag{2}$$

which can be calculated approximately as described in the monograph of [2]. In the case of a dense plasma ($n \geq 10^{19}$ cm^{-3}) with strongly broadened lines, the line absorption may reduce considerably l_R.

If $k(\varepsilon) \ll 1$ at all the photon energies, the plasma is transparent and $S = 2kS_e(\varepsilon, T)$. The total radiation energy is $2\int_0^\infty S(\varepsilon)\,d\varepsilon$ and the emissivity per unit volume $q = \frac{2\int S(\varepsilon)\,d\varepsilon}{L}$ is given by the expression $q = 4\pi \int_0^\infty \varkappa'(\varepsilon) I_e dt$ (I_e is the equilibrium intensity of the radiation). Since, in this case, the emissivity is governed by the maxima of $\varkappa'(\varepsilon)$, the contribution of the line emission and details of the behavior of the photoionization cross sections are much important than in the case of an opaque plasma.

Estimates show that two-thirds of all the energy q is emitted in the form of photons of energies $\varepsilon > I_{\alpha+1}$, where $I_{\alpha+1}$ is the ionization potential of an ion of charge $\alpha + 1$ and $I_{\alpha+1} \approx$ (5-10)T. This radiation is generated as a result of the capture of electrons by the ground state of atoms or ions. Thus, an optically transparent plasma composed of many-electron atoms emits in a wide photon range. In the case of atoms with a large difference between the successive ionization potentials, we find that at a temperature comparable with the first ionization potential, the atoms are located mainly in highly excited states and the absorption coefficient depends on the photon energy, like the absorption coefficient in the inverse bremsstrahlung process. This is very desirable for a radiation source because such a plasma can emit as a black body at low photon energies and still lose a small amount of energy (compared with a black body) due to the emission of radiation at shorter wavelengths. This possibility is considered in [9] for the specific case of a lithium plasma. This is physically due to the strong dependence of the inverse bremsstrahlung absorption coefficient on the photon energy. In this case, a plasma with a suitable number of particles has an emission spectrum narrower than the spectrum of a black body: the width of the spectrum may be approximately $2\varepsilon_1 \approx 2T$ (since $\varepsilon_1 \approx T$), compared with ~10T for a black body. Properties of this kind should be exhibited by plasmas of alkali elements [10] (for example, in the case of Li, the first ionization potential is 5.4 eV and the second 75 eV), beryllium plasma (second and third ionization potentials are 18.2 and 153.8 eV), as well as plasmas of inert gases, hydrogen, and halogens at temperatures below 0.1I (I is the ionization potential).

The absorption coefficients of a lithium plasma were computed by Nikiforov and Uvarov [11] using one of the variants of the semiempirical method for the calculation of wave functions and photoionization cross sections proposed by L. Vainshtein. The reduction of the ionization potential of a plasma was calculated on the assumption that the ions were at rest and the bremsstrahlung emission was calculated in the Born-Elwert approximation. A comparison of this approximation with the exact Sommerfeld solution is made in [12]. In the range of photon energies (plasma temperatures) of interest to us, the difference does not exceed 20-30%. The photoionization cross section of the ground state of Li obtained in [11] is in agreement with the experimental results and calculations reported by other workers [13-15].

A comparison of the electronic and radiative thermal conductivities shows that, right up to densities of 10^{10} cm^{-3}, the radiative transfer is 2-3 orders of magnitude faster than the transfer due to electronic heat conduction.

We shall conclude this brief discussion of the principal parameters of a low-temperature dense plasma by pointing out that the description of the dynamics of plasmas formed by high-current radiation-emitting discharges can be given using magnetohydrodynamics equations supplemented by equations of radiative transfer and electric circuit. The range of validity of this description is limited to $\omega_B T < 1$, where ω_B is the Larmor frequency. The limits are plotted in Fig. 1. Moreover, it is assumed that a plasma is in a local thermodynamic equilibrium which is established because of electron collisions, i.e., that the electron density is sufficiently high.

The complete system of equations which we shall use later is given below (the terms representing viscosity are omitted).

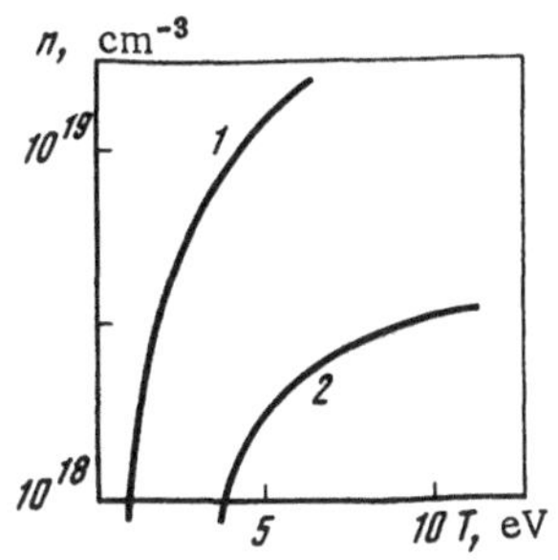

Fig. 1. Ranges of validity of the magnetohydrodynamic description of plasmas: 1) $B > 10^6$ G; 2) $B > 10^5$ G. The region between curves 1 and 2 corresponds to $m > (6 \cdot 10^{12} B T^{3/2}) \times (Z^2 \ln \Lambda)^{-1}$.

The system includes the equations of continuity, momentum, and energy:

$$\left.\begin{aligned}&\frac{\partial\rho}{\partial t}+\operatorname{div}\rho\mathbf{V}=0,\\&\rho\frac{d\mathbf{V}}{dt}=-\nabla p+\frac{1}{c}[\mathbf{j}\times\mathbf{B}],\qquad \frac{d}{dt}=\frac{\partial}{\partial t}+\mathbf{V}\frac{\partial}{\partial r},\\&\frac{\partial}{\partial t}\left(\frac{\rho V^2}{2}+\rho\varepsilon\right)+\operatorname{div}\left\{\rho\mathbf{V}\left(\frac{V^2}{2}+\varepsilon+\frac{p}{\rho}\right)\right\}=jE-\operatorname{div}S.\end{aligned}\right\}\tag{3}$$

The derivation of the above equations can be found, for example, in [16], and a system including viscosity is given in [7].

The Maxwell equations are

$$-c\operatorname{rot}E=\frac{\partial\mathbf{B}}{\partial t},\qquad \operatorname{rot}\mathbf{B}=\frac{4\pi}{c}\mathbf{j},\qquad \mathbf{j}=\sigma\left\{\mathbf{E}+\frac{1}{e}[\mathbf{V}\times\mathbf{B}]\right\}.\tag{4}$$

The radiative transfer equations (in our problems, scattering is unimportant) are given by

$$\left.\begin{aligned}&\varepsilon\nabla I=\varkappa'(\varepsilon,\ \rho,\ T)(I_e-I),\\&I_e=\frac{2k^3}{h^3e^2}\frac{\varepsilon^3}{e^{\varepsilon/T}-1}\ [\mathrm{erg\cdot cm^{-2}\cdot sec^{-1}\cdot sr^{-1}\cdot eV^{-1}}].\end{aligned}\right\}\tag{5}$$

ε is the photon energy in electron-volts.

The energy flux carried by radiation and electrons is

$$S=\int\varepsilon I d\varepsilon-\varkappa_{el}\frac{dT}{dx}.\tag{6}$$

The electric circuit (Kirchhoff) equations are:

$$B_\varphi(t)=\frac{2\mathcal{J}_p(t)}{cr}.\qquad \mathcal{J}_p=-c\frac{dU}{dt};\tag{7}$$

at the boundary of the chamber, we have

$$lE=U-\frac{1}{C^2}L_0\frac{d\mathcal{J}_p}{dt}.$$

It is assumed that the system is cylindrically symmetric; $\mathcal{J}_p$ is the total plasma (discharge) current; C and U are the capacitance of and the voltage across the capacitor bank supplying the discharge; l is the length of the discharge chamber.

The displacement current in the Maxwell equations is ignored (compared with the conduction current) because the conductivity is 10^{14} sec^{-1} and the frequency of the external field is ~10^5-10^6 sec^{-1}. It should be noted that the whole problem of radiation-emitting discharges is more complex than the corresponding problem of the thermonuclear Z pinch because of the presence of the term $\int\varepsilon I d\varepsilon$ in the expression for the energy flux and because of Eq. (5) [17].

The system (3)-(7), supplemented by the thermodynamic and radiative characteristics of the plasma, can be solved completely only by numerical methods and even then only for a specific symmetry of the problem. Analytic solutions can be obtained only in several special cases such as the steady-state conditions.

Most of the investigations of dense plasmas in high-current radiation-emitting discharges have been concerned with some aspect of the process and comprehensive experimental and theoretical studies are still lacking. This is due to the fact that theoretical treatments of the system (3)-(7) are quite difficult and comprehensive experimental studies require measurements of many parameters. In the case of radiation-emitting discharges, a comprehensive experimental investigation should include measurements of the shape of the discharge, energy balance therein, emission spectrum, and spatial distributions of various plasma properties.

CHAPTER I

THEORETICAL INVESTIGATIONS OF HIGH-CURRENT DISCHARGES IN BOUNDED PLASMAS

§ 1. Steady-State Conditions

We shall consider a discharge in which the ohmic heating energy is emitted as radiation, i.e., the thermal and kinetic energies of a plasma are low compared with the energy stored in the source. The number of particles in the plasma N is assumed to be known and this is possible only when the plasma is separated by a narrow vacuum gap from the nonionized matter (discharge-chamber wall) and, consequently, the gasdynamic pressure of the plasma is balanced out by magnetic forces, i.e., when the plasma is in a steady state in vacuum under given external conditions. We shall consider a discharge in which a plasma is a coaxial layer surrounding a thin metal conductor, located on the cylinder axis and carrying a current $\mathcal{J}_0$, directed opposite to the plasma current $\mathcal{J}_p$. The steady state is governed entirely by three quantities: N, $\mathcal{J}_0$, and E, where the last one is the electric field applied to the plasma; the plasma current is approximately 2 $\mathcal{J}_0$. This system has been considered for a thermonuclear plasma [18] and it is of interest to us because the Z pinch and planar discharge are the limiting cases corresponding to $\mathcal{J}_0=0$ and $\mathcal{J}_0=\infty$.

The equations describing the steady-state problem include the equation of continuity (the law of conservation of the number of particles per unit length of the plasma cylinder), Euler equation, Maxwell equations, and energy equation which will be formulated later. The central conductor is assumed to be infinitely thin and the plasma conductivity to be independent of its density. It is also postulated that, in accordance with the above estimates, all the Joule heat is carried away in the form of radiation and the electronic thermal conductivity is unimportant.

The dependence of the absorption coefficient of the radiation on the photon energy is assumed to be the same as in the inverse bremsstrahlung case. Equating the emissivity of the plasma to the Joule heat power, we obtain the following energy equation

$$jE = Q_0 \frac{8\pi k^4}{h^3 c^2 a} n^2 \sqrt{T}$$

[here $a = {}^3/_4 (3mk/2\pi)^{1/2} (mck/e^6 \hbar^2) = 4 \cdot 10^{36}$], which can be conveniently expressed in the form

$$p = B_0 j, \qquad B_0 = \frac{p_0}{Q_0 q_0 n_0 T_0^{1/2}}. \tag{8}$$

Equation (8) is valid if the emission of radiation from the plasma surface is negligible compared with the emission from its interior. This is true of an optically thin plasma of temperature $T \gg \varepsilon_1$, where ε_1 is the photon energy for which the plasma is of unit optical thickness $L\varkappa\varepsilon_1 \approx 1$ (L is the characteristic size of the plasma).

In the opposite case when most of the photons are emitted from the plasma surface as a result of multiple absorption and reemission events, the radiative heat conduction mechanism is important and the energy equation is

$$\sigma_0 T^{3/2} E^2 = -\frac{1}{r}\frac{d}{dr} r l_0 \varkappa_0 \frac{T^{m+3}}{\rho^n}\frac{dT}{dr}. \tag{9}$$

Here, $\varkappa_0 = 32\pi^5 k^4 / 45\hbar^3 c^2$ and the Rosseland range (2) is given by $l_R = l_0 T^m \rho^{-n}$.

We shall consider the steady-state problem assuming that the temperature distribution is homogeneous since, in this case, it is possible to obtain analytic solutions both for optically transparent and dense plasmas. In this case, the law of conservation of energy can be expressed conveniently in the integral form:

$$\sigma_0 T^{3/2} E^2 (r_2 - r_1) = 2 r_2 S(r_2). \tag{10}$$

Here, r_1 and r_2 are the boundaries of the plasma layer (they are found in the solution of the problem); $S(r_2)$ is the radiation flux at the outer boundary of the plasma.

In the three-dimensional case, the boundary conditions for the field $U = 2B$ are $U(0) = 0$ and $U(\infty) = 2\mathcal{J}_p/c$. If the temperature is independent of the coordinate, it is necessary to assume explicitly that the gasdynamic pressure vanishes at the boundaries of the plasma layer: $p(r_1) = p(r_2) = 0$. Moreover, on the inner boundary with $r = r_1$, the conditions of continuity of the radiation flux should be satisfied, whereas, on the outer boundary with $r = r_2$, it is assumed that there is no radiation flux arriving from vacuum. In the radiative heat conduction problem, the pressure satisfies corresponding conditions and the continuity of the radiation flux at the boundary is equivalent to the condition $(T^{m+3}/\rho^n)(dT/dr) = 0$; finally, the absence of the radiation flux arriving from vacuum can be formulated as follows (at the boundary $r = r_2$):

$$\frac{8}{3} l_0 \frac{T^{m+3}}{\rho^n}\frac{dT}{dr} = -T^4.$$

Equilibrium conditions in a plasma in the case of three-dimensional emission of radiation are considered in [9, 19]; the case involving radiative heat transfer is discussed in [20, 21] and the case with a homogeneous temperature distribution is analyzed in [22, 23].

Optically Transparent Discharge under Three-Dimensional Emission Conditions

Solving the system of equations discussed above, we can obtain the dependences $p(r)$, $n(r)$, and $T(r)$. These and other quantities [such as R and the optical thickness $\tau(\varepsilon) = 2\int_0^\infty \varkappa'(\varepsilon, r)\,dr$] are determined entirely by three parameters: N, $\mathcal{J}_0$, and E, which can usually be varied independently by choosing the experimental conditions. The final formulas are cumbersome and, therefore, we shall only give the relationships which define the radius

$$R^2 = \frac{\mathcal{J}_0(\mathcal{J}_0 + 2B_0 c^2)}{2\pi c^2 p(R)} \tag{11}$$

and the pressure

$$p(R) = \frac{2B_0\sigma_0}{(p_0 c^2)^{3/2}} F^{3/2}(\mathcal{J}_0)\frac{E}{N}. \tag{12}$$

The function F depends only on $\mathcal{J}_0$:

$$F(\mathcal{J}_0)=\mathcal{J}_0^{1/3}(\mathcal{J}_0+B_0c^2)^{3/2}(\mathcal{J}_0+2B_0c^2)\int\limits_0^\infty \frac{x^{\frac{\alpha+1}{3}}\,dx}{(x^2+1+2B_0c^2)^{7/3}}, \tag{13}$$

where $\alpha = 2 + (2\mathcal{J}_0/B_0c^2)$.

It should be noted that the integral (13) converges for $\alpha > 4$, i.e., for $\mathcal{J}_0 > B_0c^2$.

If $\mathcal{J}_0=0$, we obtain the following formulas for a nonhollow cylindrical filament

$$U=4B_0c\frac{r^2}{r^2+r_0}, \qquad r_0=\frac{2B_0^2c^2}{\pi p(0)}=r_2^2, \tag{14}$$

where p(0) is the maxima pressure on the filament axis and r_0 is the characteristic radius of the filament. The solution for the total number of particles in such a filament diverges, whereas the plasma current is finite ($\mathcal{J}_p=2B_0c^2$) and is independent of N and E. In this case of an isothermal finite-radius filament, the corresponding result is known as the Braginskii limiting current [24], which is $\sqrt{3}B_0c^2$. The other limiting case $\mathcal{J}_0 \gg B_0c^2$ describes a discharge whose geometric thickness is small compared with the radius R, at which the plasma pressure, temperature, and the density are maximal (the limit $\mathcal{J}_0 \to \infty$ corresponds to a planar discharge).

The maximum values of the plasma temperature and density T(R) and n(R) can be expressed quite simply in terms of N, E, and $\mathcal{J}_0$; if $\mathcal{J}_0 \gg B_0c^2$, then

$$T(R)=\frac{4.2}{(Q_0q_0\sigma_0)^{1/2}}\frac{\mathcal{J}_0}{N}\ \text{eV}, \qquad n(R)=\frac{2.05\sigma_0}{(Q_0q_0\sigma_0)^{3/4}}\frac{E\mathcal{J}_0^{1/2}}{N^{1/2}}\ \text{cm}^{-3}, \tag{15}$$

and the position of this layer is given by the formula

$$R=\frac{0.24}{\sqrt{\pi}}\frac{(Q_0q_0\sigma_0)^{5/8}}{(p_0c^2\sigma_0)^{1/2}}\frac{\mathcal{J}_0^{1/4}N^{3/4}}{E^{1/2}}\ \text{cm}.$$

The optical thickness of the discharge is

$$\tau(\varepsilon)=\frac{4.15}{\sqrt{\pi}}\frac{Q(\varepsilon)}{a\varepsilon^3}\frac{(p_0c^2\sigma_0)^{1/2}}{(Q_0q_0\sigma_0)^{9/8}}\frac{E^{3/2}N^{1/4}}{\mathcal{J}_0^{1/4}}. \tag{16}$$

The formulas (1) and (15)-(16) allow us to determine the radiation spectrum and energy. The spatial distributions of the plasma properties and the form of the spectrum are shown qualitatively in Fig. 2.

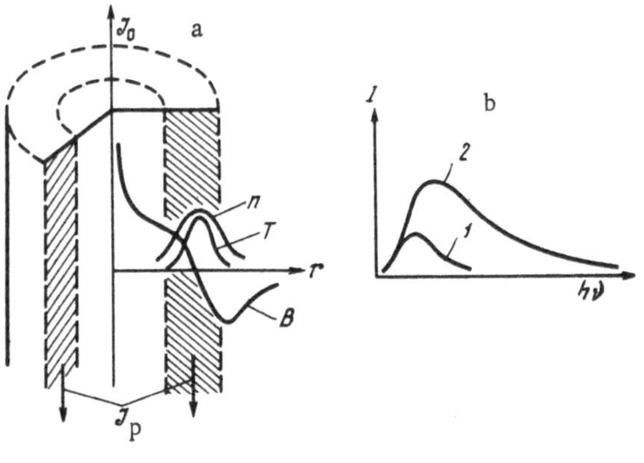

Fig. 2. Radial distributions of the plasma temperature and density in the case of three-dimensional emission (a) and the corresponding emission spectrum (b): 1) spectrum of an Li plasma; 2) spectrum of a black body.

Discharge with Homogeneous Temperature Distribution

We shall consider a discharge on the assumption that the temperature is distributed homogeneously in space.

As in the three-dimensional emission case, all the plasma quantities can be expressed in terms of N, E, and $\mathcal{J}_0$; the solution for arbitrary values of these parameters requires that the dependence of the absorption coefficient on the photon energy be specified and then numerical calculations have to be made using the radiative transfer equation. We shall confine ourselves to two special cases of optically transparent and opaque Z pinches.

In the case of an optically transparent Z pinch [$\mathcal{J}_0=0$, $\tau(\varepsilon) \ll 1$], the energy is obtained basically from the whole of the plasma interior and the homogeneity of the temperature distribution is ensured by a relatively small proportion of photons whose range is short compared with the discharge diameter. Equation (10) reduces to

$$\sigma_0 T^{3/2} E^2 r_2^2 = 2 \int_0^{r_2} q r dr, \tag{17}$$

where q is the plasma emissivity. This case has been discussed in detail for a thermonuclear plasma in which a homogeneous temperature distribution is achieved because of electronic heat conduction.

In this case, the spatial distribution of the plasma is parabolic and the magnetic field increases linearly with the radius:

$$p(r) = p_0\left(1 - \frac{r^2}{r_2^2}\right), \qquad B = \sqrt{4\pi p(0)}\frac{r}{r_2}, \tag{18}$$

$$p(0) = \frac{9}{2\sqrt{2}} \frac{p_0 c^2}{\sigma_0 (Q_0 q_0)^2} \frac{E}{N^{3/2}}, \qquad r_2 = \left(\frac{2\sqrt{2}}{3\pi}\right)^{1/2} \frac{\sqrt{Q_0 q_0}}{(p(0)c^2)^{1/4}} \frac{N^{3/4}}{E^{1/2}}. \tag{19}$$

The formulas in Eq. (18) were first obtained in [25]. For an optically transparent discharge with a homogeneous temperature distribution, the plasma current is finite, as in the three-dimensional emission case, and independent of N and E [24] [the current is $\mathcal{J}_0 = \sqrt{3}p_0c^2 \times (Q_0q_0\sigma_0)^{-1/2}$], and the plasma boundary is abrupt [this property differs from the three-dimensional emission case: see Eq. (14)]. Since the number of particles in a discharge is finite, Eqs. (18) and (19) and the formulas for the temperature and density

$$T = \frac{3}{2} \frac{p_0 c^2}{Q_0 q_0 \sigma_0} \frac{1}{N}, \qquad n(0) = \frac{3}{\sqrt{2}} \frac{(p_0 c^2)^{1/2}}{Q_0 q_0} \frac{E}{\sqrt{N}} \tag{20}$$

and for the optical thickness τ can be compared with the experimental results.

In the other case of an optically opaque Z pinch [$\mathcal{J}_0 = 0$, $\tau(\varepsilon) \gg 1$], the radiation flux from the surface of a discharge is equal to the flux emitted by a black body and the energy equation (10) can be expressed as follows:

$$\sigma_0 T^{3/2} E^2 r_2 = 2bT^4, \qquad b = \frac{2\pi^5 k^4}{15 h^2 c^2} = 10^{12}\ \text{erg} \cdot \text{cm}^{-2} \cdot \text{sec}^{-1} \cdot \text{eV}^{-4}. \tag{21}$$

The plasma properties depend on N and E in a manner different from that in the preceding transparent discharge case:

$$p(0) = \frac{\sqrt{2}}{(8\pi^2)^{1/6}} \frac{p_0 \sigma_0^{4/3}}{(p_0 c^2)^{1/3}} \frac{N^{2/3} E^2}{b^{2/3}}, \qquad r = \frac{2^{1/4} (8\pi^2)^{1/24}}{\pi} \frac{(p_0 c^2)^{5/24} b^{1/6} N^{5/24} E^{-3/2}}{\sigma^{7/12}}, \tag{22}$$

$$T = \frac{1}{(8\pi^2)^{1/12}} \frac{\sigma_0^{1/6} (p_0 c^2)^{1/12}}{b^{1/3}} N^{1/12} E^{1/3}, \qquad n(0) = \frac{\sqrt{2}}{(8\pi^2)^{1/12}} \frac{\sigma_0^{7/6} N^{7/12} E^{5/3}}{(p_0 c^2)^{5/12} b^{1/3}}. \tag{23}$$

Finally, the total plasma current depends on N and E as follows:

$$\mathcal{J}_{\mathrm{p}}=\frac{\sqrt{2}}{(8\pi^2)^{1/24}}\frac{(p_0c^2)^{3/24}\sigma_0^{1/12}}{b^{1/6}}N^{13/24}E^{1/4}. \tag{24}$$

For a cylindrical discharge, all the quantities can be expressed also in terms of the plasma current if the field E is eliminated.

Applying Eqs. (20) and (23), we can readily calculate the optical thickness of the discharge $\tau(\varepsilon)$ for photons which are of interest at this temperature and then determine the change of values of N and E in which an optically transparent discharge becomes opaque. If the photon energy is $\varepsilon = 2T$, we have

$$\tau=\frac{0.073Q_0Z^3}{a}\frac{(Q_0q_0)^2\sigma_0^{7/2}}{(p_0c^2)^{11/4}}E^{3/2}N^{3/4}\ll 1, \tag{25}$$

$$\tau=\frac{0.16Q_0Z^3}{a}\frac{\sigma_0^{7/6}b^{2/3}}{(p_0c^2)^{11/2}}E^{1/2}N^{13/12}\gg 1. \tag{26}$$

The change from a transparent to an opaque discharge occurs for

$$EN^{13/6}=\frac{4.6}{Q^2}\frac{(p_0c^2)^{11/6}b^{2/3}}{\sigma_0^{7/3}q_0^2}. \tag{27}$$

A general feature of these steady-state solutions is that, when the actual number of particles N is increased, a discharge changes from optically transparent to opaque and this is accompanied by considerable changes in the various dependences; for example, at low values of N, the plasma temperature is inversely proportional to N, whereas, at high values of N, it rises as a power of this number. The relationships for an optically transparent Z pinch demonstrate that the radiated power $\mathcal{J}_{\mathrm{p}}E \propto B_0c^2E$ is proportional to the electric field. Since the optical thickness increases with the field as $\propto E^{3/2}$ but the plasma temperature is not affected, an increase in the radiation power is due to a change in the spectrum, which becomes closer to the spectrum of a black body.

Discharge under Radiative Heat Transfer Conditions

In this case, the energy transfer is radiative and the plasma layer is not fully isothermal although the temperature gradient is not steep. It follows from Eq. (9) for the temperature profile in a cylindrical discharge that, when the dependence of T on r is weak,

$$T(r)=T(r_2)\left[1+\frac{m+5/2}{4(n+1)}\frac{\sigma_0E^2\rho_0^2r_2^2}{\varkappa_0T^{m+5/2}(r_2)}\left(1-\frac{r^2}{r_2^2}\right)^{n+1}\right]^{\frac{1}{m+5/2}}. \tag{28}$$

The pressure and density are distributed parabolically, whereas the plasma properties ρ_0 and r_2, and temperature $T(r_2)$ at the boundary are given by Eqs. (22) and (23) with slighly different numerical coefficients. The temperature on the discharge axis is

$$T_0=T(r_2)(1+\tau_R)^{\frac{1}{m+5/2}},\quad \tau_R=\frac{3(m+5/2)}{16(n+1)}\frac{r_2}{l_R(\rho_0,T_2)}. \tag{29}$$

Here, τ_R is the optical thickness of the discharge expressed in terms of the Rosseland range and calculated from the maximum density (n_0) and minimum temperature (T_2). Usually, m = 3, n = 2, and $\tau_R = r_2/2l_R$ for $r_2/l_R = 10$, $T_0/T_2 = 1.4$. It should be noted that the optical thickness calculated in terms of the Rosseland range for bremsstrahlung (m = 7/2, n = 2) differs only by a numerical factor from that given by Eq. (26).

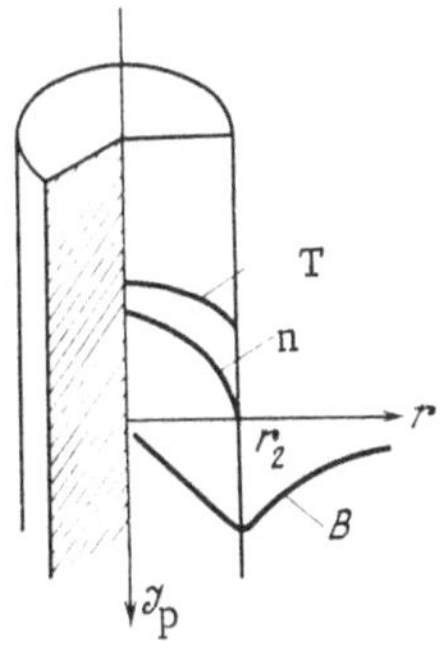

Fig. 3. Radial distributions of the plasma temperature and density for a homogeneous temperature distribution and radiative heat transfer.

Figure 3 shows qualitatively the spatial distributions of the plasma properties in the homogeneous temperature and radiative heat transfer cases.

§ 2. Discharge Stability

Numerous investigations of the high-temperature pinch effect have demonstrated that self-compressed plasma discharges are subject to hydrodynamic instabilities [18, 26], which are axially symmetric plasma constrictions or twisted configurations of the plasma column (flexural and screw instability modes). The development of these instabilities stops the current flow or forces the plasma to the chamber walls in a time of the order of several microseconds. These instabilities are due to fluctuations in the magnetic field which are not compensated by fluctuations of the kinetic pressure in the plasma. The existence of overheating instabilities in an optically transparent high-temperature plasma is also possible [26]. These instabilities may also appear under the conditions of interest to us. We shall now present briefly the results of the theory of stability of a low-temperature dense plasma, developed on the basis of magnetohydrodynamic equations in [20, 27-30]. Allowance for radiative heat transfer in the plasma and for the finite conductivity distinguishes this theory from the treatment developed for thermonuclear discharges.

In the situation under consideration, there is no need for the current to rise rapidly: quite the opposite—it is desirable to achieve quasisteady conditions under which the magnetic energy stored in the plasma as well as the thermal and kinetic energies vary slowly compared with the Joule that power and with the radiation flux integrated over the spectrum. This situation occurs if the current (or equivalent external parameter) varies so slowly that the field in the region occupied by the plasma (characteristic size L) becomes completely equalized, i.e.,

$$\mathcal{I}\Big/\left|\frac{d\mathcal{I}}{dt}\right| > \frac{4\pi\sigma L^2}{c^2}. \tag{30}$$

Discharge in the Presence of Radiative Heat Transfer

We shall use the linear approximation to consider the stability in the presence of small perturbations of the type $f(r)\exp(-i\omega t + im\varphi + ik_z)$ experienced by the steady-state distributions of a plasma under radiative heat transfer conditions (the general theory of the linear approach is presented in [26]). In this case, we can ignore the influence of changes in temperature on the processes occurring at a frequency ω if

$$\frac{l_R}{L}\frac{c^2k^2}{\sigma\omega} > 1. \tag{31}$$

Since $l_R \leqslant L$ and $\omega \sim v_s/L$ ($v_s = (p_0T/M)^{1/2}$ is the velocity of sound in the plasma), the temperature fluctuations are unimportant only in the case of a low-temperature plasma when $c^2 \gg \sigma v_s L$.

In this case, we can use equations obtained from the system (3)-(4) by linearizing it on the assumption that there are no temperature fluctuations.

The wavelength of the fluctuations in question has a lower limit set by the requirement $\lambda_{min} > l_R$, which is the criterion of validity of the radiative heat transfer approximation in the oscillation theory.

In the case of a nonhollow cylindrical discharge, the most "dangerous" force instabilities are the axially symmetric fluctuations for which m = 0 (plasma-column constrictions). The development of these instabilities is likely to stop the flow of the discharge current. In the plasma instability region $v_R > v_s$ ($v_R = (B^2/4\pi\rho)^{1/2}$ is the Alfven velocity) the growth increment of the fundamental long-wavelength constriction mode ($k_2 r_2 < 1$) is [20]

$$\gamma = \sqrt{2\sqrt{3}\frac{|k_z| v_s^2}{r_2}} < \frac{v_s}{r_2}. \quad (32)$$

The eigenvalue of (32) can be obtained by the exact solution of linearized equations for a plasma of finite conductivity subject to the boundary conditions governed by the actual nature of equilibrium quantities and by the requirement that the perturbations be finite in the range $(0, r_2)$. The same value of the growth increment is obtained for the $\sigma \to \infty$ case in [31]. A new nontrivial result obtained in [20] is the proof that the growth increment is independent of the conductivity.

Higher long-wavelength constriction modes ($\lambda \ll r_2$, $k_z r_2 < 1$) can be obtained in the geometric-optics approximation; their growth increments are given by the following expressions: in the case of a low-conductivity plasma ($\sigma \to 0$), we have

$$\gamma = \frac{4k_z^2 r_2^2}{\pi^4 (n + 1/2)^4} 4\pi\sigma \frac{v_s^2}{c^2}, \quad (33)$$

whereas, in the case of a high-conductivity plasma ($\sigma \to \infty$), the result is identical with that obtained in [32]:

$$\gamma = \frac{2|k_z| v_s}{\pi (n + 1/2)}. \quad (34)$$

Short-wavelength fluctuations $|k_z| r_2 \gg 1$ in a low-conductivity plasma have a growth increment

$$\gamma = \sqrt{\frac{2|k_z| v_s^2}{r_2}}, \quad (35)$$

which may become large compared with the reciprocal hydrodynamic time v_s/r_2. However, these fluctuations are localized in a narrow region near the plasma surface and mainfested by a "fine ripple." It follows from Eqs. (32)-(35) that the largest growth increments of the "dangerous" long-wavelength perturbations are of the order of v_s/r_2, whereas the corresponding increments of the short-wavelength fluctuations are $\sim(|k_z| r_2)^{1/2}(v_s/r_2)$. In the case of a plasma whose temperature is 3-5 eV and which occupies a region of 2-5 cm in size, the growth increments of the force instabilities are of the order of 10^5-10^6 sec^{-1}.

Discharge in an Optically Transparent Plasma

In considering the stability of an optically transparent plasma, we must bear in mind the fact that the plasma temperature fluctuates like the other plasma properties. We can show [29, 30] that, in the case of totally symmetric modes with m = k = 0, the zeroth approximation of the geometric optics (i.e., $k_z L \gg 1$) predicts overheating plasma instabilities in the wavelength range $\lambda \gtrsim v_s/\gamma$. The growth increment of such instabilities is independent of the coordinate and it is of the order of

$$\gamma \approx \frac{j^2}{\sigma p} = \frac{E}{B_0}. \tag{36}$$

These instabilities are physically due to the fact that the radiation emitted from the interior cannot compensate the temperature rise due to the Joule heating. The stability in the wavelength range $\lambda > v_s/\gamma$ can be achieved if the radiation emitted from the interior rises with increasing temperature faster than the Joule heating, i.e.,

$$\frac{\partial q}{\partial T} > \frac{\partial}{\partial T}\frac{j^2}{\sigma}. \tag{37}$$

However, in the case of a fully ionized plasma, the opposite is true ($j^2\sigma \propto T^{3/2}$, $q \propto T^{1/2}$).

Stabilization of the short-wavelength instability is obtained if a more stringent condition is satisfied:

$$\left(T\frac{\partial q}{\partial T} - \rho\frac{\partial q}{\partial \rho}\right) > T\frac{\partial}{\partial T}\frac{j^2}{\sigma}. \tag{38}$$

This is evident from the physical point of view because a small overheated region can expand (the pressure can become equalized) in a time $1/\gamma$ and this reduces the density and radiative losses. In addition to the violation of the inequalities (37) and (38), the development of an overheating instability requires that the plasma be transparent, in accordance with the condition formulated in Eq. (27). The dimensions of an overheating instability are limited from above by the skin effect; there is formally no lower limit but such a limit may appear if we allow for the electronic heat conduction.

Apart from overheating instabilities, a transparent discharge may suffer from force instabilities [30], and the absence of a sharp plasma boundary may destroy the fundamental constriction mode which is independent of the conductivity. This absence of constriction can be understood qualitatively because the discharge current flows through a cross section extending to infinity (in the range 0-$\sqrt{r_0}$, the current is $\mathcal{J}_\rho/2$ [29]).

An investigation of the stability in the geometric-optics approximation yields the following growth increments: in the case of a low-conductivity plasma ($c^2k_z^2 > 4\pi\sigma\omega$), we have

$$\gamma = \frac{k_z^2 r_2^2}{12(n+{}^1/_2)^3}\frac{\pi\sigma v_s^2}{c^2}, \qquad r_2 = \sqrt{r}; \tag{39}$$

in the case of a high-conductivity plasma ($c^2k_z^2 < 4\sigma\omega$), we obtain

$$\gamma = \frac{\sqrt{2}}{\pi}\frac{k_z v_s}{n+{}^1/_2}. \tag{40}$$

The spectra (39) and (40) can be extended to instabilities of arbitrary wavelength ($|k_z|r_2 \lesssim 1$) and low modes ($n \sim 1$). The analogs of these increments for an optically opaque discharge are the quantities given by Eqs. (33) and (34).

Numerical calculations of some increments indicate that short-wavelength force instabilities develop more rapidly than overheating instabilities. Therefore, a transparent discharge is likely to split into overheated current-carrying filaments which pulsate independently of one another.

§ 3. Numerical Calculations of Discharge Dynamics in a Lithium Plasma

It has already been pointed out that the complex hydrodynamic and radiative processes described by the system (3)-(7) cannot be investigated analytically. Therefore, the development of a discharge time was investigated by numerical calculations on a computer [32-34]. In these calculations, lithium plasma was considered and the coefficients of absorption of the radiation emitted by this plasma were postulated to depend on the frequency in the same way as in the inverse bremsstrahlung effect. The system (3)-(7) was solved for an axial discharge using thermodynamic functions, conductivity, and absorption coefficients discussed in the Introduction. The initial conditions in these calculations were certain plasma temperature and density distributions and the metal–plasma phase transition was ignored; the parameters of the power supply source (capacitance, inductance, and initial voltage) were taken into account.

Figure 4 shows computer-calculated (r, t) diagrams for linear and coaxial discharges; the caption of this figure gives the parameters of the power sources and the ratio of the Joule heating power to the radiation flux. The attainment of a steady state by a discharge is illustrated in Fig. 5, which gives the (r, t) trajectories of the various Lagrange points and the steady-state temperature and density profiles.

It follows from these figures that, in the case of radiative heat transfer and for the conditions stated above, a steady state is established in a time of the order of (10-15) · 10^{-6} sec and the characteristics of this state are practically identical with those deduced analytically.

The spectral characteristics were calculated in the five-group approximation using the transfer equations for photons and taking the absorption coefficients of the lithium plasma radiation from [35]. In the case of thin wires (small amounts of plasma), a homogeneous temperature was established at 6-7 eV and this process was affected considerably by recombination radiation involving the ground state of a doubly ionized ion; in the case of thick wires the plasma had two zones with different temperatures: the inner zone was cold and the outer zone

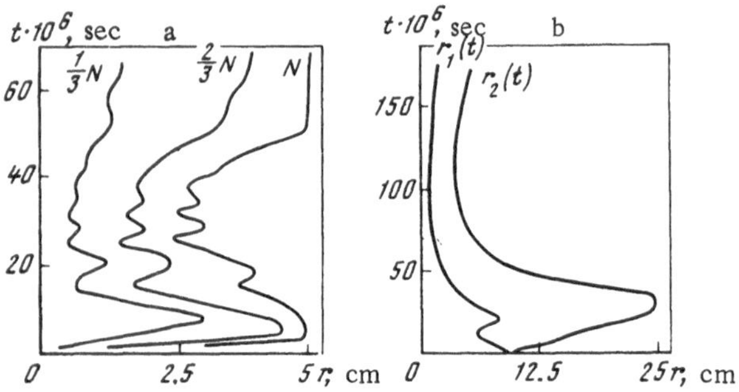

Fig. 4. The (r, t) diagrams for various Lagrange points in a linear discharge (a) and for boundaries of a coaxial discharge (b): a) C = 1800 μF, U_0 = 5 kV, L = 0.25 μH, R = 2.25 · 10^{-4} Ω, l = 15 cm, N = 5.6 · 10^{19} cm^{-1}, dE_J/dt = 3 · 10^8 W, S_{rad} = 2.7 · 10^8 W; b) C = 40,000 μF, U_0 = 10 kV, L = 0.25 μH, R = 10^{-3} Ω, l = 50 cm, N = 10^{20} cm^{-1}, J_0 = 5 · 10^5 A, dE_J/dt = 7.9 · 10^9 W, S_{rad} = 7.9 · 10^9 W.

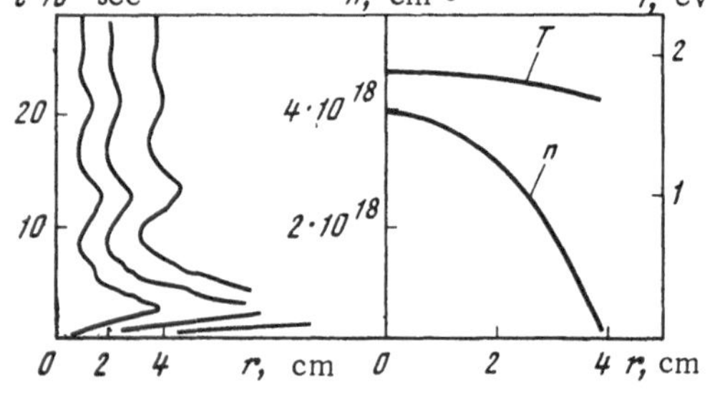

Fig. 5. Attainment of steady-state conditions in the case of a constant current.

Fig. 6. Radial distributions of the plasma temperature and density obtained by a five-group calculation without allowance for the line emission: a) $N = 5.6 \cdot 10^{19}$ cm^{-1}; b) $N = 1.2 \cdot 10^{19}$ cm^{-1}.

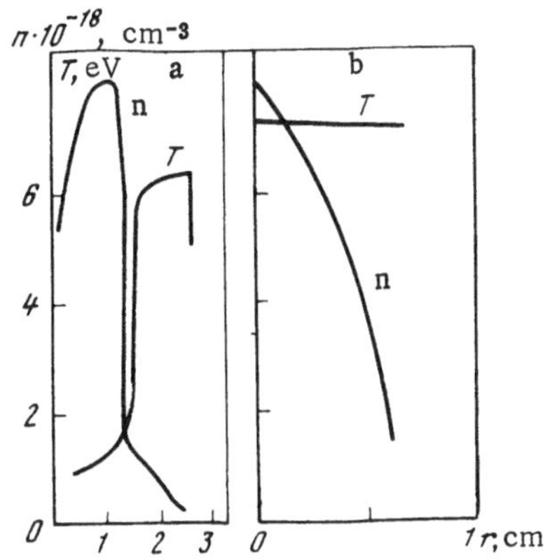

hot. The temperatures in these zones were 0.5-1 and 5-6 eV, respectively, the pressure was homogeneous, but the temperature jump was accompanied by a density jump. The energy was radiated mainly from the hot zone.

The presence of line emission due to impurities could give rise to a stronger temperature dependence of the plasma emissivity and to a closer coupling between the plasma regions with different temperatures. A calculation similar to that reported in Fig. 6 was carried out but, in this case, an effective line, 0.75 eV wide, was selected in the photon spectrum. The absorption coefficients used in this calculation were the average values taken from [35] and their absolute values were increased fivefold. This procedure was applied on the basis of a comparison of the calculations with the experimental results (see Chap. II) and corresponded to the plasma emissivity approximately 20 times higher than the emissivity in the bremstrahlung emission case. For comparison, a calculation was made in which the emission was taken into account in the radiative heat transfer approximation and the Rosseland range was assumed to be $l_R = 4.5 \cdot 10^{37} T^{7/2}/n^2$ cm; this value was 20 times smaller than in the bremsstrahlung case. The results of these calculations are presented in Figs. 4a, 7, and 8. The time dependences in these calculations [(r, t) diagrams, total current, electric field, Joule heating power, radiation flux, ohmic resistance of the plasma] were found to be very similar. The most typical feature of these dependences was the existence of a quasisteady state and reemission of all the energy supplied to the plasma.

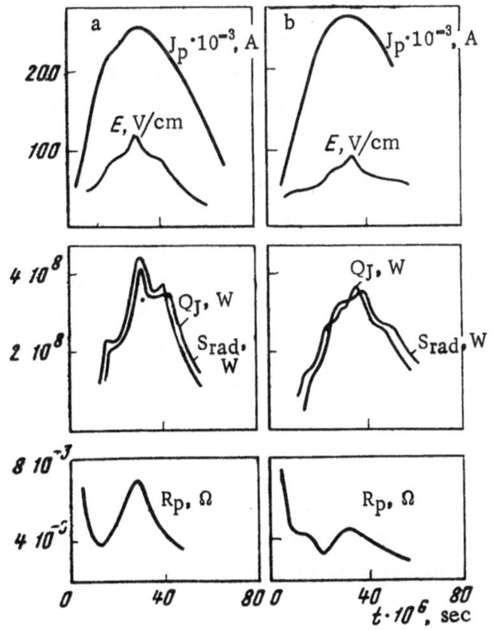

Fig. 7. Comparative results of calculations carried out using the radiative heat transfer approximation and transfer equation with an effective allowance for the line emission. Here, J_p is the current through the plasma; E is the electric field intensity; Q_J is the Joule heating power; S_{rad} is the radiation flux; R_p is the ohmic resistance of the plasma. a) Radiative heat transfer with $l_R = (1/20) l_{Rb}$ (l_{Rb} is the Rosseland radius for bremsstrahlung); b) transfer equation for five groups and a line ($\varkappa = 20\varkappa_b$, C = 1800 μF, $U_0 =$ 5 kV, $N = 5.6 \cdot 10^{19}$ cm^{-1}, $l = 15$ cm; the subscript "b" refers to bremsstrahlung).

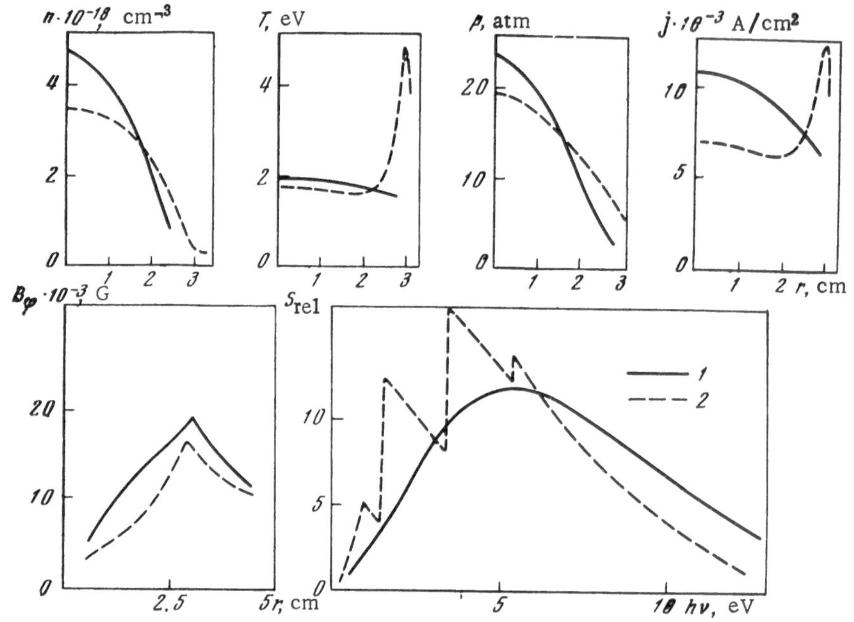

Fig. 8. Comparative results of calculations carried out using the radiative heat transfer approximation (1) and transfer equation with an effective allowance for the line emission (wire of $d_0 = 0.31$ mm diameter, $N = 5.6 \cdot 10^{19}$ cm^{-1}). Here, n is the plasma density, T is the plasma temperature, p is the plasma pressure, j is the current density, B_φ is the φ component of the magnetic field; S(hν) is the emission spectrum.

It is clear from Fig. 8 that a dense opaque plasma is surrounded by a high-temperature zone which is opaque to photons emitted as lines but is optically thin elsewhere in the spectrum. In this region, the energy is carried away from the surface by the line radiation. The distribution of the energy over the spectrum at a moment close to the current maximum, calculated allowing for the transfer processes, is shifted toward longer wavelengths, in accordance with the dependence of the emission coefficient of lithium on the photon energy.

The sensitivity of the results to the initial data, method of specification and variations of the physical constants (emissivity, conductivity), and to small perturbations is demonstrated in [33, 34]; some of the results given in these two papers will be discussed later in an analysis of the experimental data.

A strong dependence of the results on the method of description of radiative transfer in a plasma (see Figs. 5, 6 and 8) is due to the fact that the energy emitted is 10-100 times higher than the thermal energy of the plasma. This dependence (sensitivity) requires that, in the case of optically transparent discharges, the absorption coefficients of the radiation be specified in detail at different frequencies and this also applies to the line emission. Since calculations of equilibrium conditions in a plasma which is optically transparent in one part of the spectrum and opaque in another are not based on any specific emissivities of matter, they should be regarded only as estimates [36].

This discussion is based on plasma properties which are idealized to some extent. The question arises whether real discharges have these properties.

We may expect some discharge characteristics to be stable. For example, a quasisteady conversion of the electrical energy into radiation occurs always because this process is governed solely by the low specific heat of the plasma. A dynamic quasisteady state (confinement of a plasma by a magnetic field), which depends on the ratio of the magnetic and gas kinetic pressures, can also be regarded as one of the stable characteristics. Force instabilities may disturb the dynamic equilibrium but the above analysis shows that, under certain conditions and at certain times, these instabilities are not important.

The emission spectrum is sensitive to the presence of impurities. We shall now estimate the number of impurities which may halve the energy emitted in the soft part of the spectrum (ε < 5.4 eV). This occurs when the presence of impurities causes the emissivity in the hard part of the spectrum to rise twofold.

In the discharges under discussion, we may expect the presence of iron, oxygen, silicon, and carbon ions in the plasma. These impurities are not easily ionizable and, therefore, at temperatures of 3-4 eV, we may expect the degree of ionization to be $\alpha = 2$ and the excitation potentials of the impurities I_{imp} to be in the range 20-30 eV. Then, the expression for the emissivity ($4\pi \int \varkappa' I d\varepsilon$, where I is the equilibrium intensity of the radiation) yields the condition $n_{Li}^2 + 20\alpha^3 n_{imp} < 2n_{Li}^2$, which shows that the number of impurities should not exceed 5-10% of the host atoms. This estimate is based on the photorecombination contribution of impurities to the radiation. The line radiation may appear at much lower impurity concentrations but we can show that, since the total line width is small compared with the width of the whole spectrum, such line radiation does not alter significantly the emission spectrum of the discharge but plays an important role in the stability because of its strong temperature dependence.

Force instabilities (constrictions and flexural modes) still occur because they are not related to the specific properties of a plasma. However, the fundamental constriction mode, which may stop the current flow, is likely to be absent because optical transparency exists for any properties of the plasma at the periphery of a discharge. In this case, the distribution of the current density does not have a sharp boundary and, therefore, it is unlikely that the current will stop flowing.

The following comment should be made about the stability: if a high-current discharge plasma is to be used as a radiation source, the processes which alter its properties should be regarded as instabilities. These properties include the output power, emission spectrum, and geometric dimensions of the source. Force instabilities (particularly those with short wavelengths) do not alter these properties. Overheating instabilities may affect the spectrum and, therefore, they are undesirable. The presence and influence of the latter instabilities should be studied experimentally because a theoretical analysis is difficult due to the complex and, at present, insufficiently well known spectral dependence of the emissivity.

CHAPTER II

EXPERIMENTAL INVESTIGATIONS OF DISCHARGES IN LITHIUM PLASMAS

The experiments were carried out in a typical Z-pinch geometry. We employed the apparatus shown schematically in Fig. 9. A capacitor bank was discharged through a lithium wire placed in an evacuated chamber with quartz walls. The diameter of the chamber was 10 cm and the length of the discharge was varied from 15 to 100 cm. The capacitance of the bank was 1800-4200 μF and it was charged to 6 kV. A lithium (or indium) wire was extruded directly into the vacuum chamber and this was done using a special press. The wire thickness was governed by the diameter of an interchangeable die and in our experiments this diameter ranged from 10^{-2} to 0.1 cm.

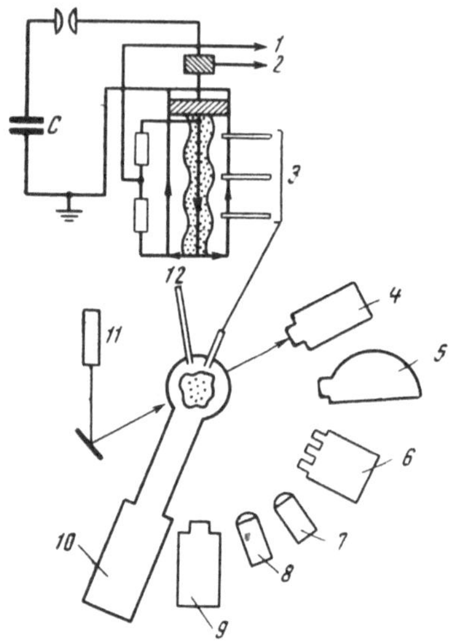

Fig. 9. Block diagram of the apparatus: 1) voltage divider; 2) Rogowski loop; 3) magnetic probes; 4) monochromator; 5) SFR-2M streak camera; 6) calibrating photocells; 7) calorimeter; 8) pyroelectric radiometer; 9) ISP-30 quartz spectrograph; 10) DFS-29 vacuum spectrograph; 11) He–Ne laser; 12) piezoelectric transducer.

We determined the electrical and energy characteristics of a discharge (current, voltage, energy balance) [10, 37, 38], studied the discharge dynamics and stability by high-speed photography [39], and measured the radiation flux and spectrum in a wide photon-energy range [10, 37, 38, 40] ($h\nu$ = 1.5–12 eV). Magnetic probes, piezoelectric transducers, and illumination of a plasma with laser radiation (transmission was measured) were used in a study of the large-scale structure of the discharge [41, 42].

§ 1. Energy Balance in Discharges

A standard method was employed in measurements of the electrical characteristics of a discharge (current and voltage): we used a Rogowski loop and divided the voltage using active resistors; the precision of these measurements was ~10%. A current pulse usually consisted of one or two half-periods of 70-100 μsec duration and it was strongly damped. The active component of the voltage across the chamber and the energy supplied to the discharge were deduced from the current and voltage oscillograms. The chamber inductance was determined from the current measurements by short-circuiting the electrodes with metal rods of different diameter.

Table 1 gives the results of measurements of the current, electric field, and ohmic resistance of different discharges. Figure 10 shows the time dependences of the ohmic resistance.

TABLE 1

Discharge	Time corresponding to current maximum, μsec	Maximum current, kA	Electric field, V/cm	Ohmic resistance of plasma near current maximum, 10^{-2} Ω
l=92 cm, V_0=6 kV, C=4200 μF	80	230	28	1.1–1.3
l=92 cm, V_0=5 kV, C=4200 μF	80	175	20	1.0–1.2
l=15 cm, V_0= 5 kV, C=1800 μF	35	220	70	0,5

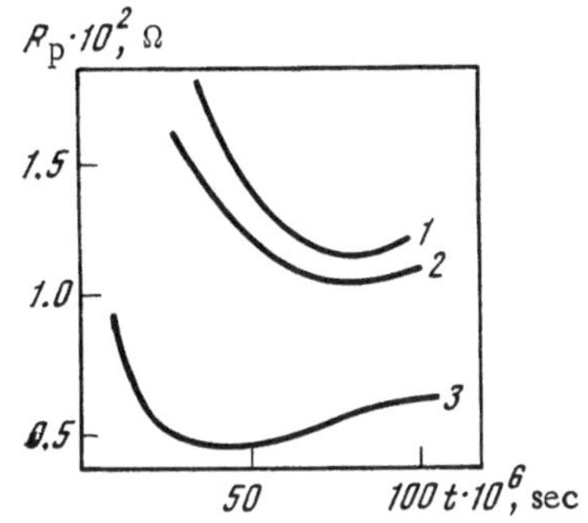

Fig. 10. Ohmic resistance of the plasma: 1) C = 4200 μF, U_0 = 5 kV, l = 92 cm; 2) C = 4200 μF, U_0 = 5 kV, l = 92 cm; 3) C = 1800 μF, U_0 = 5 kV, l = 15 cm.

The energy emitted as radiation was measured using several methods: the radiation energy in the transparency region of quartz was determined with a calorimeter and the radiation power with a pyroelectric radiometer [38]; in the region where quartz did not transmit we studied heating of the quartz bulb and the radiation fluxes in different parts of the spectrum were measured with calibrated photocells. The energy emitted by a discharge in the transparency range of quartz was determined using a cone calorimeter.

The apparatus employed enabled us to determine the radiation energy in the transparency range of quartz ($\lambda > 2200$ Å) with an accuracy of ~15%. The transparency edge was not known exactly since the transmission of quartz varied with the intensity of the radiation flux [43]. The energy deduced from the heating of a bulb was determined to within 15-20%. The time dependences of the radiometer signal, whose resolution was $5 \cdot 10^{-6}$ sec, and of the power supplied to a discharge were determined for l = 92 and 15 cm (Fig. 11). The results of measurements of the energy supplied to a discharge and emitted in the transparency range of quartz, and also of the energy absorbed by the quartz bulb, are given for various discharges in Table 2.

The measured radiation energy (two last columns of Table 2) was 60-75% of the energy supplied by the capacitor bank. The energy lost as a result of heating of the electrodes by the radiation, deduced from the difference between the temperatures before and after the discharge, ranged from 2 to 4 kJ. Thus, it was clear that all the energy supplied to the discharge was radiated by the plasma, which was to be expected because the thermal energy of the plasma was ~20 J/cm (for $n = 10^{19}$ cm^{-3}) and the magnetic field energy did not exceed 20-80 J/cm, so that the corresponding derivatives with respect to time ($< 8 \cdot 10^5$ W/sec) were almost an

Fig. 11. Discharge power Q: 1) Joule heating power for C = 4200 μF, U_0 = 5 kV, l = 92 cm; 2) total radiation flux measured in the transparency range of quartz by a pyroelectric radiometer (C = 4200 μF, U_0 = 5 kV, l = 92 cm); 3) Joule heating power (C = 1800 μF, U_0 = 5 kV, l = 15 cm).

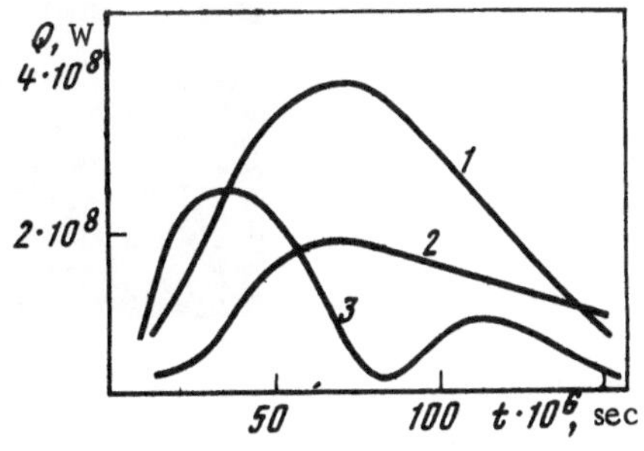

TABLE 2

Length of discharge, cm	Stored energy, kJ	Energy supplied to discharge, kJ	Energy in transmission band of quartz, kJ	Energy absorbed in quartz, kJ
92	76	60	28	28
92	52	39	19	18
15	22.5	17.5	6.7	6.6

order of magnitude lower than the Joule and radiation powers. Consequently, the experimental conditions were quasisteady: the Joule heating, equal to the radiation flux, was much greater than the time derivative of the magnetic and thermal energies. This condition could be violated only for wires of 0.1 cm diameter for which the thermal energy at a temperature of 1 eV reached 0.4 kJ/cm.

§ 2. Dynamics and Stability of Discharges

The space–time evolution of a discharge filament was investigated using a high-speed streak camera of the SFR-2M type. This camera was operated in the frame-by-frame mode

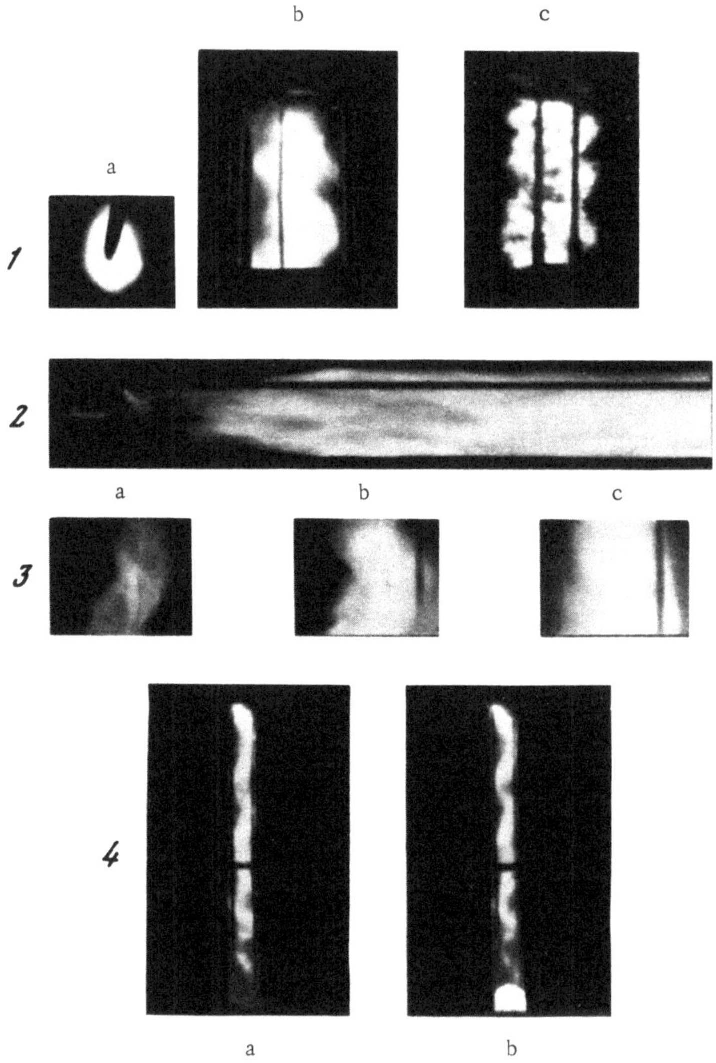

Fig. 12. Lithium discharge: 1) frame-by-frame photographs of a 15-cm-long discharge (frame exposure 1 μsec) (a – view from above, b – view from side for d_0 = 0.17 mm, c – view from side for d_0 = 1.0 mm); 2) continuous scan of a transverse section of the discharge (scanning rate 3.3 cm/μsec); 3) photographs obtained using a Kerr cell (frame exposure 20 nsec) (a – t = 12 μsec, b – 17.5 μsec, c – 25 μsec); 4) frame-by-frame photographs of a 92-cm-long discharge (d_0 = 0.17 mm) (a – 15 μsec, b – 30 μsec).

at a rate of 10^6 frames/sec as well as continuously (scanning rate 3.3 cm/sec); at certain moments we took photographs using a Kerr switch and the exposure time of these photographs was ~2 · 10^{-8} sec.

In some experiments we placed two filters with transmission bands 3700-4500 Å ("blue") and 6200-7600 A ("red") in front of the objective of the camera; these filters were located in such a way that the two upper rows of the pattern produced by the camera were illuminated through the blue filter and the two lower rows through the red filter. The results of high-speed photography of discharges l = 15 and 92 cm long are shown in Fig. 12.

It is clear from these photographs that a discharge filament with indistinct boundaries formed at the center of the chamber; in the case of a discharge l = 15 cm long, this filament expanded at a rate of 1.3 km/sec. The relatively low rate of expansion (compared with the velocity of sound $v_s \approx 10$ km/sec) of the discharge filament was due to the proximity of the gaskinetic and magnetic pressures, i.e., throughout almost the whole of the first half-period of the current pulse the discharge was compressed by its own magnetic and it did not reach the chamber walls. The discharge channel was twisted and bent into a complex structure; a "cloud" structure and jets were clearly visible inside the discharge; bright regions alternated with dark regions. The overall discharge pattern resembled a turbulent jet. The structure in question (for wires of 0.01 cm diameter) indicated that the optical transparency of the discharge was high and the complex structure of the filament was due to some inhomogeneities induced during the initial stage of the development of the discharge and due to instabilities growing in the plasma. By the time the current maximum was reached, the structure became smoother and the discharge more homogeneous. The average rate of expansion of the discharge channel was reproducible from one experiment to another. Figure 13 shows the positions of the boundary in a discharge at various moments. The boundary obtained using the blue filter represented deeper layers of the plasma and it expanded at a lower velocity.

Photographs obtained during the initial stage of a discharge revealed typical constriction and flexural instabilities of the optically opaque discharge (corresponding to the initial wire diameter $d_0 \geq 0.3$ mm) and flexural instabilities of the optically transparent discharge ($d_0 < 0.3$ mm). The experimental values of the growth increment deduced from these photographs, $\gamma \approx (2\text{-}5) \cdot 10^5$ sec^{-1}, were in good agreement with theoretical estimates (Chap. I, §2).

It was more difficult to identify the overheating instability in the photographs because the individual pulsations of the radiation of 1-5 μsec duration disappeared against a background of the whole discharge.

Experiments carried out on a longer discharge (l = 92 cm) revealed long-lived regions of different brightness (Fig. 12), but by the time the current maximum was reached, the discharge became fairly homogeneous along its length. This smoothing out by the time the maximum was reached and the subsequent retention of a homogeneous structure, revealed by the streak patterns, were typical of all the investigated discharges.

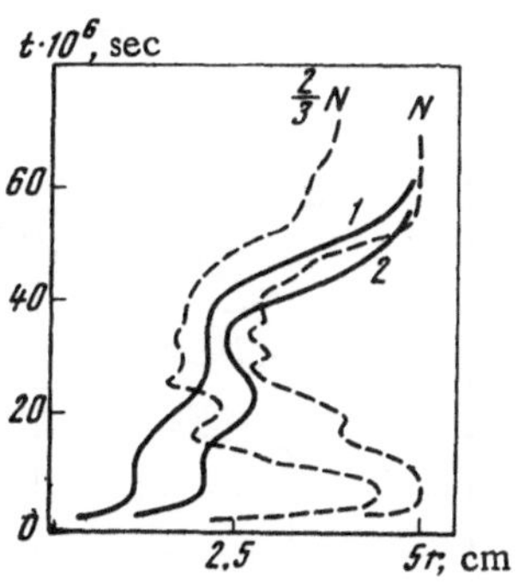

Fig. 13. Visible radius of a discharge filament (l = 15 cm, d_0 = 0.37 mm): 1) recorded in the spectral range λ = 3700-4500 Å; 2) recorded in the spectral range λ = 6200-7600 Å. The dashed curves are the calculated trajectories of various Lagrange points.

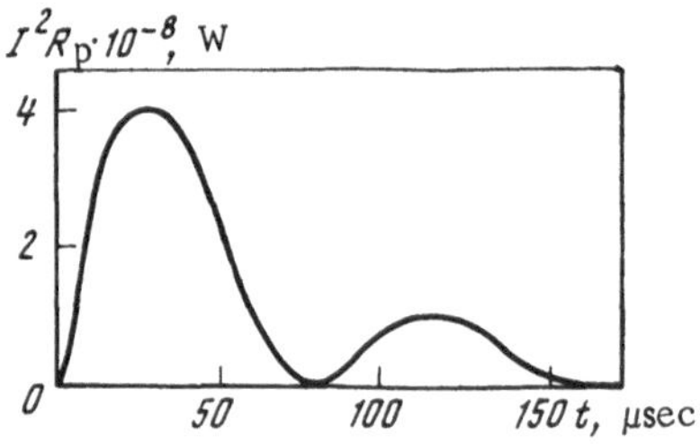

Fig. 14. Time dependence of the Joule heating power (l = 15 cm).

§ 3. Measurements of Radiation Fluxes and Spectra

The flux and spectrum of the radiation emitted by a discharge were determined using photocells calibrated in different spectral intervals. We also carried out a photometric analysis of spectrograms obtained with an ISP-30 quartz spectrograph capable of providing a time scan (this was done in the range 2.0 eV < $h\nu$ < 5.7 eV) and of time-integrated spectrograms obtained with a DFS-29 vacuum spectrograph (in the range 3 eV < $h\nu$ < 12 eV). Photocells in combination with various filters covered the spectral range from 1.2 to 5.5 eV (FÉK-15 photocell and KS filter, F1 photocell and ZhS4 filter, F1 photocell and UFS2 filter, F7 photocell and UFS1 filter). A standard light source of the ÉV-45 type was used in the absolute calibration of the photocells and of the ISP-30 spectrograph.

The time dependences of the radiation intensity in the central zone agreed with the time dependences of the Joule heating power (Fig. 14) and some of the intensity peaks during the initial stage of the discharge corresponding to the regions of localization of plasma bunches. The amplitude of irregular intensity fluctuations was about 40% of the average value.

During the initial stage of the discharge (5-10 μsec), corresponding to the breakdown on the wire surface, we observed separate bright broadened C and H lines as well as a weak continuous spectrum. During this stage the plasma temperature was estimated to be 30,000-40,000°K but the plasma density was low. The spectral composition of the emitted radiation indicated that the breakdown occurred at impurities on the wire surface. At the current maximum the emitted radiation was basically continuous with superimposed lines due to impurities picked up from the chamber walls (Si and O). During the later stages ($t \geq 70$ μsec) the importance of the line emission increased and the observed lines were due to impurities picked up from the electrodes and chamber walls (Fe, Si, O). The lithium atomic lines were hardly ever observed in the discharge spectra, which indicated that the lithium plasma was practically completely ionized. An analysis of the spectra indicated that the lithium plasma was surrounded by a low-density layer of ionized silicon and a thin layer of silicon vapor, manifested by the absorption lines, was present near the chamber walls.

The spectral distribution of the radiation flux emitted by a plasma bunch S($h\nu$), calculated per unit area, was determined by photometric analysis of the ISP-30 spectrograms (Fig. 15). In the photon energy range $h\nu$ < 2.7 eV the spectrum was identical with the emission

Fig. 15. Spectral distributions of the radiation flux (obtained by photometry) from a unit surface area of plasma bunches generated by exploding wires of different diameter d_0 (mm): 1) 0.1; 2) 0.37.

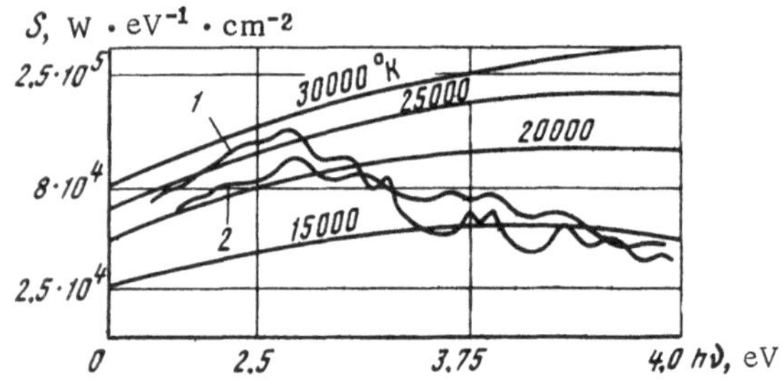

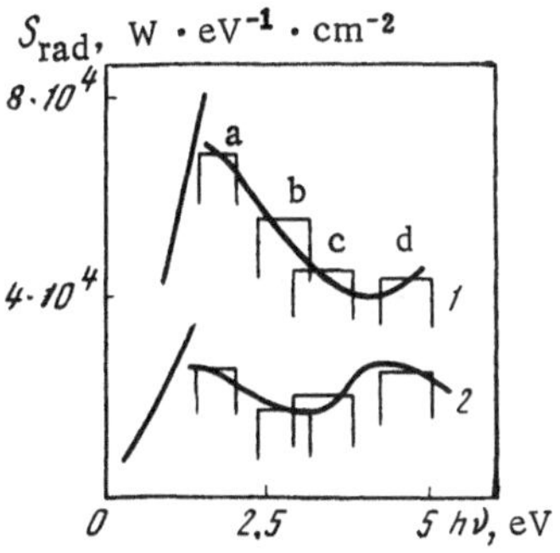

Fig. 16. Spectral distributions of the radiation flux from a unit surface of a discharge, recorded at the current maximum using calibrated photocells: 1) l = 15 cm, d_0 = 0.17 mm; 2) l = 92 cm, d_0 = 0.2 mm; a) 1.4-2 eV; b) 2.25-3.1 eV; c) 2.8-3.8 eV; d) 4.4-5.1 eV.

curve of a black body at a temperature of 25,000-30,000°K. In the range $h\nu > 2.7$ eV the brightness temperature of the discharge fell with increasing photon energy. The rate of this fall depended on the plasma density; a more rapid fall was observed for wires with an initially smaller diameter. The difference between the radiation emitted at a given frequency and the black-body radiation was used to estimate the optical thickness of the plasma. The same thickness was also estimated from the transmission of its own radiation by the plasma. These two methods gave similar results: for $h\nu = 2.5$ eV, we found that $\varkappa l \approx 1.5$, whereas for $h\nu = 4$ eV, we obtained $\varkappa l \approx 0.3$ (for a wire of 0.3 mm diameter).

The time dependences of the total radiation flux measured with photocells were reported in [37]; oscillograms of the signals produced by the F7 photocell were of the type shown in Fig. 25. They corresponded to the Joule heating power (Fig. 17). The spectral distribution of the radiation emitted by discharges 15 and 92 cm long (for U_0 = 5 kV) were of the type shown in Fig. 16, illustrating a typical selectivity of the radiation emitted by a lithium plasma and associated with a reduction in the optical thickness as shorter wavelengths.

In the vacuum part of the spectrum [40] the measurements were carried out in a special chamber connected by a tube to the DFS-29 spectrograph. The spectra were recorded on a photographic film sensitized with sodium salicylate [44]. The relative distribution of the radiation energy was determined in the range 4-12 eV. The influence of the film sensitivity in the hard-photon range was eliminated by depositing sodium salicylate on the substrate side and it was this side that faced the radiation emerging from the diffraction grating of the spectrograph. In the energy range 2.8-4.1 eV the spectrum was recorded simultaneously by the quartz spectrograph in the usual way and by the vacuum spectrograph on a photographic film with the emulsion facing the diffraction grating (in this case sodium salicylate was not used). The calibration of this part of the spectrum was carried out using the ÉV-45 standard light source.

The absolute calibration was carried out in two ways: 1) using the total radiation energy which was equated to the energy supplied to the discharge; 2) using the radiation energy emitted in the range 4.4-5 eV, where the total flux was measured with a calibrated photocell; the dependence of the reflectivity of the diffraction grating on the photon energy was allowed for in accordance with [44]. The results were found to be similar (Fig. 17). The precision of the

Fig. 17. Spectral distribution of the radiation energy emitted by a discharge and recorded with a vacuum spectrograph.

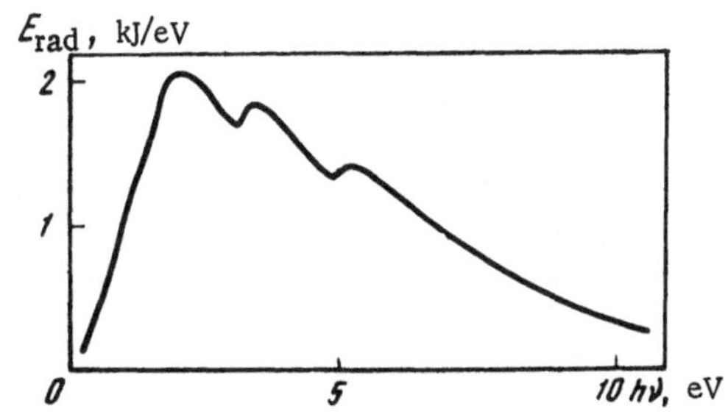

measurements in the vacuum part of the spectrum was low because of the many corrections needed and the approximate nature of the calibration method.

The energy distribution had two maxima in the transparency range of the plasma (3.5 and 5.5 eV). These maxima could be attributed to the photorecombination of electrons in the 2p and the ground 2s states of the lithium atom. The energy of the emitted radiation was approximately the same between 0 and 4 eV and between 4 and 10 eV, which was in agreement with the measurements of the radiation energy in the ranges of transparency and opaqueness of quartz described above (calorimetric and bulb heating methods).

In attempts to calculate the emission spectrum of a lithium plasma we encountered the difficulty that the measured optical thickness was several times greater than that predicted on the basis of measurements of the discharge diameter, electron density, and concentration of neutral atoms in the low-energy states (calculated using the Saha and Boltzmann formulas) and on the basis of the calculated photoabsorption cross sections.

The measured values of the optical thickness and radiation flux could be explained by assuming that the emissivity of the plasma in our experiments was approximately 5 times greater than the calculated value. It should be noted that the absorption of He−Ne laser radiation in a lithium plasma was determined under similar conditions in [45]. It was also mentioned there that the absorption was strong and that the presence of a considerable number of neutral atoms had to be postulated in order to explain this absorption.

The contribution of the line emission to the total radiation energy was small. For example, in the case of discharges with low plasma densities (wires with an initial diameter $d_0 < 0.3$ mm) we found that in the spectral range covered by the ISP-30 spectrograph the line emission represented ~15% of the total energy. This result was obtained by photometry of the various parts of the spectrum [37]. The contribution of the line emission was somewhat higher (~20-30%) in the vacuum region.

§ 4. Investigations of Large-Scale Structure of Discharges

An investigation was made of the large-scale structure of a discharge 15 cm long supplied with an energy of 17.5 kJ: this investigation involved the determination of the axial symmetry and longitudinal homogeneity, as well as of the radial distributions of the pressure, temperature, and density. We used magnetic probes to measure the local magnetic field and piezoelectric transducers to measure the pressure [41]; we also illuminated the plasma with He−Ne laser radiation [42].

Measurements of the distribution of the magnetic field along the radius and length of the chamber indicated that by the time the current maximum was reached ($t = 35$ μsec), the radial and axial components of the magnetic field B_r and B_z vanished, rapid oscillations of the depen-

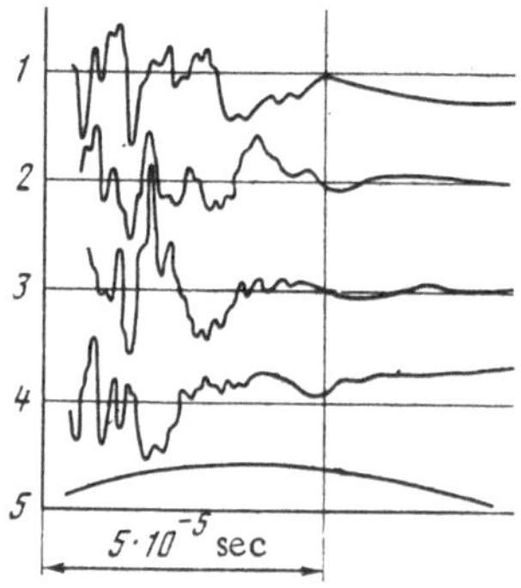

Fig. 18. Oscillograms of signals produced by magnetic probes: 1, 4) dB_φ/dt; 2) dB_r/dt; 3) dB_z/dt; 5) total current.

Fig. 19. Distributions of the magnetic field in a discharge chamber (l = 15 cm, d_0 = 0.17 mm): 1) 35 μsec; 2) 53 μsec; 3) 70 μsec.

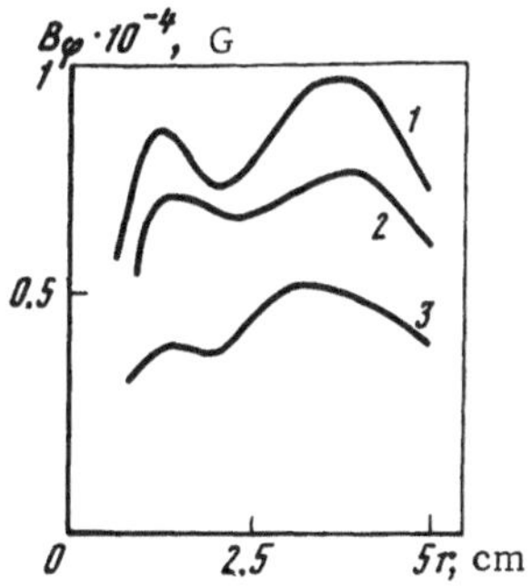

dence $B_\varphi(t)$ were damped out, and this dependence began to repeat the shape of the total current pulse. Oscillograms of B'(t) were of the type shown in Fig. 18. After passing through the current maximum the results obtained for a given radius but different lengths were identical, i.e., the discharge assumed an axial symmetry. The radial distribution of the magnetic field was determined at different moments (Fig. 19). These results were analyzed as follows. The Maxwell and motion equations for an axially symmetric quasisteady plasma were used to determine the distribution of the current density and pressure along the radius r (R is the chamber radius):

$$j_z(r) = \frac{c}{4\pi r}\frac{\partial}{\partial r}(rB_\varphi), \tag{41}$$

$$p(r) = p(R) + \frac{B_\varphi^2(R)}{8\pi} - \frac{B_\varphi^2(r)}{8\pi} + \frac{1}{4\pi}\int_r^R \frac{B_\varphi^2}{r}\,dr; \tag{42}$$

the distribution p(r) was measured with a piezoelectric transducer. These quantities, equation of state $p(n, T) = (1 + \alpha)nT$, and calculations of the plasma conductivity $\sigma(n, T)$ [5] were used in the determination of the radial distributions of the plasma temperature and density:

$$\sigma(n, T) = \frac{j(r)}{E(r)}. \tag{43}$$

Here, $E(r) = E_z(R) - \frac{1}{c}\frac{\partial}{\partial t}\int_r^R B_\varphi dr$ is the electric field intensity; the eddy component of E(r) does not exceed 15%; $E_z(R) \approx 70$ V/cm for t = 35 μsec; α is the degree of ionization of lithium.

The radial distributions j(r), p(r), n(r), and T(r) obtained for different initial diameters of the wire and at different moments are plotted in Figs. 20 and 21. The maximum temperature of the discharge occurred in the central part when the thinner wires were exploded (this temperature was ~50,000°K). For the sake of comparison, measurements were made also in an indium plasma and in this case the temperature distribution exhibited a typical (for an opaque plasma) dependence of the type shown in Fig. 8. The spectrum of the radiation emitted by an indium discharge was the same as that for a black body. The pressure at the center of a lithium discharge was 10-15 atm, whereas the pressure at the chamber walls was much smaller (0.5-2 atm); clearly, the discharge was compressed by the magnetic field. The total number of particles in the discharge (determined to within a factor of 1.5-2) agreed with the initial number of particles in the wire and it decreased with time because particles escaped to the cold region near the walls.

We also studied the large-scale structure by illuminating a plasma with laser radiation. A laser beam passed through the plasma along a chord located at a distance r from the center (typical oscillograms are shown in Fig. 22). In each experiment we determined the optical thickness of the plasma at just one distance but since the results were reproducible from one

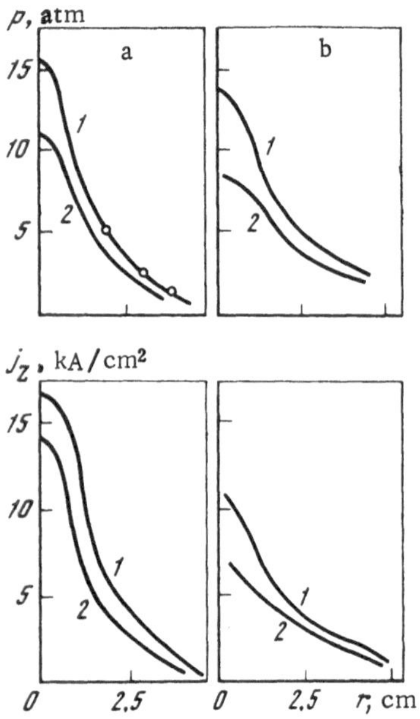

Fig. 20. Dependences p(r) and j_z(r) in a discharge at a different moments: 1) 35 μsec; 2) 53 μsec; a) d_0 = 0.17 mm; b) d_0 = 0.31 mm. The circles denote pressure measurements made with a piezoelectric transducer.

experiment to another, we could use such results to plot the distribution of the optical thickness across the discharge diameter. Measurements along two mutually perpendicular directions confirmed the axial symmetry of the discharge.

The distribution of the optical thickness along the discharge diameter was of the type shown in Fig. 23. The quantity $\varkappa l$, related to the absorption coefficient $\varkappa(r)$ by the integral

$$\bar{\varkappa}l = 2\int_r^R \frac{\varkappa(x)\,x\,dx}{\sqrt{r^2 - x^2}}, \tag{44}$$

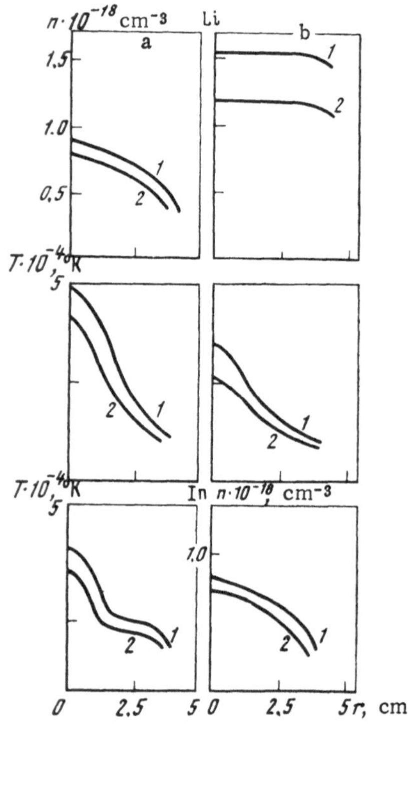

Fig. 21. Dependences n(r) and T(r) for Li and In plasmas (d_0 = 0.17 mm) obtained at different moments: 1) 35 μsec; 2) 53 μsec; a) d_0 = 0.17 mm; b) 0.31 mm.

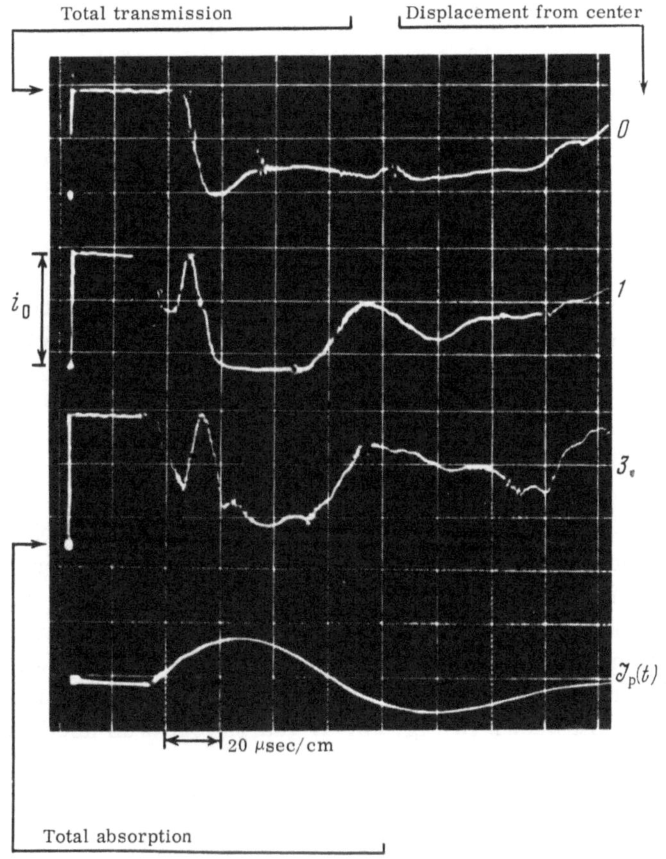

Fig. 22. Oscillograms produced by a photodetector recording laser radiation transmitted through a discharge plasma (J_{max} = 220 kA).

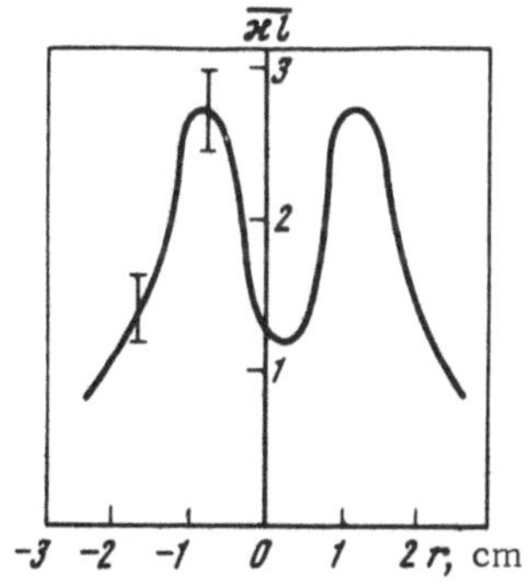

Fig. 23. Dependence of the optical thickness $\overline{\varkappa l} = f(r)$ of a Li plasma on the distance from the center of the chamber at the moment corresponding to the current maximum (d_0 = 0.17 mm).

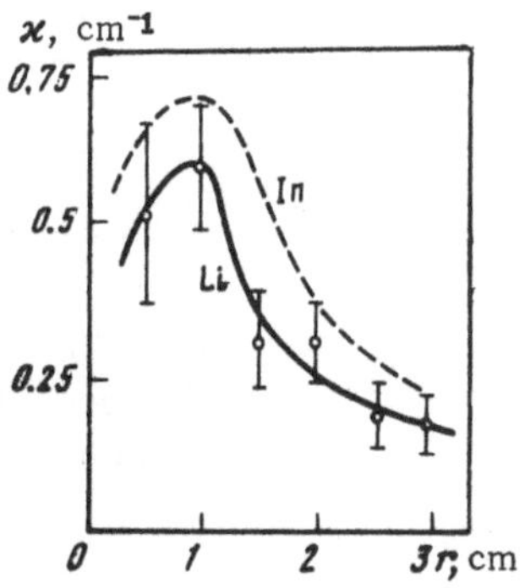

Fig. 24. Radial distributions of the absorption coefficient of the radiation emitted by indium and lithium plasmas.

was subjected to an inverse transformation by the Abel method [46] on the assumption that the symmetry was axial and this yielded the radial distribution of the absorption coefficient of the plasma at the laser radiation frequency (λ = 6138 Å, $h\nu \approx 2$ eV), as shown in Fig. 24. A reduction in the absorption coefficient was observed at the center of the discharge and this reduction corresponded to the temperature maximum deduced from the conductivity. The optical thickness and the absorption coefficient of a lithium plasma deduced from the laser radiation absorption were in agreement with those found from measurements of the radiation emitted by the plasma.

The measured radial distributions n(r) and T(r) of Fig. 21 were used to calculate the absorption coefficients in [11] and the values of these coefficients were found to be 5 times smaller than those determined experimentally (Fig. 24).

§ 5. Investigations of Small-Scale Structure of Discharges

Measurements of the magnetic field (Fig. 19) and high-speed photography of discharges (Fig. 12) indicated that during the first quarter of the discharge-current period (corresponding to 0-35 μsec) the discharge did not have a regular structure and the plasma exhibited inhomogeneities of the structure and emission, whose positions and size varied with time [47]. We carried out an additional experiment which confirmed the presence of small-scale inhomogeneities in the plasma parameters [48]. The experimental arrangement used and typical oscillograms obtained are shown in Fig. 25. A quartz lens was used to project a cross section

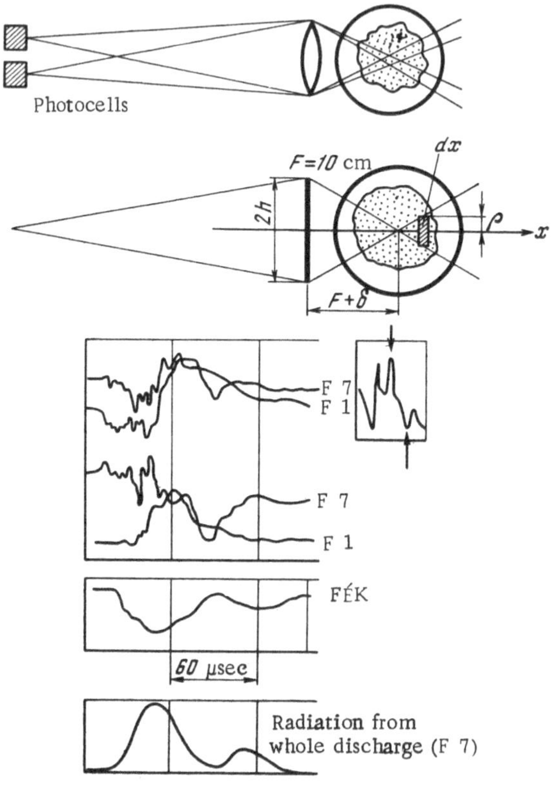

Fig. 25. Experimental arrangement used to observe radiation fluctuations and oscillograms produced by different photodetectors.

of a plasma filament on a plane in which there were cathodes of three pairs of photocells which recorded the radiation in different spectral intervals. The photocell signal obtained for the optically transparent plasma represented the radiation emitted from the interior within cones identified in Fig. 25. Two identical photocells were focused inside the plasma at points separated by 0-3 cm and in recording the signal of one photocell was subtracted from the signal of the other. We could easily show that the plasma layers x and x + dx lying within the region of the sharp focusing of the cells made the same contributions to the radiation. However, when the plasma emitted inhomogeneously, the energies reaching the photocells should depend strongly on whether the inhomogeneity was close to the region of sharp focusing or outside this region.

Since the signals produced by identical photocells were subtracted, the resultant signal differed significantly from zero only when an inhomogeneity was located inside the sharp-focusing region. In particular, when an inhomogeneity moved across the sharp-focusing regions of both photocells, an oscillogram exhibited characteristic peaks of opposite polarity. A signal of this type is shown in Fig. 25. The time τ separating such peaks enabled us to determine the velocity of an inhomogeneity $v = l/\tau$, where l was the known distance between the sharp-focusing regions. The duration of each peak $\Delta\tau$ gave the size of the inhomogeneity: $\Delta x = \Delta\tau l/\tau$, These relationships were valid for $\Delta x < l$, whereas for $\Delta x > l$, we should use $\tau = \Delta x/v$ and $\Delta\tau = l/v$. An analysis of the oscillograms obtained for different values of l indicated that the dimensions of all the inhomogeneities were below 1.0 cm and the former case applied for $l > 1.0$ cm. We employed calibrated photocells and this enabled us to use the amplitude of the signals in estimating the emissivity of the plasma and, finally, the scale of fluctuations of the plasma density.

Table 3 gives the calculated emissivity of a lithium plasma as a function of the temperature and density [35]. The emissivity ε_ν was obtained in a photon energy interval $\Delta h\nu = 1$ eV at $h\nu = 4.8$ eV. It is clear from Table 3 that in the temperature range ~3.16 eV the emissivity depended quite weakly on the temperature and it varied approximately quadratically with the density. Thus, the amplitude of the signal $\mathcal{J}$ was related to n^2 or, more exactly, to optical thickness of the inhomogeneity $\mathcal{J} \approx \int n^2 dx$.

Typical dimensions of the inhomogeneities, their velocities, and amplitudes of the density fluctuations calculated from the oscillograms (Fig. 25) were as follows:

Size of inhomogeneity	0.1—0.6 cm,
Velocity	1.5—10 km/sec,
Density	$2\cdot10^{17}$—10^{18} cm^{-3}.

Beyond the current maximum the plasma became quiescent: oscillograms assumed a smooth shape and the same results were obtained using magnetic probes (Fig. 19).

We shall now consider the results obtained with magnetic probes. Oscillograms produced by these probes showed random pulsations of different amplitudes and periods, and we observed all the components of the magnetic field H_φ, H_z, and H_r. Since the magnetic fields in

TABLE 3. Emissivity of Lithium Plasma ε_ν ($ergs \cdot eV^{-1} \cdot cm^{-3} \cdot sec^{-1}$)

n, cm^{-3}	T, eV		
	1.77	3,16	5,6
10^{17}	$1,05\cdot10^{9}$	$8.3\cdot10^{8}$	$2.2\cdot10^{9}$
10^{18}	$1.1\cdot10^{10}$	$7.9\cdot10^{10}$	$1.05\cdot10^{11}$
10^{9}	$6.2\cdot10^{12}$	$6.8\cdot10^{12}$	$7\cdot10^{12}$

a plasma were related to the hydrodynamic quantities p, v and T, these pulsations of the magnetic field corresponded to fluctuations of the hydrodynamic quantities and the latter were manifested by fluctuations of the radiation emitted from small volumes of a plasma. Beyond the current maximum the plasma became quiescent and then the hydrodynamic quantities at different space–time points were related by a fully determinate functional dependence, considered in the preceding sections.

The initial stage of the discharge, when physical quantities fluctuated, was turbulent and it was dealt with by statistical methods of the kind used in investigations of low-density plasmas [49]. In practical calculations one should replace the "averages over an ensemble" with averages during one realization time. In the case of a transient process this replacement was valid if

1) the averaging interval t was sufficiently long compared with the correlation time t_{corr};

2) the averaging interval was short compared with the characteristic time of changes in the external conditions t_{av} (in our case these were the current and voltage), so that the average characteristics of the plasma such as $\bar{n}$ and $\bar{T}$ did not change greatly during this time [50].

The averaging interval t = 15 μsec satisfied these two conflicting requirements because $t_{corr} \approx$ 3-5 μsec and $t_{av} \approx$ 150 μsec. Moreover, one should satisfy also the condition of turbulence homogeneity. This implied that the dimensions of the inhomogeneity should be considerably smaller than the characteristic dimensions of the discharge (D = 10 cm).

In the analysis of the experimental results the averaging time t was selected approximately in the middle of the first quarter of the discharge period. When t was changed by ±5 μsec, the autocorrelation coefficient was practically unaffected, which indicated that the interval t was selected reasonably. The correlation functions for identical experiments (identical initial wire diameter, same energy, same wire material), i.e., for different realizations of a random process, were identical. Hence, we found that not only the average discharge parameters ($\mathcal{J}_p$, V, $\bar{T}$, $\bar{n}$) were reproduced from one experiment to another, but also the statistical properties of fluctuations of the initial phase of the discharge as well as the nature of the turbulent motion remained constant.

Miniature magnetic probes were used to measure local magnetic fields (or their time derivatives); a typical oscillogram is shown in Fig. 18. Various components of the fields could be determined at four different points in a discharge. The space and time resolutions of these magnetic probes and the recording apparatus were 0.3 cm and 10^{-7} sec, respectively. The experimental data were tabulated in steps governed by the fluctuation frequency and by the pass band of the apparatus—the step was (2-3) · 10^{-7} sec—and were introduced into a computer together with suitable calculation programs. The results of a correlation analysis are plotted in Figs. 26-28.

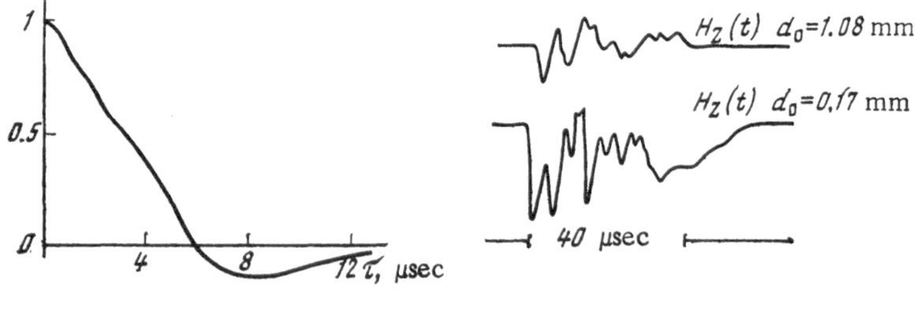

Fig. 26. Autocorrelation coefficient $R_{xx}(\tau)$ and oscillograms of $H_z(t)$ ($x = dH_\varphi/dt$).

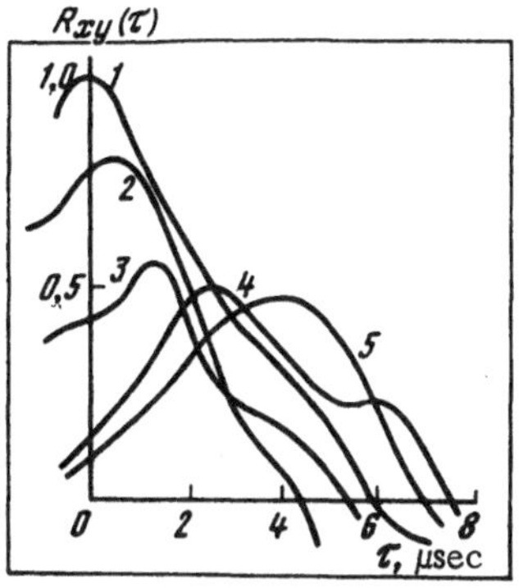

Fig. 27. Mutual correlation coefficient for different distances between probes: 1) r = 0; 2) 0.5 cm; 3) 1.5 cm; 4) 3.0 cm; 5) 3.5 cm; $x = dH_{\varphi}(r_1)/dt$, $y = dH_{\varphi}(r_2)/dt$, $r = |r_1 - r_2|$.

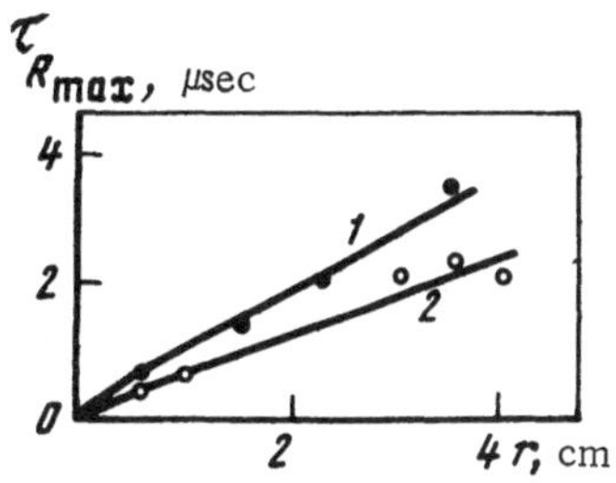

Fig. 28. Dependence $\tau_{Rmax} = f(r)$: 1) d_0 = 0.31 mm; 2) d_0 = 0.17 mm.

The autocorrelation coefficient of the derivative of the azimuthal component of the field $\partial H_{\varphi}/\partial t$ for wires of 0.17 mm initial diameter (Fig. 26) allowed us to determine the correlation time t_{corr} = 3 μsec. The correlation functions were similar for the axial and radial components of the magnetic field. It was interesting to note that although the average values of $\bar{H}_r$ and $\bar{H}_z$ were close to zero, the average values of the squares of the amplitudes of the different components of the field were approximately equal, $\bar{H}_r^2 \approx \bar{H}_z^2 \approx \bar{H}_{\varphi}^2$, and the fields H_r and H_z decayed in 40 μsec. Oscillograms of $H_z(t)$ for wires of d_0 = 1.08 and 0.17 mm initial diameter are also included in Fig. 26. It is clear from these oscillograms that the pulsation amplitude increased and the pulsation spectrum became wider when the number of particles in a discharge was increased.

A family of the mutual correlation functions of signals of two magnetic probes separated by different distances (Fig. 27) made it possible to determine the velocity of inhomogeneities in a plasma. The strongest time correlation between two probes corresponded to the time needed for an inhomogeneity to travel from one probe to the other. We could thus find the time τ_{Rmax} when the correlation function reached maximum and this could be done as a function of the distance between the probes. The function was linear (Fig. 28) and differentiation of this function yielded the velocity of an inhomogeneity $v = r/\tau_{Rmax}$, which was ~10 km/sec for a wire of d_0 = 0.17 mm diameter and ~7 km/sec for a wire of d_0 = 0.31 mm diameter. It was interesting to note that these velocities were equal to the velocity of isothermal sound in a plasma, calculated using the average parameters of the plasma.

The spatial correlation coefficient $R_{xy}(r)|_{\tau=0}$ (x and y denote magnetic fields measured at different points in space at the same time τ = 0) is plotted in Fig. 29. When the distance

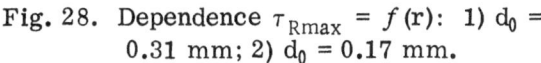

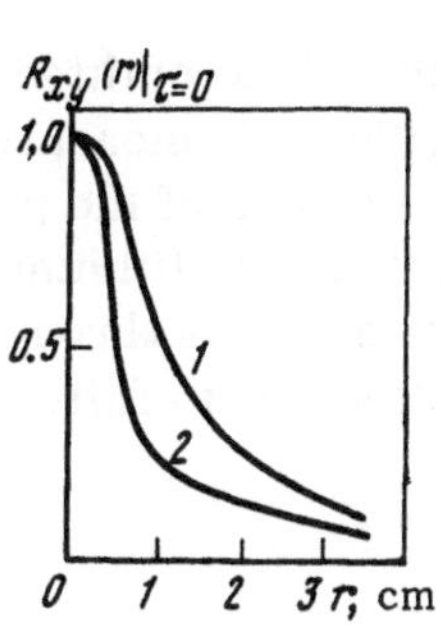

Fig. 29. Mutual correlation coefficient $R_{xy}(r)|_{\tau=0}$: 1) d_0 = 0.31 mm; 2) d_0 = 0.17 mm.

between the probes was increased, the correlation coefficient decreased. The size of plasma inhomogeneities could be defined as the distance at which there was still a significant correlation between the values of the field at two points [50]

$$L = \int_0^\infty R_{xy}(r)\,|_{\tau=0}\, dr. \tag{45}$$

In the case of a plasma formed by exploding a wire of $d_0 = 0.17$ mm diameter, we found that L = 0.9 cm, whereas for $d_0 = 0.31$ mm the corresponding size was L = 1.3 cm (Fig. 29). When the wire mass and, consequently, the optical thickness of the discharge were increased, the dimensions of inhomogeneities became greater and larger volumes of a plasma were linked by a functional dependence so that turbulence became less marked. The suppression of turbulence in the explosions of "thick" wires could be related to the weakening of the overheating instability mechanism, which developed in an optically transparent plasma but not in an optically dense one.

The dimensions and velocities of inhomogeneities estimated by optical methods (0.1-0.6 cm and 1.5-10 km/sec) were in agreement with the results of a correlation analysis of the magnetic measurements (0.9 cm and 10 km/sec).

§ 6. Discussion of Results and Comparison with Theory

Experiments revealed clearly a discharge stage which was quasisteady in the energy sense (all the injected energy was emitted as radiation) and in the dynamic sense (the gasdynamic pressure was balanced by the magnetic forces, the boundaries were at rest, and the profiles of plasma properties did not vary greatly with time). This conclusion was confirmed by various independent measurements. The quasisteady stage was also found in numerical calculations concerned with such discharges (Chap. I) and its characteristics (duration, typical discharge size, current, resistance, plasma temperature, and magnetic field) predicted by these calculations were close to the experimental values. However, this agreement could be obtained only by increasing the emissivity of a lithium plasma compared with that predicted by calculations.

High-speed photography of discharges made it possible to observe and follow (i.e., to determine growth increments) the development of various types of force instabilities. The theoretical representations (Chap. I) were again confirmed experimentally. It was interesting to note that the developed instabilities ceased to vary and retained their shape and dimensions for a time considerably longer than $1/\gamma$, where γ was the corresponding increment. A full explanation of this phenomenon was not available but we could assume that under real conditions the distribution of the current at the periphery of a discharge differed considerably from that predicted theoretically and it fell more rapidly away from the discharge center. In principle, this difference could stabilize the developed force instabilities in a discharge. The discharge expanding slowly after this "confinement" stage was found to be stable so that the plasma configuration could be retained for a long time.

The selective nature of the emitted radiation was revealed by the results of a photometric analysis of the spectra (Fig. 14), measurements with calibrated photocells (Fig. 15), and determination of the optical thickness of the plasma in different parts of the spectrum: these results indicated that an optically opaque lithium plasma of ~2.5 eV temperature and a corresponding energy flux emitted 5.5 times less throughout the spectrum than a black body. The experimental results could be described quite fully by the theoretical ideas developed in [9, 19].

Measurements of the magnetic field distribution in a discharge plasma (Fig. 19) yielded the pressure and current density distributions with maxima on the axis of the system (Figs. 20 and 21). These results excluded reliably the distributions predicted by calculations for a developed overheating instability and characterized by a high-temperature zone on the discharge periphery (Fig. 6). In these calculations the ohmic resistance of the plasma and the total radiation power were approximately three times smaller than the experimental values and the emission spectrum was found to be much "harder" than that actually observed.

The dependences p(r) and j(r) yielded information on the plasma emissivity. In the case of a steady-state discharge and an optically thin plasma, when the dominant process was the bremsstrahlung, $q = Qq_0n^2(T)^{1/2}$ and $\sigma = \sigma_0 T^{3/2}$, the following relationships were obeyed:

$$p(r) = B_0 j(r), \quad Q = \frac{q}{q_{\text{brems}}} = \frac{p_0^2 j^2(r)}{\sigma_0 q_0 p^2(r)}. \tag{46}$$

It was readily found that the ratio of the experimentally determined quantities was $j(r)/p(r) \approx 1$ (for a wire of 0.17 mm diameter) and, consequently, the corresponding coefficient was Q = 16, i.e., the value of Q was approximately 4-5 times greater than the calculated value q/q_{brems} (for temperatures of 1.7-4 eV and concentrations $n = 10^{17}$–10^{18} cm^{-3}, the calculated value was $Q_{\text{calc}} \approx$ 3-4 [35]). The dependences n(r) and T(r) of Fig 21 could be used, employing the expression for the inverse bremsstrahlung absorption coefficient, to calculate $\varkappa(r)$ at the He–Ne laser frequency $h\nu \approx 2$ eV. Such a calculations has demonstrated that $\varkappa(r)$ had a minimum on a discharge axis and a maximum at r = 1 cm, and it fell toward the discharge periphery. This dependence was in agreement with the results obtained by illumination of a lithium plasma with He–Ne laser radiation. The absolute values of $Q\varkappa_{\text{brems}}(r)$ for Q = 16 were approximately half the experimental results plotted in Fig. 24.

Thus, the plasma emissivity, its optical thickness, absorption coefficient, and magnetic measurements can be made to agree if we assume that the plasma emissivity is actually 4-5 times higher than the calculated value. However, if we assume that for some reason the plasma density is twice as high as calculated, then the optical thickness agrees with the experiments for calculated values of $\varkappa$ but then the temperature should be halved because the pressure can be determined quite reliably. However, a fall of the temperature by a factor of two would have been noticed in measurements of radiation fluxes with photocells. The results of a numerical calculation of a radiation flux emitted by a cylindrical plasma column with a specified (experimental) absorption coefficient at $h\nu = 2$ eV (Fig. 24) and with a dependence $\varkappa(r)$ obtained from magnetic measurements (Fig. 24) gave $7 \cdot 10^4$ W·V^{-1}·cm^{-2}, in full agreement with the value measured directly. The results of calculations of the emission spectrum of a cylindrical plasma column with specified T(r) and n(r) are plotted in Fig. 21. These results are based on calculated values of $\varkappa(n, t)$ multiplied by a factor of 5.

In Chap. I we have reported a calculation of the dynamics of a discharge in which $\varkappa$ is increased by a factor of 5 and the stability against overheating is ensured by an "effective spectral line."* Comparing the results of calculations (Fig. 7) with the experimental data we can show that they agree to within a factor of 1.5-2.

The plasma emissivity differs from the calculated value probably because of incomplete equilibrium between the direct and reverse processes in the investigated plasma. In fact, the ionization (excitation) occurs only because of electron collisions since the radiation density is

*Recent measurements of the emission spectrum of discharges in the $h\nu > 10$ eV range demonstrated the existence of a recombination peak at $h\nu = 75$ eV. The energy carried away by photons of this kind was ~0.8 kJ.

below the equilibrium value and the recombination (deactivation) is due to electron collisions and results in photon emission. If we consider only the transitions between the ground states of atoms and free electrons, we find that deviations from equilibrium should be small. These deviations can be considerable if the probabilities of radiative transitions of bound electrons from upper levels are compared with the probabilities of excitation by electron impact. These conditions are realized for the lower excited states of atoms. An experimental check of this hypothesis would require complex diagnostic experiments. One would have to measure the number of atoms in certain states and the proportion of these atoms is 10^{-2}-10^{-3} of the total density of atoms and ions.

The increase in the emissivity may be partly due to the emission in the far wings of broadened lines, particularly near the series limits. Finally, calculations of the photoionization cross sections may not be accurate. A theoretical analysis of all these possibilities is not a reliable process because of the complex conditions in a plasma and we have not attempted it.

As shown earlier, the initial stage of a discharge is characterized by turbulence. Individual realizations of the turbulent motion fields satisfy magnetohydrodynamic equations but the equations of turbulent motion found by separating the regular and pulsating parts are never closed (i.e., there are more variables than equations) and the problem in the turbulence theory cannot be reduced to finding a unique solution governed by the known initial and boundary conditions. Therefore, the average values of pulsations of plasma quantities have to be found invoking additional considerations. It is possible to use dimensional analysis in which physical parameters affecting steady-state turbulence can be identified. In the problem under consideration these dimensional parameters are the electric field intensity E, $\sigma_0 = \sigma/T^{3/2}$, and $Q = Q_0 q_0 = q/\rho^2(T)^{1/2}$. The quantities E and σ_0 determine the energy supplied to a plasma with given values of ρ and T, and Q represents the loss of energy because of emission of radiation. Then, magnetohydrodynamic equations (3)-(5) can be used in the three-dimensional radiation approximation to find the following relationships for the parameters of pulsations of the hydrodynamic quantities:

$$\tau \approx A\sigma_0^{-1/2}E^{-1}Q^{-1/2}\ \text{sec}, \qquad x \approx c^{2/5}A^{11/10}\sigma_0^{-3/5}E^{-4/5}Q^{2/5}\ \text{cm},$$
$$v \approx c^{2/5}A^{1/10}\sigma_0^{-1/10}E^{1/5}Q^{1/10}\ \text{cm/sec}, \qquad T \approx c^{4/5}A^{-4/5}\sigma_0^{-1/5}E^{2/5}Q^{1/5}\ \text{eV},$$
$$\rho \approx c^{2/5}A^{-2/5}\sigma_0^{2/5}E^{6/5}Q^{-2/5}\ \text{g/cm}^3, \qquad p \approx c^{6/5}A^{-1/5}\sigma_0^{1/5}E^{8/5}Q^{-1/5}\ \text{erg/cm}^3.$$

Here, $A = 5 \cdot 10^{11}\ \text{cm}^2 \cdot \text{eV}^{-1} \cdot \text{sec}^{-2}$ is the specific heat of the plasma; $c = 3 \cdot 10^{10}$ cm/sec is the velocity of light. In numerical estimates we can use E = 70 V/cm and $Q = 24 \cdot 10^{24}\ \text{cm}^5 \cdot \text{sec}^{-3} \cdot \text{g}^{-1} \cdot \text{eV}^{-1/2}$, found experimentally for the investigated conditions (t = 25 μsec, initial wire diameter d_0 = 0.17 mm), and $\sigma_0 = 4 \cdot 10^{13}\ \text{sec}^{-1} \cdot \text{eV}^{-3/2}$ [5]. Substituting these values into Eq. (47), we find that

$$\tau = 2 \cdot 10^{-6}\ \text{sec}, \quad x = 3\ \text{cm}, \quad v = 23\ \text{km/sec}, \quad T = 3.1\ \text{eV},$$
$$\rho = 1.7 \cdot 10^{-5}\ \text{g/cm}^3\ (n = 1.7 \cdot 10^{18}\ \text{cm}^{-3}).$$

It follows from the above results that the dimensional analysis gives results which are close to the experimental values: the average instability lifetime is ~3 μsec, whereas the characteristic dimensions determined by optical methods are up to 0.6 cm (magnetic probes give 0.9 cm), and the velocity is 10 km/sec. Therefore, we may assume that the above relationships give approximately correct values of the amplitudes of the pulsations of ρ, T, and p. These amplitudes are close to the average plasma parameters. Therefore, turbulence does not alter basically the properties of a discharge (i.e., it does not alter significantly the average charge of the ions and produces no anomalous changes in the emissivity).

This means that the general characteristics of a discharge (total current, voltage applied to the electrodes, total light flux) vary smoothly.

The development of a discharge can be described generally as follows. The explosion of a wire in the center of a chamber produces an inhomogeneous conducting channel. The electric field E applied to the plasma heats the channel and expands it. Instabilities appear in the dense optically transparent plasma and these instabilities give rise to and maintain turbulence in the discharge. After 40 μsec the turbulence decays because instabilities are damped out. We may assume that the damping of the overheating instability is due to the line emission from the plasma.

The following qualitative analogy may be quoted to account for the development of turbulence in the plasma under consideration. In hydrodynamics and magnetohydrodynamics turbulence appears when the ratio of the parameters representing local and overall motion becomes large. Ratios of this kind are the Reynolds number, which is the ratio of the inertial and viscosity forces, and the magnetic Reynolds number, which is the ratio of the velocity of a magnetic field together with a plasma to the velocity of plasma diffusion. In the presence of radiation the local tendency affects the energy of the radiation emitted from the interior (each separate regions emits independently) whereas the overall tendency is reflected in the energy emitted from the plasma surface (photons are repeatedly absorbed and reemitted, tending to equalize the conditions in the plasma interior). Essentially, the ratio of the energy emitted from the interior to the energy emitted from the surface (including line emission) is a dimensionless parameter representing the appearance of turbulence in a plasma because of the overheating instability. The instabilities are damped out when the current passes through the maximum and a steady-state optically transparent (for wires with $d_0 < 0.3$ mm) Z pinch is observed: this pinch has axial symmetry and it is fairly homogeneous along the discharge. A discharge of this type is a high-power selective radiation source: under the experimental conditions the radiation fluxes emitted in a wide part of the spectrum indicate that the surface temperature is 2-3 eV. The use of more powerful sources should make it possible to raise the temperature to 4-5 eV and to produce a coaxial discharge; experimental studies of such discharges could be of great interest because of their great stability.

LITERATURE CITED

1. W. Lochte-Holtgreven, Rep. Prog. Phys., 21:312 (1958).
2. Ya. B. Zel'dovich and Yu. P. Raizev, Physics of Shock Waves and High-Temperature Hydronamic Phenomena, 2 vols., Academic Press, New York (1966-7).
3. L. Spitzer, Jr., Physics of fully Ionized Gases, 2nd ed., Wiley-Interscience, New York (1962).
4. L. Landau and E. M. Lifshitz, Statistical Physics, 2nd ed., Pergamon Press, Oxford (1969).
5. V. S. Imshennik and V. F. D'yachenko, Preprint No. 960 [in Russian], Institute of Atomic Energy, Moscow (1965).
6. L. D. Landau and E. M. Lifshitz, Electrodynamics of Continuous Media, Pergamon Press, Oxford (1960).
7. N. M. Kuznetsov, Thermodynamic Functions and Shock Adiabats of Air at High Temperatures [in Russian], Mashinostroenie, Moscow (1965).
8. P. N. Kalitkin, Teplofiz. Vys. Temp., 6:801 (1968); V. S. Rogov, Preprint No. 47 [in Russian], Institute of Applied Mathematics, Academy of Sciences of the USSR, Moscow (1969).
9. V. B. Rozanov, Dokl Akad. Nauk SSSR, 182:320 (1968).

10. A. A. Klementov, G. V. Mikhailov, F. A. Nikolaev, et al., Teplofiz. Vys. Temp., 8:736 (1960).
11. A. F. Nikiforov and V. B. Uvarov, Preprint No. 36 [in Russian], Institute of Applied Mathematics, Academy of Sciences of the USSR, Moscow (1969).
12. J. Berger, Astrophys. J., 124:550 (1956).
13. A. Burgess and M. Seaton, Mon. Not. R. Astron. Soc., 120:121 (1961).
14. R. D. Hudson and V. L. Carter, Phys Rev, 137:A1648 (1965).
15. Kh. B. Gezalov and A. V. Ivanova, Teplofiz. Vys. Temp., 6:416 (1968).
16. V. Ferraro, Nuovo Cimento Suppl., 13:9 (1959).
17. S. I. Braginskii, I. M. Gel'fand, and R. N. Fedorenko, in: Physics of Plasma and Problems in Controlled Thermonuclear Reactions [in Russian], Vol. 4, Izd. AN SSSR, Moscow (1958), p. 201.
18. L. A. Artsimovich, Controlled Thermonuclear Reactions, Gordon and Breach, New York (1964).
19. V. B.Rozanov, Zh. Tekh. Fiz., 40:62 (1970).
20. A. A. Rukhadze and S. A. Triger, Zh. Prikl. Mekh. Tekh. Fiz., No. 3, 11 (1968); Preprint No. 168 [in Russian], Lebedev Physics Institute, Academy of Sciences of the USSR, Moscow (1968).
21. V. B. Rozanov and A. A. Rukhadze, Preprint No. 132 [in Russian], Lebedev Physics Institute, Academy of Sciences of the USSR, Moscow (1969).
22. V. V. Pustovalov and V. B. Rozanov, Proc. Second All-Union Conf. on Physics of Low-Temperature Plasma [in Russian], Nauka i Tekhnika, Minsk (1968).
23. V. V. Pustovalov and V. B. Rozanov, Preprint No. 52 [in Russian], Lebedev Physics Institute, Academy of Sciences of the USSR, Moscow (1969).
24. S. I. Braginskii, in: Physics of Plasma and Problems in Controlled Thermonuclear Reactions [in Russian], Vol. 1, Izd. AN SSSR (1958), p. 115.
25. A. von Schlüter, Z. Naturforsch. a, 5:72 (1950).
26. B. B. Kadomtsev, Vopr. Teor. Plasma, 2:132 (1963).
27. A. A. Rukhadze and S. A. Triger, Zh. Eksp. Teor. Fiz., 56:1029 (1969).
28. A. F. Aleksandrov, A. A. Rukhadze, and S. A. Triger, Proc. Ninth Conf. on Phenomena in Ionized Gases, 1969, Bucharest, publ. Editura Akademiei Republici, Bucharest, (1969), p. 379.
29. V. B. Rozanov, A. A. Rukhadze, and S. A. Triger, Zh. Prikl. Mekh. Tekh. Fiz., No. 5, 18 (1968).
30. A. A. Rukhadze and S. A. Triger, Preprint No. 26 [in Russian], Lebedev Physics Institute, Academy of Sciences of the USSR, Moscow (1968).
31. B. A. Trubnikov, in: Physics of Plasma and Problems in Controlled Thermonulcear Reactions [in Russian], Vol. 1, Izd. AN SSSR (1958), p. 289.
32. P. P. Volosevich, V. Ya. Gol'din, N. N. Kalitkin, and S. P. Popov, Proc. Ninth Conf. on Phenomena in Ionized Gases, 1969, Bucharest, publ. Editura Akademiei Republici, Bucharest (1969), p. 348.
33. V. Ya. Gol'din, D. A. Gol'dina, G. V. Danilova, et al., Preprint No. 36 [in Russian], Institute of Applied Mathematics, Academy of Sciences of the USSR, Moscow (1971).
34. P. P. Volosevich, V. Ya. Gol'din, N. N. Kalitkin, et al., Preprint No. 40 [in Russian], Institute of Applied Mathematics, Academy of Sciences of the USSR, Moscow (1971).
35. A. F. Nikiforov and V. B. Uvarov, Preprint No. 36 [in Russian], Institute of Applied Mathematics, Academy of Sciences of the USSR, Moscow (1969).
36. A. F. Aleksandrov, V. V. Zosimov, A. A. Rukhadze, et al., Preprint No. 72 [in Russian], Lebedev Physics Institute, Academy of Sciences of the USSR, Moscow (1971).
37. A. D. Klementov, G. V. Mikhailov, F. A. Nikolaev, et al., Proc. Second All-Union Conf. on Low-Temperature Plasma [in Russian], Nauka i Tekhnika, Minsk (1968), p. 269.

38. A. A. Vekhov, A. D. Klementov, F. A. Nikolaev, et al., Kratk. Soobshch. Fiz., No. 10, 53 (1970).
39. A. D. Klementov G. V. Mikhaikov, F. A. Nikolaev, et al., Preprint No. 126 [in Russian], Lebedev Physics Institute, Academy of Sciences of the USSR, Moscow (1969); Kratk. Soobshch. Fiz., No. 4, 69 (1971).
40. A. D. Klementov, F. A. Nikolaev, and V. B. Rozanov, Proc. Tenth Intern. Conf. of Phenomena in Ionized Gases, Oxford, 1971, publ. by Donald Parsons, Oxford (1971), p. 389.
41. F. A. Nikolaev, V. B. Rozanov, and Yu. P. Sviridenko, Preprint No. 99 [in Russian], Lebedev Institute, Academy of Sciences of the USSR, Moscow (1971); Teplofiz. Vys. Temp., 10:486 (1972).
42. A. A. Vekhov, F. A. Nikolaev, and V. B. Rozanov, Preprint No. 79 [in Russian], Lebedev Physics Institute, Academy of Sciences of the USSR, Moscow (1971); Teplofiz. Vys. Temp., 10:728 (1972).
43. S. N. Belov, M. I. Demidov, N. N. Ogurtsova, I. V. Podmoshenskii, P. N. Rogovtsev, and V. M. Shelemina, Zh. Prikl. Spektrosk., 10:408 (1969).
44. A. N. Zaidel' and B. Ya. Shraider, Spectroscopy of Vacuum Ultraviolet [in Russian], Nauka, Moscow (1967).
45. E. Oktay and D. R. Bach, J. Appl. Phys., 41:1716 (1970).
46. H. Griem, Plasma Spectroscopy, McGraw-Hill, New York (1964).
47. F. A. Nilolaev, Yu. V. Novitskii, V. B. Rozanov, and Yu. P. Sviridenko, Preprint No. 21 [in Russian], Lebedev Physics Institute, Academy of Sciences of the USSR, Moscow (1972); Zh. Eksp. Teor. Fiz., 63:884 (1972).
48. A. D. Klementov, and V. B. Rozanov, Kratk. Soobshch. Fiz., No. 9, 42 (1972).
49. V. A. Rodichkin, G. A. Serebryannyi, and A. M. Timoshin, in: Plasma Diagnostics [in Russian], No. 2, Atomizdat, Moscow (1968).
50. A. S. Monin and A. M. Yaglom, Statistical Hydromechanics [in Russian], 2 vols., Nauka, Moscow (1967).

GENERATION AND AMPLIFICATION OF LIGHT BY STIMULATED SCATTERING

A. Z. Grasyuk

A theoretical and experimental study was made of the dynamics of Raman lasers emitting two Stokes components. A study was made of transient processes and of the saturation effect in stimulated scattering amplification. It was found experimentally that, under saturation conditions, the quantum efficiency of the first Stokes component could be close to 100%. The influence of the pump spectrum on the stimulated scattering amplification was manifested by the spatial asymmetry of the gain and by the threshold nature of the dependence of the gain on the excitation intensity.

INTRODUCTION

The stimulated scattering of light was one of the first nonlinear-optics phenomena whose observation became possible after the appearance of high-power lasers, although it was predicted theoretically in the nineteen-thirties. Stimulated Raman scattering (SRS) [1] is described by the general Placzek equation [2]. If follows from the very physical nature of the scattering of light that each type of spontaneous scattering* has a stimulated analog involving the amplification of the scattered radiation. The number of types of stimulated scattering discovered so far and investigated intensively is large (Table 1). Papers reporting investigations of stimulated scattering are very numerous. It is not even possible to list a significant part of these papers and it would be pointless because there are already several reviews and monographs dealing with SRS [3-5], stimulated Mandelshtam-Brillouin scattering referred to in Western literature as the stimulated Brillouin scattering (SBS) [13, 14], and stimulated thermal scattering (STS) [14, 31].

The physical phenomena responsible for a given type of scattering can differ greatly. However, they can be classified in accordance with the nature of the excitation transferred to the material system as a result of scattering (Table 1).†

* The term "spontaneous scattering" is used for that part of the scattered light whose power is proportional to the intensity of the exciting radiation (the stimulated scattering power is proportional to the product of the intensities of the exciting and scattered light).

† Stimulated scattering is possible in which four rather than two photons participate in each elementary event [39]. An event of this kind leaves the material system in the unexcited state. The probability of such events is much lower and will not be considered here.

TABLE 1. Types of Stimulated Scattering

Type of stimulated scattering	Cause of spontaneous scattering	Spectral range of pumping	Frequency shift, cm^{-1}	Applications
Stimulated Raman scattering (SRS) [1-11]	Molecular vibrations [1-6] or rotations [7, 8]	From ultraviolet [11] to infrared [9]	$2 \cdot 10^2$-$4 \cdot 10^3$	1. Stimulated emission (Table 2) 2. Formation of pulses (Table 3)
Stimulated Brillouin scattering (SBS) [12-15]	Fluctuations of density of medium	Visible	10^{-2}-10^{-1}	1. Stimulated emission (Table 2) 2. Formation of pulses (Q switching) (Table 3)
Stimulated spin-flip Raman scattering (SSFRS) by spin Landau sublevels in semiconductor in magnetic field [16-28]	Spin flip of conduction electrons in magnetic field	Far infrared (5 and 10.6 μ)	10-200; controlled by magnetic field	Tunable-wavelength stimulated emission (Table 2)
Stimulated Raman scattering by polaritons (SPRS) [29, 30]	Simultaneous fluctuations of density and charge in ionic crystals	Visible	10^2-$5 \cdot 10^2$; controllable	Tunable-wavelength stimulated emission (Table 2)
Stimulated thermal scattering (STS) [31-36]	Entropy fluctuations due to: 1) absorption of light [32, 33] 2) electrocaloric effect [34]	Visible	–	Q switching (Table 3)
Stimulated concentration scattering (SCS) [37, 38]	Fluctuations in concentrations of components of mixtures	Visible	1-10	Stimulated emission (Table 2)
Stimulated Rayleigh line wing scattering (SRLWS) [13, 14]	Fluctuations in molecular anisotropy	Visible	1-10	Stimulated emission (Table 2)

If light of intensity exceeding a certain threshold value is directed into an active medium (gas, liquid, or solid), the beam emerging from this medium in the forward and/or backward directions has one or more new spectral components of longer (Stokes) or shorter wavelengths (anti-Stokes); these components may carry a considerable proportion of the pump energy. The frequency shift resulting from the scattering depends on the nature of the latter and the type of active substance. This shift may range from several thousandths to several thousands of reciprocal centimeters. This provides extensive opportunities for frequency conversion of laser radiation which are now being utilized in a wide spectral range from ultraviolet to far infrared. For example, pump radiation of wavelength λ_p = 1.0787-1.064 μ was converted efficiently into radiation with λ = 10.52-9.26 μ by SRS in compressed hydrogen [40] and this made it possible to generate pulses with optimal characteristics in the spectral range of the CO_2 laser.

In practical applications such as spectroscopy, communications, plasma diagnostics, etc., it is necessary to perform efficient frequency conversion of the exciting radiation into a light beam of minimum divergence containing just one Stokes stimulated-scattering component. This cannot be achieved by amplifying spontaneous scattering in one pass through the active medium. In fact, such amplification follows the exponential law $\exp(gI_pL)$, where g (cm/W) is the gain of the active medium, L (cm) is the length of this medium, and I_p (W/cm^2) is the intensity of the pump radiation; significant conversion begins when the gain increment $b = gI_pL$ reaches ~30. When the intensity of the first Stokes component becomes comparable with the intensity of the pump (exciting) radiation, this component amplifies the second component and such a process weakens the first component. Thus, a continuous increase in the pump intensity produces successfully Stokes components of increasing orders (second, third, etc.) and 100% quantum efficiency of the pump radiation conversion into the first component can never be achieved.*

Considerable advantages are gained by the use of stimulated scattering for the generation of coherent radiation (Table 2). If an active medium excited by a laser beam is placed in, for example, a Fabry–Perot resonator formed by mirrors whose reflection coefficients are r_1 and r_2, pump radiation of sufficiently high intensity I_p ensuring the necessary gain increment $b = gI_pL$ per pass generates the first Stokes component.† The self-excitation condition of such a Raman laser has the usual form, applicable to any laser:

$$r_1 r_2 \exp(2gI_pL) = 1.$$

A Raman laser may be advantageous in respect of the efficiency of frequency conversion and of the intensity of the converted light compared with a system using single-pass amplification of the spontaneous scattering in an active medium. In fact, the reflection coefficients r_1 and r_2 can be made different for different Stokes components; for example, they can be made optimal for the first component and minimal for the second. This is particularly easy to achieve in the stimulated Raman scattering because, in this case, the frequency interval between the neighboring components is 10^2-10^3 cm^{-1}. This selection of r_1 and r_2 makes it possible to increase the threshold intensity of the first component at which the second component is excited. Nevertheless, this method does not allow us to eliminate completely the generation of the second component because in practice it is very difficult to suppress the feedback to the second component if this feedback is strong for the first component. As the intensity of the pump radiation is increased, the second component is generated sooner or later and this reduces the quantum efficiency of generation of the first component.

* Moreover, if this approach is employed, the diffraction divergence of the converted radiation cannot be achieved either.

† We shall call a system of this kind a Raman laser or oscillator.

TABLE 2.

Type of stimulated scattering	Active medium					Pumping
	Active substance	Aggregate state	T, °K	Line width, cm^{-1}	Gain, 10^{-2} cm/MW	System
Stimulated Raman scattering	Nitrobenzene [41-43]	Liquid	300	1.5	—	Longitudinal axial [41] Longitudinal non-axial [42] Transverse [43]
	Benzene [44-48]	Liquid	300	2.15 [49]	0.28 [50]	Longitudinal axial [44] Longitudinal non-axial [47, 48] Transverse [44]
	Hydrogen [51-56]	Gas, 50 atm	300	0.2	0.15	Longitudinal axial [51, 53, 54] Transverse [52]
	Nitrogen [57-59]	Liquid	77	0.067 [49]	1.6 [50] 1.0	Longitudinal non-axial [57, 58]
					1.6	Transverse [59]
	Diamond [60]	Solid	300	—	0.6	Longitudinal axial
	Carbon disulfide [61]	Liquid	300	0.5 [49]	—	The same
	Glass [61]	Solid	300	—	0.005	" "
	Methane [63]	Gas	300	0.2	—	" "
Stimulated Brillouin scattering	Benzene [64-68]	Liquid	300	0.0082 [67]	2 [68]	Transverse [66] Longitudinal non-axial [64, 65]
	Methane [69]	Gas, 130 atm	300	—	9	The same
	Ether [70]	Liquid	300	0.0039	3.4	Transverse
	Carbon disulfide [70]	Liquid	300	0.0013	17	The same
Stimulated spin-flip Raman scattering	InSb [16—20]	Solid	4	—	1670	" "
	InSb [24]	Solid	30	—	—	Longitudinal axial
Stimulated polariton Raman scattering	$LiNbO_4$ [30]	Solid	300	—	—	Longitudinal nonaxial

Raman Lasers

Pumping		Stimulated emission	
Wave-length, μ	Regime	Wavelength, μ	Resonator
0.694	Single nanosecond pulses	0.767	Fabry-Perot (FP) with plane mirrors (PM)
0.694	The same	0.756	The same
		0.975	The same [51, 53, 55, 56]; F-P with spherical mirrors (SM) [54]
0.694	" "	0.975; 0.725	FP-PM [52]
0.694 [57]	" "	0.828	The same
1.06 [58]	" "	1.41	" "
0.694	" "	0.828	" "
0.694	" "	—	FP (polished crystal faces)
0.48	Periodic microsecond pulses	0.51	Light guide between plane mirrors
0.53	Continuous	0.54	The same
0.694	Single nanosecond pulses	0.87	FP-SM
0.694	The same	0.694	FP-PM
0.694	Single 200 nsec pulses	0.694	Unidirectional ring
0.694	Single nanosecond pulses	0.694	FP-PM
0.694	Single 0.8 μsec pulses	0,694	FP-PM
10.6	Periodic nanosecond pulses	Rearrangement by magnetic fields: 11–13	FP (polished crystal faces)
5.32	Continuous	Rearrangement by magnetic fields: 5.6–6.4	The same
0.694	Single nanosecond pulses	0.7–0.74	FP-SM

The basic difference between a Raman laser and a single-pass spontaneous scattering amplifier is the ability of the former to concentrate, with the aid of a resonator, the transformed light in a narrow solid angle, which can be close to the diffraction limit. It is important to note that such concentration is also possible in multimode pumping, when the exciting radiation is spread over an angle much greater than the diffraction value. An important consequence of this property is the possibility of increasing the brightness of the output radiation ($W \cdot cm^{-2} \cdot sr^{-1}$) [51, 53, 57]. Such an increase in brightness has already been achieved for several types of stimulated scattering, including Raman (SRS) [51, 53, 57], Brillouin (SBS) [70], and spin-flip Raman (SSFRS) [24] (Table 2). The methods of excitation of Raman lasers can vary considerably. For example, pump radiation may travel along the resonator axis (longitudinal pumping). Such pumping can be axial (the pump beam is parallel to the resonator axis) or nonaxial (the pump beam is inclined at a small angle to the resonator axis). If the angle of inclination is 90°, the pumping configuration is known as transverse [43, 52, 59, 66, 71]. The active medium may be gaseous [51-55], liquid [42, 44-46, 58, 59], or solid [30, 60]. The nonselectivity of the excitation makes it possible to perform pumping with laser radiation ranging from ultraviolet [11] to far-infrared (Table 2 and [16-28]).

The dynamics of Raman lasers has several features which distinguishes it from the dynamics of conventional lasers. The gain of a Raman laser is proportional to the instantaneous pump intensity and it "follows" this intensity with a delay ranging from several picoseconds to several nanoseconds [72]. Consequently, stimulated emission is obtained earlier because the duration of the transient stage is governed by the longer of two characteristic times: 1) the development time of a "photon avalanche," which is determined by the number of passes in the resonator needed to amplify spontaneous scattering to the pump level; 2) the formation time of the diffraction-limit angular distribution of the output radiation [70].

In spite of the important advantages which Raman lasers have over single-pass spontaneous scattering amplifiers, it is, nevertheless, difficult to achieve efficient generation of one Stokes component in a wide range of pump intensities. The difficulty is due to the fact that when the pump intensity is increased, the second Stokes component is generated sooner or later [57-59] because it is practically impossible to suppress feedback completely. This slows down the rise of the first component and then even reduces its intensity [57-59, 73] and the situation then resembles one obtained by single-pass amplification.

This difficulty can be avoided using a Raman amplifier,* in which the amplification of a sufficiently intense input signal results in saturation [76]. Such an amplifier can operate at a low gain increment ($5 \leqslant b \leqslant 10$), when spontaneous scattering is not enhanced to a level comparable with the intensity of the pump radiation. In this situation, the pump radiation may be weak enough to avoid such undesirable phenomena as self-focusing, stimulated emission due to a nonresonant feedback, etc. It should be pointed out that, in the case of multimode pumping, a sufficiently strong saturable signal in the form of the first Stokes component with the diffraction-limit divergence can be obtained only using a Raman laser of high efficiency and brightness close to the maximum value. The construction of such a laser is a task in itself and the difficulty of this task has been discussed on several occasions [77-79]. This is probably why, up to the time of the experiments carried out by the present author [76], there was only one paper [54] describing an attempt to achieve efficient conversion in an amplifier whose input signal was generated by a Raman laser (SRS in compressed hydrogen pumped with a ruby laser). Saturation in a Raman amplifier was not achieved in [54] since the input-signal source (Raman laser) was of insufficient power.

* The basic possibility of amplification of an external signal by stimulated scattering (without saturation) was demonstrated experimentally in [74] (SRS) and later investigated in [54] (SRS in compressed hydrogen) and in [75] (SBS).

TABLE 3. Formation of Pulses

Type of stimulated scattering	Pumping source	Active medium	Aggregate state	T, °K	Main result	Special features	Pulse parameters: wavelength, μ	Pulse parameters: duration, nsec	Pulse parameters: energy, J	Reference
Raman	Ruby laser	Carbon disulfide	Liquid	300	Generation of picosecond pulses	Opposed Raman amplification	0.71	0.03	0.0002	[86, 88]
Raman	The same	Hydrogen	Gas	300	Generation of subnanosecond pulses	The same	0.975	0,3	—	[87]
Brillouin	" "	Carbon disulfide	Liquid	300	Q switching	Focusing of radiation in CS_2 cell	0.694	25	—	[89]
Brillouin	" "	Methanol	Liquid	300	Shortening of pulse duration	Preliminary Q switching	0.694	25	—	[90]
Brillouin	" "	Nitrogen	Gas (500 atm)	300	Generation of series of high-power pulses	Absence of focusing elements	0.694 0.83	4 3	0.3×n 0.2×n	[91]
Brillouin	" "	Ether, acetone	Liquid	300	The same	The same	0.694			[92]
Thermal	Neodymium laser	Ethanol	"	300	Q switching	Initial transmission 50%	0.694	15—25	0.15—0.25	[93]
Raman	Ruby laser	Nitrogen	"	77	Lengthening of pulse duration	Q switching	0.694	Up to 3000	0.9	[94]
Raman	Neodymium laser	Nitrogen	"	77	The same	The same	1.06	Up to 3000	1.5	[95]

The radiation emitted by a real pump source is nonmonochromatic. The width of the pump spectrum may be close to or even greater than the width of the spontaneous scattering line. The experiments reported in [80] and the subsequent theoretical investigations [81-84] enabled us to establish experimentally [85] the following new properties of scattering under wide-band pumping conditions.

1. The gain is asymmetric relative to the mutual orientation of the signal and pump beams.

2. The dependence of the gain increment on the pump intensity is of the threshold type: if the pump intensity exceeds a certain value, the gain becomes the same as for monochromatic pumping. The latter property is important in practical applications since it makes it possible to perform efficient frequency conversion of wide-band laser radiation [80]. We shall use the term "wide-band" for pump radiation whose spectrum is wider than the spontaneous scattering line.

Stimulated scattering of light can be used to form subnanosecond pulses [86-88] (opposed amplification in a Raman amplifier). Stimulated scattering can also be used successfully in the control of the duration of pulses of various lasers, such as Q switching [89-93], lengthening of nanosecond giant pulses to several microseconds [94, 95], and so on (Table 3).

Classifying and dividing various types of stimulated scattering, we must bear in mind the general properties of these phenomena which are manifested in the amplification and generation of coherent light. These general properties are manifested, for example, by the identical mathematical description of the various types of stimulated scattering. This allows us to consider from a unified standpoint (without specifying the nature of the scattering process) the principal characteristics and properties of stimulated scattering, including transient-processes, steady-state operation, saturation effect, influence of spectral composition of pump radiation, etc. The same generality makes it possible to analyze the processes occurring in Raman lasers or amplifiers utilizing various types of stimulated scattering.

The present paper is based mainly on the results reported in [57-59, 70, 73, 76, 80, 85, 91, 94-96] dealing with various processes and phenomena associated with the generation [57-59, 73] and amplification [76, 80, 85, 91, 96] of light by SRS [57-59, 73, 76, 80, 85, 96] and SBS [70, 91, 96]. These investigations were carried out in the Quantum Electronics Laboratory at the Lebedev Physics Institute of the Academy of Sciences of the USSR during 1967-1972.

CHAPTER I

DYNAMICS OF RAMAN LASERS

§ 1. Theoretical Analysis of the Dynamics of Emission from a Raman Laser

Derivation of Initial Equations for Slowly Varying Amplitudes

Let us assume that an active medium placed inside a Fabry–Perot resonator, formed by mirrors whose reflection coefficients are r_1 and r_2, is illuminated with a plane electromagnetic wave (pump wave of frequency ω_p) whose wave vector is directed along the resonator axis (longitudinal axial pumping). We shall assume that stimulated emission is possible at the first and second Stokes frequencies. The excitation of the resonator at the pump frequency can be ignored if we assume that the reflection coefficients of the mirrors vanish at this frequency. The electric field inside the resonator consists of the traveling pump-wave field and the fields of the standing Stokes waves:

$$\mathbf{E}(x, t) = \mathbf{E}_p(x, t) + \sum_k E_k^S(t)\,\mu_k(x) + \sum_l E_l^{SS}(t)\,\mu_l(x). \tag{1.1}$$

Here, $E_k^S(t)$ is the field of the k-th mode at the frequency of the first Stokes component; $E_l^S(t)$ is the field of the l-th mode at the frequency of the second Stokes component; $\mu_{kl}(x) = (2/LA)^{1/2} \times \sin(\pi kx/L)$ are the axial resonator modes with eigenfrequencies within the width of the spontaneous scattering line. We recall that $\int \mu_k \mu_i dV = \delta_{ik}$; $E_p(xt)$ is the pump field.

The equations describing the behavior of the standing Stokes waves in the resonator are

$$\left.\begin{aligned} \frac{d^2E_k^S}{dt^2} + \frac{2}{\tau_k^S}\frac{dE_k^S}{dt} + \omega_k^2 E_k^S &= -\frac{4\pi}{n_c^2}\int \mu_k \frac{\partial^2 P_k^{S(nl)}}{\partial t^2}\, dV, \\ \frac{d^2E_l^{SS}}{dt^2} + \frac{2}{\tau_l^{SS}}\frac{dE_l^{SS}}{dt} + \omega_l^2 E_l^{SS} &= \frac{4\pi}{n_{cc}^2}\int \mu_l \frac{\partial^2 P_l^{SS(nl)}}{\partial t^2}\, dV, \end{aligned}\right\} \tag{1.2}$$

where τ_k^S and τ_l^S are the photon lifetimes in the resonator at the first and second Stokes frequencies, respectively; n_S and n_{SS} are the refractive indices of the medium at the same frequencies; $P_k^{S(nl)}$ and $P_l^{SS(nl)}$ are the nonlinear polarizations at the frequencies in question.

The nonlinear polarization in the system (1.2) is given by the expression

$$\mathbf{P}^{(nl)} = N\left(\frac{\partial \alpha}{\partial q}\right) qE. \tag{1.3}$$

Here, α is the molecular polarizability; N is the concentration of the molecules; q is the vibrational coordinate defined by the equation

$$\ddot{q} + \Delta\omega_l \dot{q} + \Omega_{1,2}^2 q = \frac{1}{m}\left(\frac{\partial \alpha}{\partial q}\right) E^2, \tag{1.4}$$

where $\Delta\omega_l$ is the width of the spontaneous Raman scattering line; $\Omega_{1,2}$ is the eigenfrequency of the molecular vibrations; E(xt) is defined by Eq. (1.1).

The pump (exciting) radiation is a traveling wave propagating along the x axis:

$$\frac{\partial^2 E_p}{\partial x^2} - \frac{n_p^2}{c^2}\frac{\partial^2 E_p}{\partial t^2} = \frac{4\pi}{c}\frac{\partial^2 P_p^{(nl)}}{\partial t^2}, \tag{1.5}$$

where $P_p^{(nl)}$ is the nonlinear polarizability at the pump frequency.

Equations (1.2)-(1.5) form a closed system which, subject to suitable boundary and initial conditions, can be used to analyze the dynamics of stimulated emission from Raman lasers. In deriving this system, it is assumed that the concentration of molecules N in the ground state is independent of the spatial coordinates and time.

We shall analyze the system (1.2)-(1.5) by the method of slowly varying amplitudes and assume that the stimulated emission does not occur at more than two Stokes components.* We

* The case of n components can be considered in a similar manner. For example, the stimulated emission of n Stokes components is considered in [97] for a system excited by a standing pump wave.

shall represent $E_p(x, t)$, $E_k^S(t)$, $E_l^{SS}(t)$, and $q(t)$ in the form

$$\left.\begin{aligned}
E_p(x, t) &= \frac{1}{2} A_p(x, t) \exp i(\omega_p t - k_p z) + \text{c.c.},\\
E_k^S(t) &= \frac{1}{2} A_k^S(t) \exp i\omega_k^S t + \text{c.c.},\\
E_l^{SS}(t) &= \frac{1}{2} A_l^{SS}(t) \exp i\omega_l^{SS} t + \text{c.c.},\\
q(t) &= q^{S\text{-}p}(t) + q^{S\text{-}SS}(t),\\
q^{S\text{-}p}(t) &= \frac{1}{2} \sum_k r_k^{S\text{-}p}(t) \exp i\Omega_k^{S\text{-}p} t + \text{c.c.},\\
q^{S\text{-}SS}(t) &= \frac{1}{2} \sum_l r^{S\text{-}SS}(t) \exp i\Omega_l^{S\text{-}p} t + \text{c.c.},\\
\Omega_k^{S\text{-}p} &= \omega_p - \omega_k^S, \quad \Omega_l^{S\text{-}SS} = \omega_k^S - \omega_l^{SS}.
\end{aligned}\right\} \tag{1.6}$$

The indices "S" and "SS" represent the first and second Stokes components, respectively, whereas "S-p" and "S-SS" represent the interaction of the first Stokes component with the pump radiation and with the second Stokes component. The vibrational coordinate has two components $q^{S\text{-}p}(t)$ and $q^{S\text{-}SS}(t)$, which oscillate in general with different amplitudes $r^{S\text{-}p}$ and $r^{S\text{-}SS}$ and different (although similar) frequencies. We shall substitute the expressions for $E_p(x, t)$, $E_k^S(t)$, $E_l^{SS}(t)$, and $q(t)$ into the system (1.2)-(1.5). Ignoring the second derivatives with respect to time and the products of the first derivatives, collecting terms with the same frequencies ω_k^S, ω_l^{SS}, ω_p, $\Omega_k^{S\text{-}p}$, and $\Omega_l^{S\text{-}SS}$, and bearing in mind $\Delta\omega_l \ll \Omega_{k,l}$, $\Delta\omega_l \sim 2/T_2 \ll \omega_{k,l}$, we obtain the following system of reduced equations for slowly varying amplitudes:

$$\dot{A}_k^S + \frac{\Delta\omega_r}{2} A_k^S = \frac{i2\pi\omega_k^S (\partial\alpha/\partial q) N}{n_S^2} \int \mu_k (A_p \overset{*}{r}{}_k^{S\text{-}p} + A_l^{SS} r_l^{S\text{-}SS} \mu_l)\, dV, \tag{1.7}$$

$$\dot{A}_l^{SS} + \frac{\Delta\omega_r}{2} A_l^{SS} = \frac{i2\pi\omega_l^{SS} (\partial\alpha/\partial q) N}{n_{SS}^2} \int \mu_k \mu_l A_k^S \overset{*}{r}{}_l^{S\text{-}SS} dV, \tag{1.8}$$

$$\dot{r}_k^{S\text{-}p} + \frac{\Omega_{12}^2 - (\Omega_k^{S\text{-}p})^2 - i\Delta\omega_l \Omega_k^{S\text{-}p}}{-2i\Omega_k^{S\text{-}p}} r_k^{S\text{-}p} = \frac{i(\partial\alpha/\partial q)}{2m\Omega_k^{S\text{-}p}} \overset{*}{A}{}_k^S A_p \mu_k, \tag{1.9}$$

$$\dot{r}_l^{S\text{-}SS} + \frac{\Omega_{12}^2 - (\Omega_l^{S\text{-}SS})^2 - i\Delta\omega_l \Omega_l^{S\text{-}SS}}{-2i\Omega_l^{S\text{-}SS}} r_l^{S\text{-}SS} = \frac{i(\partial\alpha/\partial q)}{2m\Omega_l^{S(SS)}} A_k^S \overset{*}{A}{}_l^{SS} \mu_k \mu_l, \tag{1.10}$$

$$\frac{\partial A_p}{\partial x} + \frac{n_p}{c}\frac{\partial A_p}{\partial t} = \frac{2\pi i N (\partial\alpha/\partial q)\, \omega_p^2}{c^2 k_p} \sum_k \mu_k A_k^S r_k^{S\text{-}p}. \tag{1.11}$$

In the single-mode emission case, the sum over k in Eq. (1.11) disappears.

Quasistatic Approximation. Kinetic Equations for Intensities

An analysis of the system (1.7)-(1.11) simplifies considerably if we consider only the quasistatic approximation in which the polarization of the medium follows the field quasistatically. We can then simplify Eqs. (1.9) and (1.10) by dropping $\dot{r}$ compared with $\Delta\omega_l r/2$. Then, for example, $r_k^{S\text{-}p} \approx \frac{i(\partial\alpha/\partial q) A_p A_k \mu}{2m\Delta\omega\, \Omega_k^{S\text{-}p}}$. This makes it possible to obtain equations relating the intensity. We shall substitute the values of r from Eqs. (1.9) and (1.10) into Eqs. (1.7), (1.8), and (1.11) and introduce the intensities of the first and second Stokes components emerging from a Raman laser resonator, as well as the pump intensity and gain of the active medium g:

$$I_S = \frac{n_S^2 v_S}{8\pi} |A_k^S|^2 \frac{\ln(1/r)}{LA}, \quad I_{SS} = \frac{n_S^2 v_S}{8\pi} |A^{SS}|^2 \frac{\ln(1/r_S)}{LA}, \quad I_p = \frac{n_p^2 v_p}{8\pi} |A_p|^2,$$

$$\xi_{(SS)} = [r_1^{S(SS)} r_2^{S(SS)}]^{1/2}, \qquad g = \frac{32\pi^2 \omega_S}{S}\chi = 32\pi^2 \omega_S N \left(\frac{\partial\alpha}{\partial q}\right)^2 \Big/ c^2 \hbar n_p \Delta\omega_p.$$

We shall next replace μ^2 with its average value $\mu^2 = 1/LA$, where L is the resonator length and A is the area of the transverse cross section of the resonator. The system (1.7)-(1.11) now becomes

$$\left.\begin{aligned}&\frac{\partial I_p}{\partial x}+\frac{1}{v_p}\frac{\partial I_p}{\partial t}=-\frac{\omega_p}{\omega_S}\frac{g}{\ln(1/r_S)}I_S I_p,\\&\frac{dI_S}{dt}+\frac{1}{\tau_S}I_S=\frac{v_S}{L}gI_S\int I_p(x)\,dx-\frac{gv_S}{\ln(1/r_{SS})}I_S I_{SS},\\&\frac{dI_{SS}}{dt}+\frac{1}{\tau_{SS}}I_{SS}=\frac{\omega_{SS}}{\omega_S}\frac{gv_{SS}}{\ln(1/r_S)}I_S I_{SS}.\end{aligned}\right\}\qquad(1.12)$$

The first equation in the above system can be solved immediately by replacing the variables as follows: $z = x$, $\tau = t - x/v_p$.

In terms of these new variables, the equation becomes

$$\frac{dI_p}{dz}=-\frac{\omega_p}{\omega_S}\frac{g}{\ln(1/r_S)}I_S I_p.$$

Integrating from 0 to x, we obtain

$$I_p\left(x,\ t-\frac{x}{v_p}\right)=I_p(0,\ t)\exp\left(-\frac{\omega_p}{\omega_S}\frac{g}{\ln(1/r_S)}I_S x\right).$$

We shall substitute the expression for the pump intensity into the second equation of the system (1.12) and integrate with respect to x. We shall next normalize the time t relative to the Stokes photon lifetime in the resonator

$$\xi=\frac{t}{\tau_S},\qquad \tau_S=\frac{L}{v_S\ln(1/r_S)}$$

and denote the Stokes components by $Y(\xi)=\frac{\omega_p}{\omega_S}\frac{I_S}{I_p^{max}(0,\ t)}$ and $Z(\xi)=\frac{\omega_S}{\omega_{SS}}\frac{I_{SS}}{I_p^{max}(0,\ t)}$, where $I_p^{max}(0, t)$ is the maximum pump intensity at the entry to the resonator,

$$X(0,\ \xi)=\frac{I_p(0,\ t)}{I_p^{max}(0,\ t)},\qquad b=gI_p^{max}(0,\ t)\,v_S\tau.$$

We obtain in this way the following system of dimensionless kinetic equations:

$$\left.\begin{aligned}&\frac{dY}{d\xi}+Y=[1-\exp(-bY)]\,X(\xi)-\frac{b}{\ln(1/r_{SS})}YZ,\\&\frac{dZ}{d\xi}+\tau Z=\frac{b}{\ln(1/r_{SS})}YZ.\end{aligned}\right\}\qquad(1.12a)$$

The solution of the above system subject to the initial conditions Y_0 and Z_0 and for a given shape of pump pulse $X(\xi)$ yields the shape of single-mode pulses emitted by a Raman laser at the frequencies of both Stokes components.

Figures 1 and 2 give the solutions of the system (1.12) for the Gaussian shape of pump pulses of different intensities (the Gaussian shape describes quite accurately multimode emission from a Q-switched ruby laser). It is clear from Figs. 1 and 2 that the stimulated emission of the second Stokes component reduces considerably the intensity of the first component. If we wish a Raman laser to emit just the first component, we must suppress the generation of the second component. This can be achieved by introducing additional losses at the frequency of the second component using, for example, unidirectional stimulated emission from a Bril-

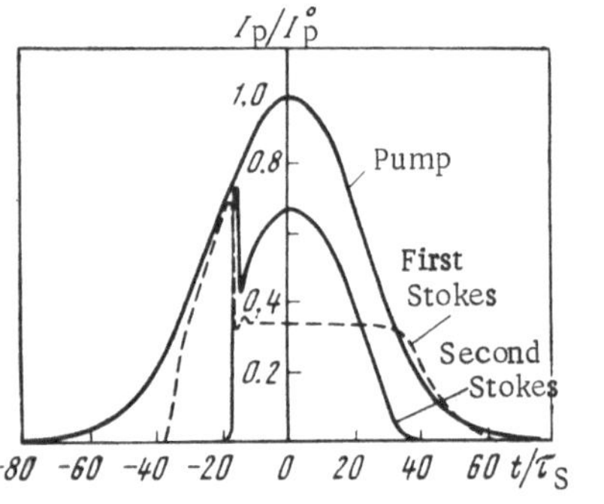

Fig. 1. Development of the emission of the second Stokes component and its influence on the first component with increasing pump intensity ($b \propto I_p^0$) in a resonator with a low Q at the frequency of the second component (the reflection coefficients of the resonator mirrors are assumed to be r_{SS} = 50 and 0.1%). Here, τ_S is the lifetime of a Stokes (first component) photon in the resonator; $I_p(t)/I_p^0 = \exp[-a \times (t/\tau_S)^2]$; $a = 10^{-3}$, b = 20, r_S = 20%.

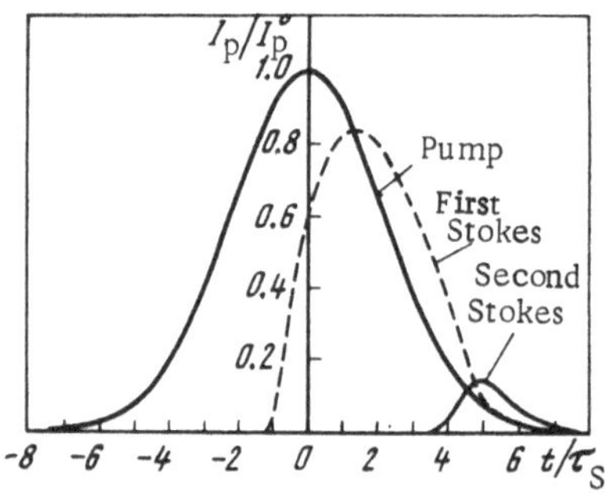

Fig. 2. Stimulated emission of two Stokes components as a result of excitation with pump pulses of duration comparable with the Stokes photon lifetime in the resonator τ_S. The time is insufficient for the development of the second component in spite of a high Q factor of the resonator and a large gain increment b ($b \propto I_p^0$); $I_p(t)/I_p^0 = \exp[-a\,(t/\tau_S)^2]$; $a = 10^{-1}$, b = 20, r_S = 20%, r_{SS} = 10%.

louin laser [69]. The second component can also be suppressed by reducing the reflection coefficient at the frequency of this component* and selecting suitably the gain increment for the first component (Fig. 1). In those cases when the duration of the pump pulses is not too great compared with the lifetime of the Stokes photons in the resonator, we can select the pump intensity (i.e., the gain increment of the first Stokes component) so that the stimulated emission at the second component does not develop (Fig. 2). This reduces the undesirable influence of the second component on the first.

§ 2. Experimental Investigations of Raman Laser Dynamics

An experimental investigation [73] was made of the shape and duration of the stimulated emission pulses at the frequency of the first Stokes component, position and duration of the leading edge of these pulses, and dependences of the energies of the first and second Stokes components on the pump intensity. The emission from the investigated Raman laser was close to the single-mode type.

A block diagram of the apparatus is shown in Fig. 3. The pump source was a Q-switched ruby laser. This laser generated almost Gaussian pulses and the width of its line was $\Delta\nu_p$ = 0.017 cm^{-1}. The duration of these pulses was 18 nsec at midamplitude. This ruby laser radiation was obtained by mode selection: the exit mirror of the resonator was a Fabry-Perot etalon (mirror M_2) in the form of two glass plane-parallel plates; the Q switching was performed by a solution of vanadium cryptocyanine in methanol, which resulted in additional narrowing of the spectrum. The width of the ruby laser line was measured in each experiment

* In practice, the value of r_{SS} cannot be reduced below 0.1% because of random reflections and nonresonant feedback.

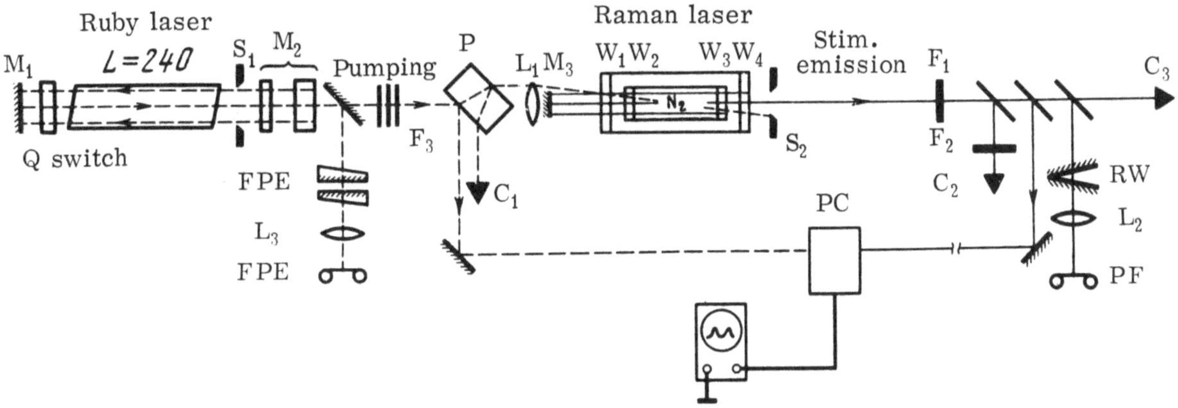

Fig. 3. Block diagram of the apparatus used in investigations of Raman lasers: M_1 and M_2 are mirrors used in the pump laser resonator; M_3 and W_4 are mirrors used in the Raman laser resonator; L_1-L_3 are lenses; P is a plane-parallel glass plate, 5 cm thick; W_1-W_4 are glass windows of a cell containing liquid nitrogen; S_1-S_2 are screens with apertures; F_1 is a selective filter transmitting the first and second Stokes components; F_2 is a filter transmitting the second Stokes component; C_1-C_3 are calorimeters; PF is a photographic film; PC is a coaxial photocell of the FÉK type; F_3 are neutral filters; RW is a reflecting wedge; FPE is a Fabry–Perot etalon.

using a Fabry-Perot etalon with mirrors 3 cm apart. The ruby laser (pump) radiation was focused by a lens with F = 1 m (lens L_1) into a 25-cm long cell filled with liquid nitrogen. A plate P (5 cm thick) acted as an auxiliary element in the system for mutual alignment of the Raman and ruby laser resonators. It was also used as a beam splitter, which avoided the influence of the interference effects on the energy and time characteristics of the ruby laser radiation (for the same reason, all the other beam splitters were 5 cm thick). The Raman laser resonator consisted of a mirror M_3 (the reflection coefficient was 98% at the frequency of the first Stokes component and 50% at the frequency of the second) and one surface of a glass wedge W_4, acting also as the exit window of the liquid-nitrogen cell. The use of this exit mirror ensured the maximum quantum efficiency and output power at the first Stokes component. The Raman laser radiation was separated by a screen S_2 and a filter F_1. Calorimeters C_1, C_2, and C_3 were used to measure the energy of the pump radiation, second Stokes component, and the sum of the first and second Stokes components. The energy of the first Stokes component was found by subtracting the readings of the calorimeter C_2 from those of the calorimeter C_3. The pump radiation and first Stokes component pulses were delayed relative to one another by an optical line and reached a coaxial photocell (of the FÉK type), which produced a current pulse recorded with an I2-7 oscillograph. The divergence of the converted radiation was measured with a reflecting wedge [98] and a lens L_2, whose focal length was 1 m.

The transverse dimensions of the amplifying region were selected so as to ensure that the Fresnel number was close to infinity; consequently, only the dominant mode was excited in the Raman laser. This ensured that the divergence of the converted radiation was $4.5 \cdot 10^{-4}$ rad, whereas the divergence corresponding to the diffraction limit ($\theta_{\text{diffr}} = 2.4\lambda/d$) was $4.1 \cdot 10^{-4}$ rad. All the subsequent measurements were carried out using the same geometry.

Figure 4 shows an oscillogram of pump and first Stokes component pulses. We found that the leading edge of the Stokes pulse was steep and had a fairly flat top: this was due to the generation of the second Stokes component (compare with Fig. 1). Figures 5 and 6 show the results of an analysis of the time and energy characteristics of the first Stokes component

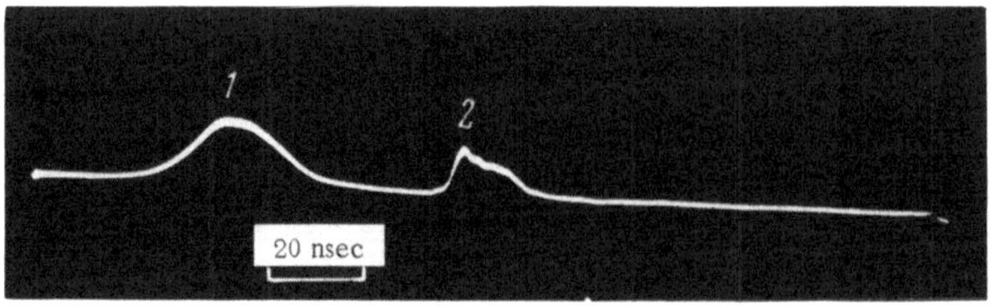

Fig. 4. Typical oscillogram of pump (1) and first Stokes (2) pulses.

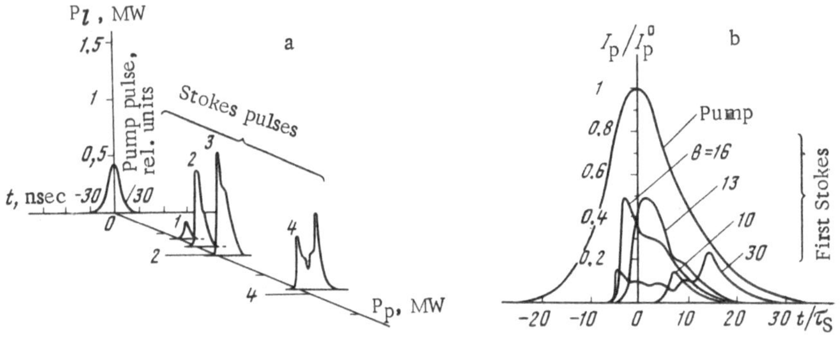

Fig. 5. a) Stimulated emission of the first Stokes component and the influence of the second component when the pump intensity is increased. b) Time dependences of the quantum efficiency of the generation of the first Stokes component and the influence of the second component when the pump intensity is increased ($b \propto I_p^0$). Here, τ_S = 1.3 nsec and τ_{SS} = 1.12 nsec are the photon lifetimes in the resonator of a Raman laser for the first and second Stokes components, respectively; P_p is the pump power; P_l is the laser output power.

using different pump intensities. When the pump rate was increased, the output power of the Raman laser first increased and then began to fall. The rise of the power of the first Stokes component with increasing pumping rate was limited by the generation of the second Stokes component, whose power was also measured. Moreover, it was found (Fig. 5) that, when the pumping rate was increased, the duration of the Stokes pulse became greater and the leading edge shifted toward the onset of stimulated emission.

The results of calorimetric measurements (Fig. 6) yielded the dependences of the energies of the first and second Stokes components on the pump energy. As in [57-59], the generation of the second Stokes component reduced the energy of the first component. Consequently, the peak power of the first component began to fall with rising pump energy (pulses 3

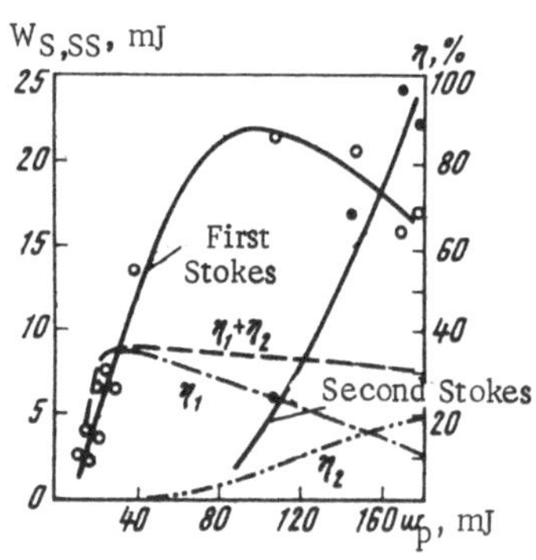

Fig. 6. Dependences of the output energy on a Raman laser (continuous curves) at the frequencies of the first and second Stokes component, and of the quantum efficiencies η_1 and η_2 of the first and second Stokes components (dashed curves) on the pump energy.

and 4 in Fig. 5). This fall in power was faster than the increase in the pulse duration and, consequently, the energy of the first component decreased.

When the pumping rate was increased, the leading edge became shorter and shifted forward (Fig. 5). The appearance of the second Stokes component deformed the top of the pulse of the first component (Figs. 4 and 5). At the same time, the quantum efficiency of the first Stokes component decreased. The maximum quantum efficiency of the first Stokes component reached 50% (Fig. 5b) when the divergence of the radiation was close to the diffraction limit.

§ 3. Conditions for Attaining Maximum Quantum Efficiency and Minimum Divergence of Raman Laser Output

Theoretical Estimates

The maximum quantum efficiency and minimum divergence of the output radiation of a Raman laser are obtained when the principal characteristics of the pump radiation, including its intensity I (W/cm^2), energy density ϵ_p (J/cm^2), and pulse duration τ_p (sec), exceed certain threshold values:

$$1)\ I_p > I_p^{th},\quad 2)\ \epsilon_p > \epsilon_p^{th},\quad 3)\ \tau_p > \tau_p^{th}. \tag{1.13}$$

In the stimulated Brillouin scattering case, there is also a fourth condition which applies to the width of the angular spectrum $\Delta\nu_p$ (divergence) of the pump radiation [56, 57]:

$$\Delta\nu_p < \varphi(\vartheta), \tag{1.14}$$

where ϑ is the angle between the pump beam and the resonator axes.

The first of the above inequalities is simply the usual self-excitation condition of a laser: the gain should exceed the losses $R_1R_2 \exp(2gI_pL) \geq 1$. Here, R_1 and R_2 are the reflection coefficients of the resonator mirrors and $b = gI_pL$ is the gain increment.

The second condition $\epsilon_p > \epsilon_p^{th}$ is the criterion for the development of a "photon avalanche": the scattered radiation should make a sufficient number of passes through the resonator for its intensity to become of the same order of magnitude as the pump radiation intensity. Therefore, a certain excess over the threshold pump intensity should be maintained for a certain minimum time so that the scattered light amplified during each pass increases in intensity until it becomes comparable with the pump intensity. We can easily show [58, 59] that, after N passes in the resonator, the power P^{sp} of the spontaneously scattered radiation, which is amplified by a factor exp b during each pass, reached the level

$$P^{sp} = \hbar\omega \frac{S}{\lambda^2} do\Delta\nu_l (\exp b - 1) \frac{[R_1R_2 \exp 2b]^{N/2} - 1}{[R_1R_2 \exp 2b]^{1/2} - 1}. \tag{1.15}$$

Here, S is the cross sectional area of the pump beam; $\Delta\nu_l$ is the width of the spontaneous scattering line; do is a solid-angle element.

We shall assume that a "photon avalanche" develops if the spontaneously scattered radiation is amplified to the power $P^{sp} \sim P_p$. Since the development time of a photon avalanche at the self-excitation threshold is infinitely long, we must ensure that the pump intensity is such that $R_1R_2 \exp 2b > 1$. Then, ignoring unity in Eq. (1.15) and equating the power of the amplified spontaneous radiation to the pump power, we find that Eq. (1.15) yields the following condition for the minimum duration of the pump pulses necessary for the development of a photon avalanche:

$$\tau_p > \frac{L}{c} N = \frac{L}{c} \frac{1}{b(1 - I_p^{th}/I_p)} \ln \left\{ \frac{(R_1 R_2)^{1/2} I_p}{\frac{\hbar\omega}{\lambda^2} do\Delta\nu_l} \right\}$$

or, since $b = gI_pL$ and $\ln \left\{ \frac{(R_1R_2)^{1/2} I_p}{\hbar\omega do\Delta\nu_l/\lambda^2} \right\} \approx 25\text{-}30$, we find that the introduction of $I_p\tau_p = \epsilon_p$ (J/cm^2) and $\varkappa = I_p^{th}$ yields

$$\epsilon_p > \frac{20-30}{cg(1-\varkappa^{-1})}. \tag{1.16}$$

If the pump intensity exceeds sufficiently the threshold value I_p^{th}, the inequality (1.16) becomes

$$\epsilon_p = I_p\tau_p > \frac{20-30}{gc} = \epsilon_0. \tag{1.17}$$

The third inequality in Eq. (1.13) governs the time necessary for the formation of an angular distribution of the output radiation close to the diffraction limit. In fact, the diffraction-limit divergence can be obtained only by the use of angular mode selectors, i.e., by ensuring that the conditions are such that the losses experienced by the angular modes are considerably higher than those experienced by the axial modes. Two types of selection are possible: 1) spatial; 2) frequency.

1) The spatial selection is achieved relatively easily by the use of selectors with focusing elements. However, there is a danger of optical breakdown or damage to the mirrors. In the absence of such selectors, the diffraction-limit angular distribution can be achieved by ensuring that the output radiation travels a distance L to the spatial coherence area [99], defined by the condition $L = d^2/\lambda$, where d and λ are, respectively, the beam diameter and output wavelength. The corresponding time is $\tau_p^{th} = L/c = d^2/c\lambda$. Thus, the diffraction-limit divergence is obtained if the pump pulse duration satisfies the inequality

$$\tau_p > \tau_p^{th} = \frac{d^2}{c\lambda}. \tag{1.18}$$

2) The frequency selection can be performed only if the duration of the Raman laser pulses is $\tau_l > 1/\Delta\nu$, where $\Delta\nu$ is the difference between the frequencies of the axial and nearest angular modes. Employing the well-known expressions for $\Delta\nu$ [100], we again obtain the condition (1.18).

Thus, to achieve the maximum quantum efficiency and minimum divergence, we must ensure that the intensity of the pump radiation, its energy density, and pulse duration satisfy simultaneously the inequalities

$$I > I_p^{th}, \quad \epsilon_p > \epsilon_p^{th}, \quad \tau_p > \tau_p^{th}.$$

It should be pointed out that these requirements apply to any laser utilizing stimulated scattering.

Experimental Results

The conditions (1.13) were checked experimentally [70] for Brillouin lasers. The apparatus of Fig. 7 was used in an investigation of an ether Brillouin laser pumped by Q-switched ruby laser pulses of nanosecond duration. The Brillouin laser resonator was formed by two plane dielectric mirrors, 1.7 cm apart. The transmission of each of these mirrors was 4% at the wavelength $\lambda = 0.7\ \mu$. The mirrors were used also as the internal walls of a cell filled with ether. The pump radiation passed through the walls which were 2 cm apart and inclined

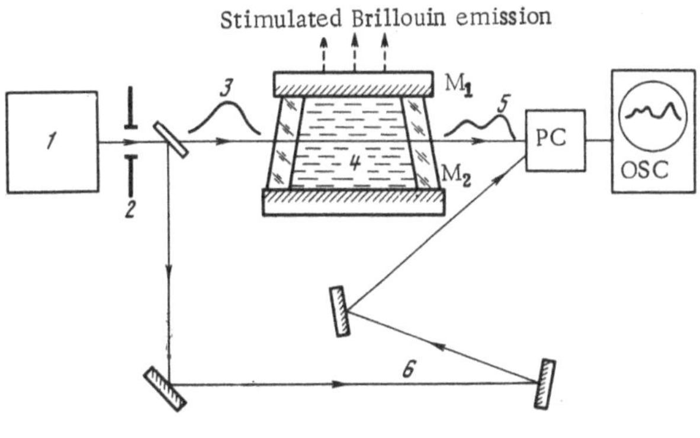

Fig. 7. Block diagram of the apparatus for investigating a transversely pumped ether Brillouin laser: 1) ruby laser (pump source); 2) screen with aperture; 3) pump pulse on entry; 4) cell with ether; 5) pump pulse on exit; 6) optical delay line; M_1 is a semitransparent mirror, M_2 is a totally reflecting mirror; PC is a coaxial photocell of the FÉK type; OSC is an I2-7 oscillograph.

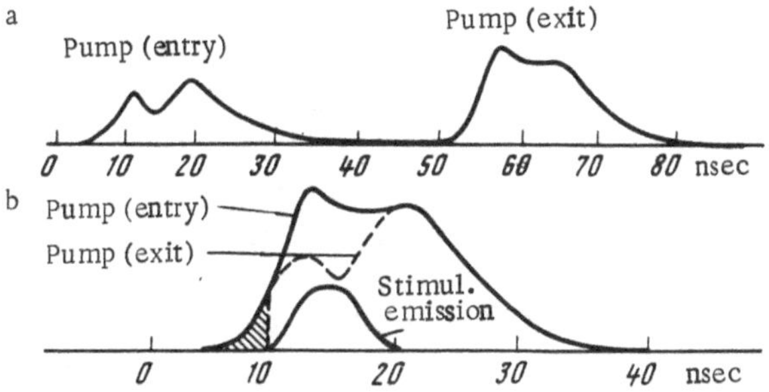

Fig. 8. Oscillograms of the pump radiation at the entry to an ether cell and at the exit from this cell (a), compared with the normalized oscillogram (b).

at 5° relative to one another. A screen with an aperture of 1.3 cm size along the resonator axis and 0.1 cm at right-angles to this axis was placed in the path of the pump beam in front of the cell. The dynamics of the stimulated emission and the time dependence of the quantum efficiency were determined by the simultaneous recording of the shape of the pump pulses entering and leaving the cell. The pump pulse at the entry to the cell was delayed (with a special optical line) by 46 nsec relative to the pulse leaving the cell.

We observed coherent radiation generated as a result of stimulated Brillouin scattering; this radiation was emitted in the form of pulses of 0.04 J energy and 5 nsec duration (Fig. 8). The divergence of this radiation was $3 \cdot 10^{-3}$ rad in the vertical plane (at right-angles to the pump beam) and $5 \cdot 10^{-2}$ rad in the horizontal plane. Thus, the divergence along one coordinate was close to the diffraction limit whereas, along the other coordinate, it was much greater than this limit.* This asymmetry was due to the fact that only the conditions $I_p > I_p^{th}$ and $\epsilon_p > \epsilon_p^{th}$ were satisfied. The third condition, relating to the duration of the pump pulses, was not satisfied, so that the diffraction limit was not reached in the horizontal plane. It is clear from Fig. 8 that the quantum efficiency was 50% because the above two conditions were satisfied.† The shaded part of the pulse in Fig. 8b corresponded to the pump energy density equal to ϵ_p^{th}. It is clear from Fig. 8b that the quantum efficiency differed from zero when the condition $\epsilon_p > \epsilon_p^{th}$ was satisfied.

* A similar asymmetry of the angular distribution was also observed in [59] for a transversely pumped liquid nitrogen Raman laser.

† In spite of the considerable excess over the threshold of the first two conditions in Eq. (1.13), the quantum efficiency rose relatively slowly because of the simultaneous generation of up to five Stokes components.

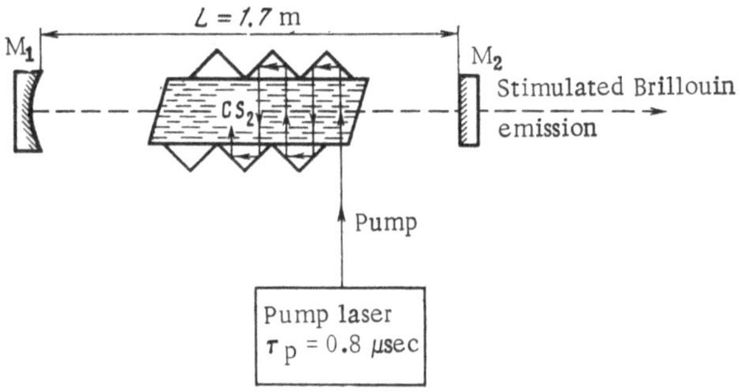

Fig. 9. Transversely pumped CS_2 Brillouin laser.

The formation of the angular spectrum in the case when all the conditions of Eq. (1.13) were satisfied was studied using the apparatus shown in Fig. 9. The resonator was formed by two dielectric mirrors, 1.7 m apart, and the reflection coefficients were $r_1 = 98\%$ and $r_2 = 80\%$, respectively. The resonator contained a rectangular screen with a 0.9 · 0.9 cm aperture. A glass cell filled with CS_2 was 24 cm long, 1 cm wide, and 2 cm high. The windows through which the output radiation passed were inclined by 6° and were antireflection-coated. A system of 17 total-internal-reflection prisms ensured repeated passage of the pump radiation at right angles to the resonator axis. A thin layer of glycerin was located between the prisms and the cell. Since the refractive indices of carbon disulfide, glass, and glycerin were similar, the reflection coefficients at the interfaces were small (≤ 0.1%). Therefore, the losses of the pump radiation were mainly due to the absorption in glass and they did not exceed 50%.

The pump source was a ruby laser emitting 0.8 μsec pulses, which satisfied the third inequality in the system (1.13). The arrangement shown in Fig. 9 generated a 0.9 · 0.5 cm beam with a $\vartheta_r = 3 \cdot 10^{-4}$ rad divergence. In contrast to the preceding case, the angular distribution was symmetric and its width exceeded the diffraction limit by a factor of 2.5-4.

CHAPTER II

FUNDAMENTALS OF THE THEORY OF AMPLIFIERS BASED ON STIMULATED SCATTERING

§ 1. Equations Describing Amplification

Initial Equations for Slowly Varying Amplitudes

The phenomena responsible for the various types of scattering are naturally different but the scattering process is common to all cases. Normally, the first Stokes component, shifted in the direction of longer wavelengths, is amplified. This amplification process can be followed readily by considering the Raman scattering, i.e., the scattering in which a photon interacts with one molecule (atom, electron) in each elementary event. By way of example, we shall consider the longitudinal pumping case and two amplification variants: parallel and opposed. In the parallel variant the signal being amplified and the pump radiation travel in the same direction, whereas in the opposed variant, they travel in opposite directions. Both variants are possible in the stimulated Raman scattering. The opposed variant is also possible in the stimulated Brillouin scattering.

Following the method in [47, 48, 52], we shall assume that there is an active medium in which a pump beam and a signal at the frequency of the first Stokes component are propagated.

The Maxwell equations describing the pump E_p and Stokes $E_S(x, t)$ waves are of the form

$$\frac{\partial^2 E_S}{\partial x^2} - \frac{n_S^2}{c^2}\frac{\partial^2 E_S}{\partial t^2} = \frac{4\pi}{c}\frac{\partial^2 P_S^{nl}}{\partial t^2}, \tag{2.1}$$

$$\frac{\partial^2 E_p}{\partial x^2} - \frac{1}{v_p^2}\frac{\partial^2 E_p}{\partial t^2} = \frac{4\pi}{c}\frac{\partial^2 P_p^{nl}}{\partial t^2}, \tag{2.2}$$

where P_p^{nl} and P_S^{nl} are the nonlinear polarizations at the pump and Stokes frequencies, respectively. The nonlinear polarization is given by Eq. (1.3), in which the field $E(x, t)$ has just two components, which are the pump radiation E_p and the first Stokes field E_S:

$$\mathbf{P}^{nl} = N\left(\frac{\partial\alpha}{\partial q}\right) q\mathbf{E}, \tag{2.3}$$

where $E(x, t) = E_p(x, t) + E_S$.

The vibrational coordinate q is given by Eq. (1.4). The system (2.1)-(2.3) and (1.4) is closed and can be analyzed by the method of slowly varying amplitudes, in which $E_p(x, t)$, $E_S(x, t)$, and q are represented in the form

$$E_p(x, t) = \frac{1}{2} A_p(x, t) \exp i(\omega_p t - k_p x) + \text{c.c.}, \tag{2.4}$$

$$E_S(x, t) = \frac{1}{2} A_S^{\pm}(x, t) \exp i(\omega_S t \mp k_S x) + \text{c.c.}, \tag{2.5}$$

$$q(t) = \frac{1}{2} r(t) \exp i\Omega t + \text{c.c.}, \tag{2.6}$$

where $A_S^+(x, t)$ and $A_S^-(x, t)$ are the amplitudes of the parallel and opposed Stokes signals.

Substituting these expressions into Eqs. (2.1)-(2.3), (1.3), and (1.4), we obtain the following system of reduced equations for the amplitudes $A_p(x, t)$, $A_S(x, t)$ and $r(t)$:

$$\frac{\partial A_S^{\pm}}{\partial x} \pm \frac{1}{u_S}\frac{\partial A_S^{\pm}}{\partial t} = \frac{i2\pi\omega_S^2\left(\frac{\partial\alpha}{\partial q}\right)N}{c^2 k_S} A_p r^*, \tag{2.7}$$

$$\frac{\partial A_p}{\partial x} + \frac{1}{v_p}\frac{\partial A_p}{\partial t} = \frac{i2\pi\omega_p^2}{c^2 k_S}\left(\frac{\partial\alpha}{\partial q}\right) A_S^{\pm} r, \tag{2.8}$$

$$\dot{r} + \frac{\Omega_{12} - \Omega^2 - i\Delta\omega_L\Omega}{-2i\Omega} r = \frac{i}{2m\Omega}\left(\frac{\partial\alpha}{\partial q}\right) \dot{A}_S^{\pm} A_p. \tag{2.9}$$

The above considerations apply to the stimulated Raman scattering (SRS) but the same procedure can be used to obtain reduced equations for slowly varying amplitudes in the cases of stimulated Brillouin and thermal scattering (SBS and STS) [68].

Given Field Approximation

Solutions of the type (2.7)-(2.9), describing SBS and STS, are obtained in [68, 101] using the given field approximation in which the pump intensity is assumed to be constant. This means that only fairly weak Stokes signals are considered so that the conversion of the pumping radiation into the Stokes component can be ignored (no saturation).

A solution of systems of the (2.7)-(2.9) type is found in [102] for SRS excited by monochromatic radiation and it is generalized in [103] to an arbitrary pumping (amplitude modulation, phase modulation, etc.). Examples of SBS and STS are used in [68] to give a detailed general theoretical analysis of transient processes and derive analytic formulas for the gain g(t) as a function of time. In [101, 104], the general solution of [68] is applied to the description of transient processes excited by pulses of the type encountered in experiments, such as pulses with exponential leading edges [104] and Gaussian pulses [101]. It follows from [101] that, in the case of a Gaussian pulse, the ratio $g(t_{th})/g_0$, considered as a function of the normalized pulse duration $\Delta\omega_l t_0$, is

$$\frac{g(t_l)}{g_0}=\frac{\Delta\omega_l t_0}{b_0 C}\left\{\left(1+\frac{4b_0 C}{\Delta\omega_l t_0}\right)^{1/2}-1-\ln\left[1+\left(1+\frac{4b_0 C}{\Delta\omega_l t_0}\right)^{1/2}\right]+\ln 2\right\}, \tag{2.10}$$

where $C \sim 1.2$. It is thus clear from Eq. (2.10) that steady-state stimulated scattering is obtained if $g_0/g(t) \sim 1$ and the duration t_{th} of a Gaussian exciting pulse, reduced to the relaxation time of the type of scattering under investigation ($T_2 \sim 2/\Delta\omega_l$), is greater than the time needed to reach the maximum (steady-state) gain increment b_0:

$$\tau_0=\frac{\Delta\omega_l t_0}{b_0}\gg 1. \tag{2.11}$$

Figure 10 shows the theoretical dependence $g_0/g(t)$, plotted on the basis of Eq. (2.10), on which we have superimposed the experimental values of the transient gain of various substances taken from [101] and [96]. It is clear from Eq. (2.10) and Fig. 10 that the duration of the transient process (i.e., the time during which a steady-state gain is established) increases with decreasing value of the parameter $\tau_0 = \Delta\omega_l t_0/b_0$. In other words, the time needed to establish a steady-state gain is directly proportional to the ratio of the pump pulse duration t_0 (measured at midamplitude) to the relaxation time $\tau_{rel} \approx 2/\Delta\omega_l$ of the scattering process in question and it is inversely proportional to the steady-state gain increment $b_0 = g_0 I_p L$.

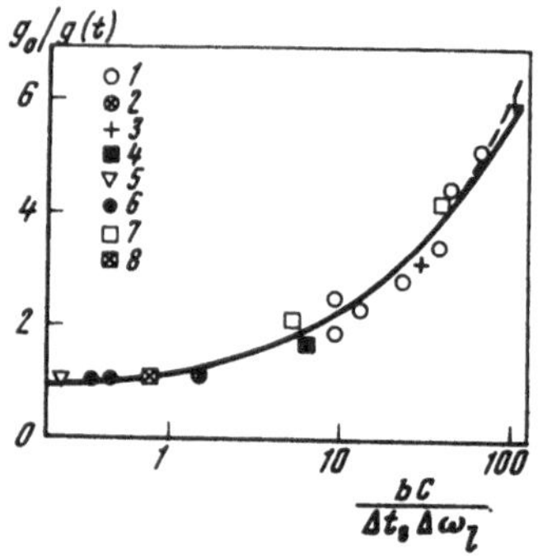

Fig. 10. Dependence of the gain on the duration of pump pulses; the continuous curve is plotted on the basis of [72, 101] and the dashed curve on the basis of [102].

Experimental points	Pulse duration, nsec	Active medium	T, °K	Type of scattering	References
6	17	Compressed hydrogen (p = 50 atm)	300	SRS	
1	1.7	The same	300	SRS	[72, 101]
3	17	Comp. nitrogen	300	SBS	
4	17	Acetone	300	SBS	
7	17	CCl_4	300	SBS	
5	500	Liquid nitrogen	77	SRS	
8		Liquid oxygen		SBS	[96]
2				STS	

Allowance for Saturation (Quasistatic Approximation)

A general solution of the system (2.7)-(2.9) can be obtained analytically by assuming that the polarization follows the field quasistatically; this is known as the quasistatic approximation (slow motion range [97]). It is assumed that all the changes in the pump and signal radiations require a time much longer than the transverse relaxation time T_2. In this case, we can ignore the time derivative of the amplitude of the vibrational coordinate of a molecule in Eq. (2.9). Substituting the value of r from Eq. (2.9) and assuming that $\Omega = \Omega_{12}$, we multiply both sides of the equations by A_S^* and A_p^*, and we introduce the following notation:

the intensity is

$$I_{pS} = v_{pS} \frac{n_{pS}^2 |A_{pS}^{\pm}|^2}{8\pi}, \qquad (2.12)$$

the velocity of propagation of signal and pump radiations is

$$v_{pS} = c/n_{pS},$$

and the gain is

$$g = \frac{32\pi^2 \omega_S}{c} \chi'' = \frac{32\pi^2 \omega_S N \left(\frac{\partial \alpha}{\partial q}\right)^2}{c^2 \hbar n_S^2 n_p \Delta\omega_l}. \qquad (2.13)$$

The system (2.7)-(2.9) now becomes

$$\frac{\pm \partial I_S^{\pm}}{\partial x} + \frac{1}{v_S} \frac{\partial I_S^{\pm}}{\partial t} = g I_S^{\pm} I_p, \qquad (2.14)$$

$$\frac{\partial I_p}{\partial x} + \frac{1}{v_p} \frac{\partial I_p}{\partial t} = -g I_S^{\pm} I_p, \qquad (2.15)$$

where $I_S^{\pm}$ and I_p are the intensities of the Stokes component and pump waves traveling at group velocities v_S and v_p. The "+" sign on the left-hand side of Eq. (2.14) and I_S^+ correspond to the parallel amplification, whereas the "−" sign and I^- correspond to the opposed amplification. The pump radiation is traveling along the positive direction of the x axis. This system of kinetic equations and the corresponding boundary and initial conditions describe both the amplification and generation of coherent radiation as a result of stimulated scattering under longitudinal pumping conditions.* In the absence of an external signal, the initial Stokes-frequency

---

* In the general case, the systems (2.7)-(2.9) and (2.14)-(2.15) include an additional equation describing the change in the concentration of excited molecules N [103]. However, in many practical cases, this equation can be ignored. In fact, it can be shown that, under steady-state conditions, when the duration of the pump pulses τ_p is much longer than the excited-state lifetime T_1, the change in N can be ignored if

$$N^{st} \approx T_1 \frac{I_p(0)}{L} \frac{1}{\hbar\omega_p} \ll N$$

or

$$I_p(0) \ll \frac{NL}{T_1} \hbar\omega_p. \qquad (2.16)$$

radiation appears because of the spontaneous scattering. This can be allowed for by introducing a random force into the system (2.14)-(2.15) and by assuming zero initial and boundary conditions, as in [103].

It should be pointed out that the quasistatic condition imposes restrictions on the pump and Stokes wave intensities, in particular,

$$I_S,\ I_p \ll \frac{1}{g v_{pS} T_2} = \frac{1}{2}\frac{2\pi\Delta\nu_l}{gv}, \tag{2.17}$$

where $\Delta\nu_l$ is the width of the spontaneous scattering line. Since we shall consider mainly the solution in the form of waves traveling without dispersion ($v_p = v_S = v$), it is sufficient to specify the boundary conditions in the form

$$I_S^+(x,\ t)\,|_{x=0} = I_0^+(t), \tag{2.18}$$

$$I_S^-(x,\ t)\,|_{x=L} = I_L^-(t), \tag{2.19}$$

$$I_p(x,\ t)\,|_{x=0} = I_p^0(t). \tag{2.20}$$

In the subsequent analysis, it is convenient to introduce new variables:

$$z = x, \qquad \tau = t - \frac{x}{v_p}. \tag{2.21}$$

The system (2.14)-(2.15) can be reduced to the form

$$\left(\frac{1}{v_S} \mp \frac{1}{v_p}\right)\frac{\partial I_S^\pm}{\partial \tau} \pm \frac{\partial I_S^\pm}{\partial z} = g I_p I_S^\pm, \qquad \frac{\partial I_p}{\partial z} = -g I_p I_S^\pm. \tag{2.22}$$

It is now preferable to consider separately the parallel and opposed amplification cases because each of these cases has its own peculiarities.

§ 2. Parallel Amplification

We shall assume, for simplicity, that there is no dispersion ($v_p = v_S = v$). The system (2.22) then reduces to the steady-state amplification conditions when $\partial/\partial t = 0$ in Eqs. (2.14)-(2.15). Equations (2.14)-(2.15) and the boundary conditions then become

$$\frac{dX}{d\xi} = -bXY^+, \qquad \frac{dY^+}{d\xi} = bXY^+, \qquad X(0) = 1, \qquad Y^+(0) = Y_0^+. \tag{2.23}$$

We have introduced here the dimensionless quantities

$$\xi = \frac{z}{L}, \qquad X = \frac{I_p(\xi)}{I_p^0(t)}, \qquad Y^+ = \frac{I_S^+(\xi)}{I_p^0(t)}. \tag{2.24}$$

The inequality (2.16) is satisfied practically always. For example, in the case of H_2 compressed to ~100 atm, we have $T_1 \approx 10^{-6}$ sec [105], $N \sim 10^{21}$ cm^{-3}, $L = 100$ cm, $\hbar\omega = 3 \cdot 10^{-19}$ J, and $I_p(0) < 10$ GW/cm^2. If the pump pulse duration τ_p is less than the relaxation time T_1 ($\tau_p \ll T_1$), the condition (2.16) becomes

$$I_p(0)\,\tau_p = \epsilon_p(0) \ll \hbar\omega_p NL,$$

where $\epsilon_p(0)$ is the pump energy density in J/cm^2. For example, if $\hbar\omega_p = 3 \cdot 10^{-19}$ J ($\lambda = 0.7\mu$), $N \sim 10^{21}$ cm^{-3}, $L = 10^2$ cm, we find that $\epsilon_p(0) \ll 3 \cdot 10^4$ J/cm^2.

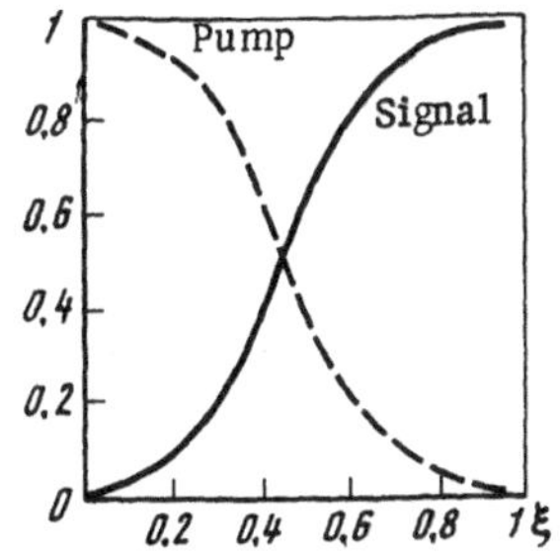

Fig. 11. Distributions of the intensity of the pump and signal waves along the normalized length ($\xi = x/L$) of an active medium in a parallel amplifier. The linear gain increment is $b = gI_p(0)L = 10$ and the input signal is $Y_0^+ = I_S(0)/I_p(0) = 0.01$.

The solution (2.23) is (see also [106] and Fig. 11)

$$Y^+(\xi) = Y_0^+ \frac{(1+Y_0^+)\exp[(1+Y_0^+)b\xi]}{1+Y_0^+\exp[(1+Y_0^+)b\xi]}. \tag{2.25}$$

The quantum efficiency at the point ξ $(0 \leq \xi \leq 1)$ is $\eta^+(\xi)$. At the amplifier exit, i.e., at $\xi = 1$, the quantum efficiency is given by

$$\eta^+(b,\ Y_0^+) = Y^+(1) - Y_0^+. \tag{2.26}$$

§ 3. Opposed Amplification

In contrast to the parallel amplification variant, the system (2.14)-(2.15) in the opposed case cannot be reduced by any change of variables to the form identical to the steady-state case. This is a manifestation of the fact that the pump and signal move in opposite directions. Consequently, it is necessary to consider separately the steady and nonsteady opposed amplification regimes, described by Eqs. (2.14)-(2.15) (with the "-" sign) subject to the boundary conditions of Eqs. (2.19) and (2.20).

Nonsteady Regime

An analysis of nonsteady opposed amplification on the assumption that the active medium is lossless and of infinite length is given in [88, 107]. The "infinity" of the active medium means, in practice, that the length of a pulse in space (i.e., the product of the duration and velocity) is less than the length of the active medium. However, the solutions obtained in [88] are unsuitable for the description of amplification in a lossy medium.* Nevertheless, we need not solve again Eqs. (2.14) and (2.15) but, substituting the variables of Eq. (2.21), we can transform these equations so that they are formally identical with the system (2.9) in [108]. This system describes the nonlinear amplification of a light pulse in a resonantly amplifying medium in which the active losses are γ (cm^{-1}); this system is

$$\frac{\partial I}{\partial t} + c\frac{\partial I}{\partial x} = c(\sigma N - \gamma)I, \qquad \frac{\partial N}{\partial t} + \frac{1}{T_1}(N - N_0) = -2\sigma I N. \tag{2.27}$$

Here, x is the coordinate; t is the time; N is the population inversion; I is the intensity of the amplified light pulse; σ is the amplification cross section.

We can readily show that Eq. (2.22) transforms into Eq. (2.27) if we substitute the variables

$$\tau = -\frac{4x}{v}, \quad z = 2vt, \quad I_p = -\frac{\hbar\omega_p N v}{2}, \quad gv = \sigma/\hbar\omega_p, \quad I_S^- = I \tag{2.28}$$

and if we assume that $N_0 = 0$.

---

* The losses experienced of the Stokes signal can be allowed for by introducing on the right-hand side of Eq. (2.14) a term γI_S^-, where γ (cm^{-1}) is the active loss coefficient.

In the resonance amplification case, a light pulse moves in a medium with a population inversion and its energy increases until the gain per unit length becomes equal to the losses per unit length. The analogy of this case with the opposed amplification in a Raman laser is physically clear. In fact, a Stokes pulse which meets pump radiation increases in energy until the energy rise per unit length is compensated by the linear losses with a coefficient γ. The formal identity of Eqs. (2.22) and (2.28), subject to Eq. (2.27), allows us to use the results obtained in [108], where the limiting shape of an amplified pulse is found for Gaussian, Lorentzian, and steplike profiles. In particular, in the resonance amplification case, the limiting energy density in a pulse is

$$\epsilon_{\mathrm{lim}}^{\mathrm{res}} = \frac{1}{2}\frac{\hbar\omega N}{\gamma}; \qquad (2.29)$$

using the notation of Eq. (2.28), we find that, in the case of a Stokes signal amplified by stimulated scattering, we have

$$\epsilon_{\mathrm{lim}} = I_{\mathrm{p}}/v\gamma. \qquad (2.30)$$

If $I_{\mathrm{p}} = 100$ MW/cm^2, $v = 2 \cdot 10^{10}$ cm/sec, and $\gamma = 10^{-4}$ cm^{-1}, we find that $\epsilon_{\mathrm{lim}}^{\mathrm{S}} = 100$ J/cm^2.

In the case of nanosecond pulses, the assumption that the medium is infinite is no longer valid: the length of the active medium does not usually exceed 1 m and the spatial length of a pulse may exceed tens of meters. Therefore, we must go back to the system (2.14)-(2.15) subject to the boundary conditions of Eqs. (2.19) and (2.20) and we have to repeat all the detailed calculations [76].

Figure 12 describes the solutions of Eqs. (2.14) and (2.15) [76] for the constant intensity of the pump radiation, $I_{\mathrm{p}}^0(t)$ = const, and an input pulse in the form of a step. It is clear from this figure that a short single pulse appears at the amplifier output. Steady amplification is achieved after a transient stage, which is in the form of damped pulsations of period

$$\tau > \tau_{\mathrm{tr}} = 2\frac{L}{v}.$$

Steady Regime

If the amplification regime is steady ($\partial/\partial\tau = 0$), the system (2.22) (with the "-" sign) and the boundary conditions of Eqs. (2.19) and (2.20) in the notation of Eq. (2.24) are

$$\frac{dX}{d\xi} = -bXY^{-}, \quad \frac{dY}{d\xi} = -bXY^{-}, \quad X(0) = 1, \quad Y(1) = Y_{1}^{-}. \qquad (2.31)$$

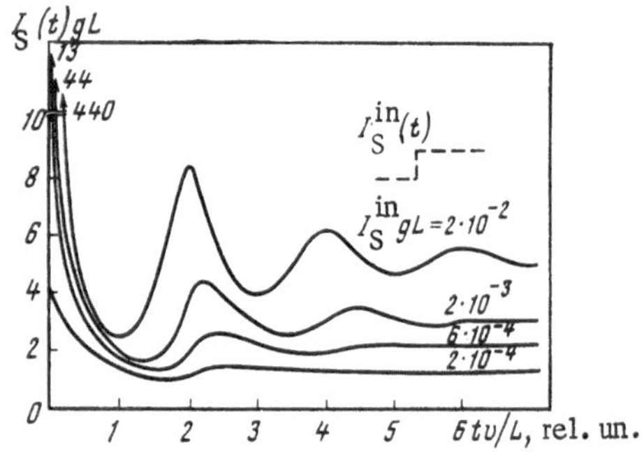

Fig. 12. Opposed amplification of a Stokes "step" pulse in a medium of finite length subjected to constant pumping $I_{\mathrm{p}}gL = 10$.

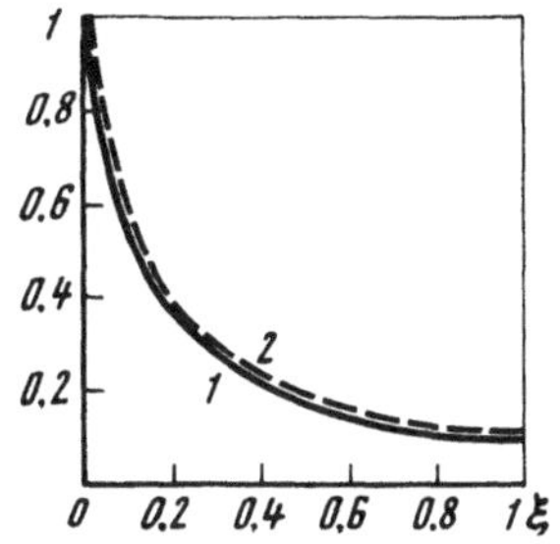

Fig. 13. Distribution of the intensity of the signal (1) and pump (2) radiations along the normalized length $\xi = x/L$ of an active medium in a steady opposed amplifier. The linear growth increment is $b = gI_p(0)L = 10$ and the input signal is $Y_1^- = I_s(1)/I_p(0) = 0.09$.

The expression for the intensity of the amplified signal $Y^-(\xi)$, normalized to the maximum pump intensity, is

$$Y^-(\xi) = \frac{[1 - Y^-(0)]\, Y^-(0)}{\exp\{b\,[1 - Y^-(0)]\,\xi\} - Y^-(0)}. \qquad (2.32)$$

The expression (2.32) is identical, to within $Y^-(\xi) = N_s(x)/N_p(0)$, with Eq. (2.30) [109] for a signal amplified by the stimulated Brillouin scattering. The function $Y^-(0)$ can be found from a transcendental equation obtained as a result of substitution in Eq. (2.32) of the boundary condition for $\xi = 1$: $Y^-(1) = Y_1^-$,

$$Y^-(1) \equiv Y_1^- = \frac{[1 - Y^-(0)]\, Y^-(0)}{\exp\{b\,[1 - Y^-(0)]\} - Y^-(0)}. \qquad (2.33)$$

Solving Eq. (2.33) for b, we obtain

$$b = \frac{1}{1 - Y^-(0)} \ln\left\{Y^-(0)\left[1 - \frac{1 - Y^-(0)}{Y_1^-}\right]\right\}.$$

Substituting this expression into Eq. (2.32), we find that the distribution of the signal intensity along the active medium is (Fig. 13):

$$Y^-(\xi) = \{[1 - Y^-(0)]\, Y^-(0)\} \Big/ \left\{\left[Y^-(0)\left(1 - \frac{1 - Y^-(0)}{Y_1^-}\right)\right]^{\xi} - Y^-(0)\right\}. \qquad (2.34)$$

Equation (2.33) can be used to find the quantum efficiency of the conversion of the pump radiation into a signal in an opposed amplifier:

$$\eta^-(b,\ Y_1) = Y^-(0) - Y_1^-. \qquad (2.35)$$

For a fixed input signal Y_1^- (or Y_0^+), the formulas (2.35) and (2.25) give the dependence of the quantum efficiency η^- (η^+) on the gain increment (Fig. 14). It is clear from Fig. 14 that, in particular, the saturation for a given gain increment is reached earlier in a parallel amplifier than in an opposed one.

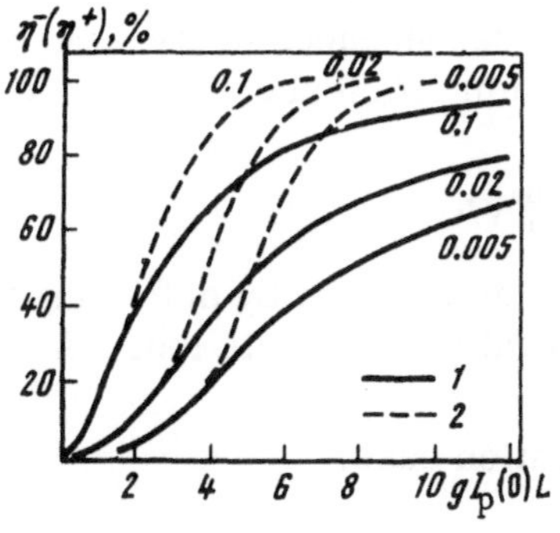

Fig. 14. Dependence of the quantum efficiency of the conversion of the pump radiation into the first Stokes component on the linear gain increment $b = gI_p(0)L$; here, 0.005-0.1 represent intensities of the input signal normalized to $I_p(0)$; 1) opposed amplifier; 2) parallel amplifier.

CHAPTER III

EXPERIMENTAL INVESTIGATION OF AMPLIFICATION DUE TO STIMULATED RAMAN SCATTERING

§ 1. Dynamics of Amplification and Saturation Effect

Apparatus and Measurement Method

We investigated parallel and opposed amplifiers using apparatus shown schematically in Fig. 15. A ruby laser system was used as the pump source: this system was composed of a Q-switched laser and a ruby amplifer (length l = 230 m, diameter 13 mm). The Q switching was performed by a solution of vanadium cryptocyanine in methanol. Part of the ruby laser radiation, deflected by a glass plate, was amplified by the ruby amplifier and used to excite a Raman amplifer. The rest of the ruby laser radiation was used to pump a Raman laser.

The shape and duration of the pulses were measured with coaxial photocells (FÉK-09) in combination with an oscillograph (I2-7). The duration of the pump pulses was varied from 15 to 30 nsec and the spectral line width was about 0.017 cm^{-1}. Liquid nitrogen was the active medium and its length was 25 and 50 cm, respectively. The ruby and Raman lasers were located 3 m from the Raman amplifier to avoid feedback. The signal at the frequency of the first Stokes component was passed through an inverted Galilean telescope with twofold magnification (to reduce the signal divergence); it was also passed through a screen with an aperture and directed by a selective mirror * to the Raman amplifier (Figs. 15b and 16). The characteristics of this mirror were as follows:

Wavelength, μ	Reflection coefficient, %	Transmission coefficient, %
0.6943	10	90
0.828	90	10

The input signal was controlled by neutral filters. The input and amplified signals were delayed by 70 and 35 nsec, respectively, using special lines and they reached the same coaxial photocell. Thus, a single 250-nsec scan performed by the I1-7 oscillograph enabled us to display all three pulses (Fig. 17). Knowing the energy, shape, and duration of the pump, input signal, and output signal pulses, as well as the transverse cross sections of the beams and radial distributions of the intensity in these beams (Fig. 16), we could find the instantaneous quantum efficiency of the conversion of the pump radiation into the first Stokes components and calculate other characteristics of the Raman amplifier (Fig. 18). In particular, the quantum efficiency (deduced from the photon flux via the signal beam cross section) was determined as follows. Oscillograms and the results of calorimetric measurements yielded the time dependences of the pump power $P_p(t)$ and the corresponding dependences of the input $P_S^{in}(t)$ and amplified (output) $P_S^{out}(t)$ signal powers (Fig. 18). The quantum efficiency $\eta(t)$ at a moment t was found from

$$\eta(t) = \omega_p/\omega_S[P_S^{out}(t) - P_S^{in}(t)]/P_p(t).$$

The divergence of the input and amplified signals was determined by deflecting them by a lens L_1 (Fig. 15) and placing a photographic film in the focal plane of this lens. A series of

* The selective mirror was needed in the opposed amplification case.

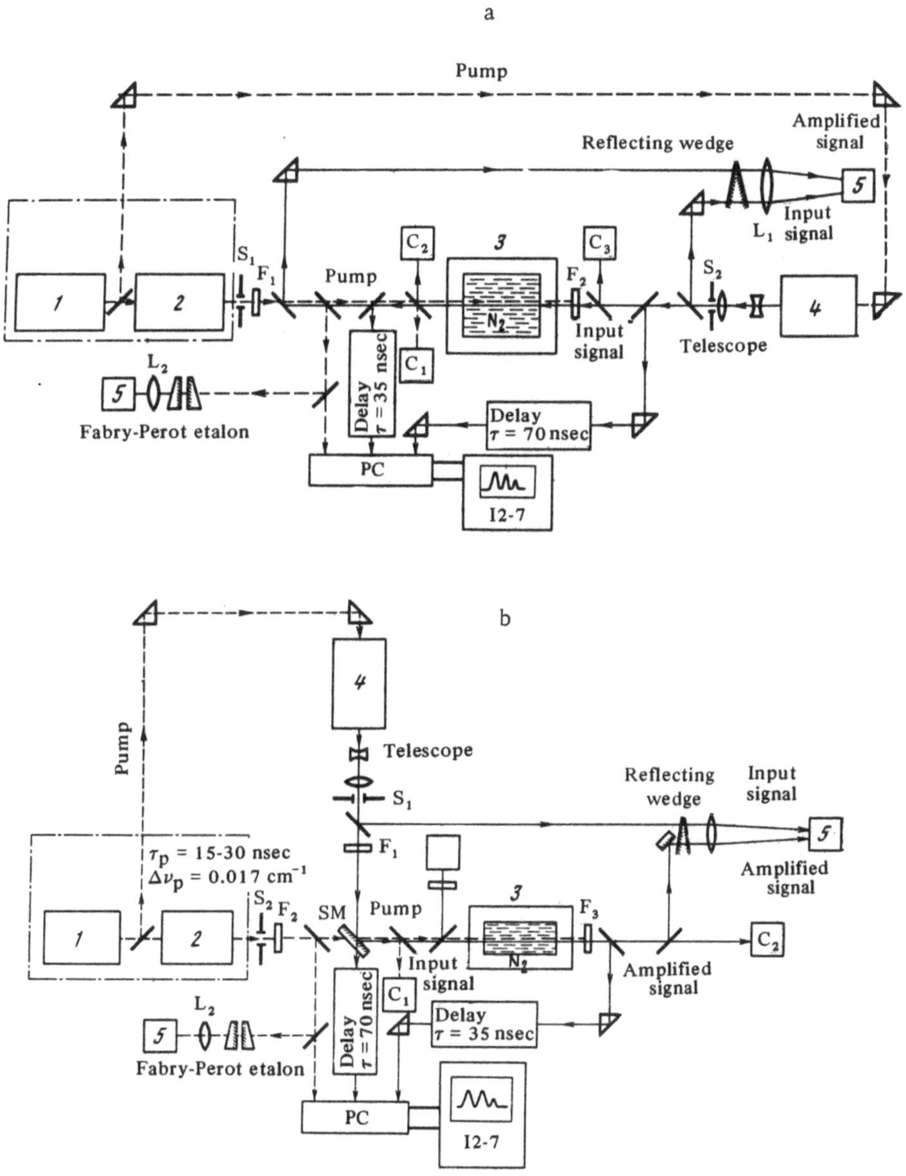

Fig. 15. Block diagram of the apparatus used in studies of opposed (a) and parallel (b) Raman amplifiers. The dashed arrows represent pump radiation of λ_p = 0.6943 μ wavelength; the continuous arrows represent a signal (first Stokes component of SRS) of λ_S = 0.828 μ wavelength (ruby laser); 2) ruby amplifier (τ_p = 15-30 nsec, $\Delta\nu$ = 0.017 cm^{-1}); 3) SRS amplifier; 4) master (SRS) laser; 5) photographic camera; C_1-C_3 are calorimeters; F_1 and F_2 are neutral filters; F_3 and F_4 are selective filters (FS-7) transmitting the signal and absorbing the pump radiation; L_1 and L_2 are lenses (focal length of L_1 is 1 m); PC is a coaxial photocell (FÉK-09); S_1 and S_2 are screens with apertures of 3 mm diameter; SM is a selective mirror.

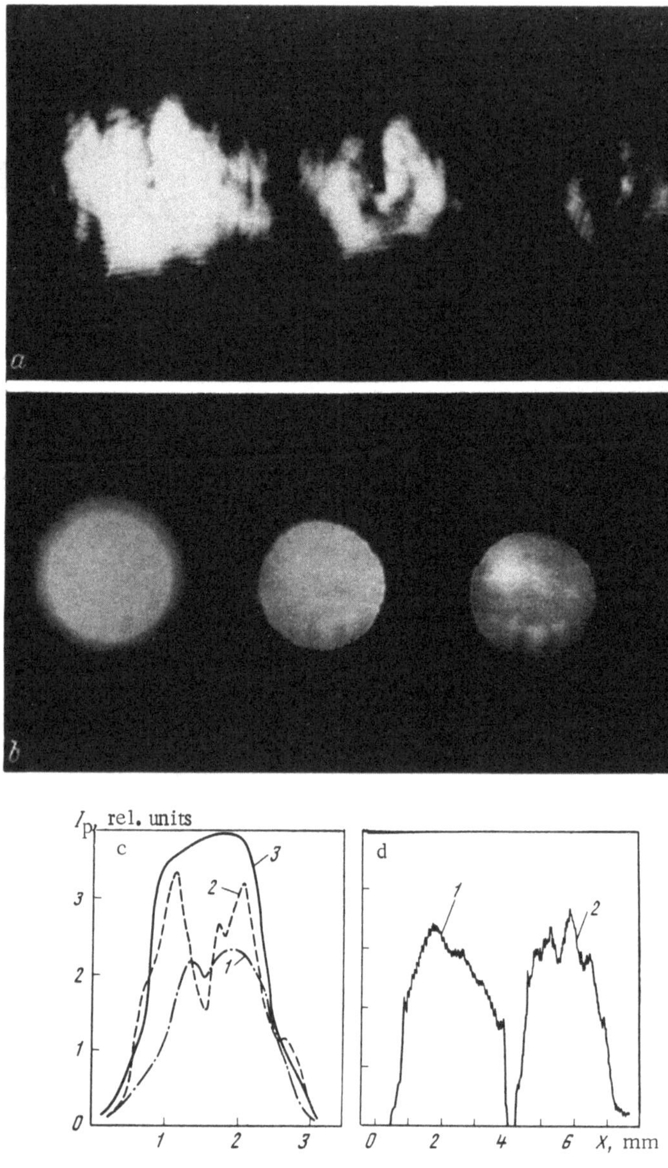

Fig. 16. Transverse distributions of the intensities of pump and input signals: a) pumping at the center of an amplifying cell (attenuation 1:2:4); b) signal (first Stokes component) at the entry to amplifier (attenuation 1:2:4); c) results of photometric analysis of pump radiation (1 – pumping along Y axis, 2 – pumping along X axis, 3 – input signal); d) results of photometric analysis of input signal along two mutually perpendicular directions (1 and 2).

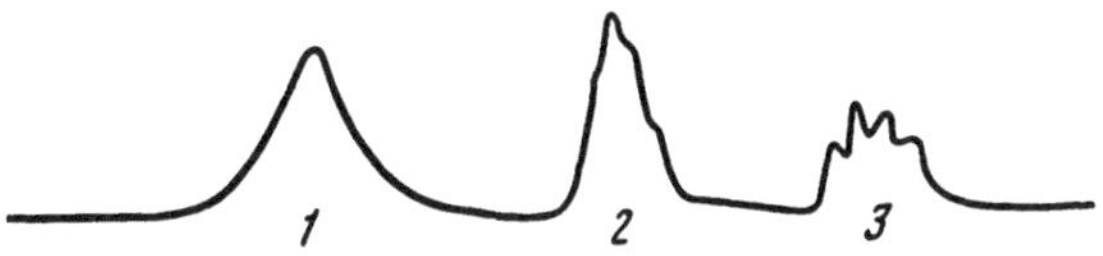

Fig. 17. Oscillograms of light pulses (duration of total scan 250 nsec): 1) pump (λ_p = 0.6943 μ); 2) signal at the exit from parallel Raman amplifier; 3) signal at the entry to the same amplifier (first Stokes component λ_S = 0.828 μ).

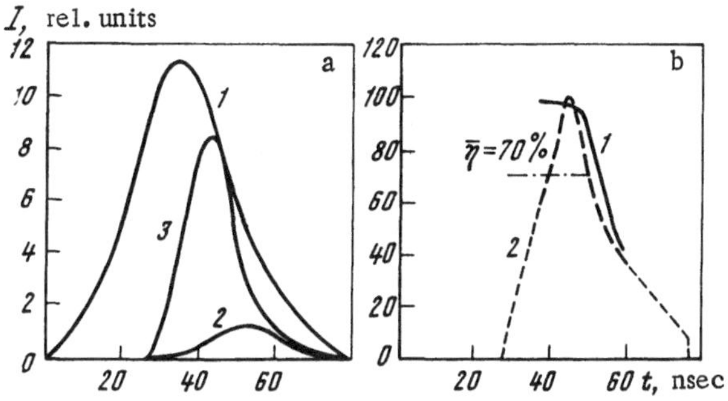

Fig. 18. Oscillograms of pump (1), input signal (2), and amplified signal (3) pulses for a parallel amplifier (b = 6.1) (a) and results of analysis of oscillograms (b) showing the theoretical (1) and experimental (2) time dependences of the instantaneous quantum efficiency (η, %); $\bar{\eta}$ is the time-average value of the quantum efficiency.

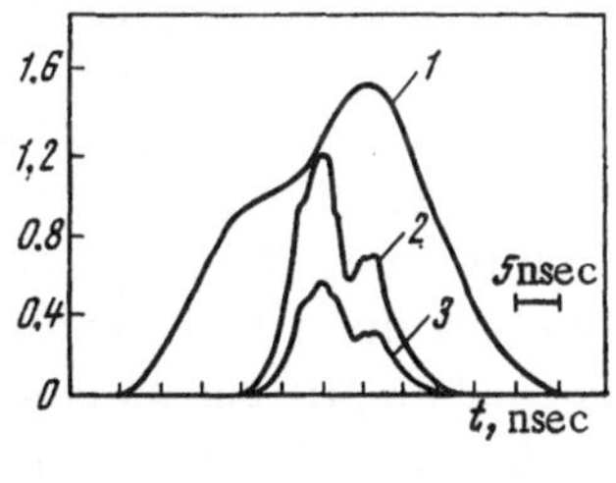

Fig. 19. Parallel amplification of a Stokes pulse in the absence of saturation: 1) pump; 2) amplified signal; 3) input signal (curves 2 and 3 are multiplied by 20).

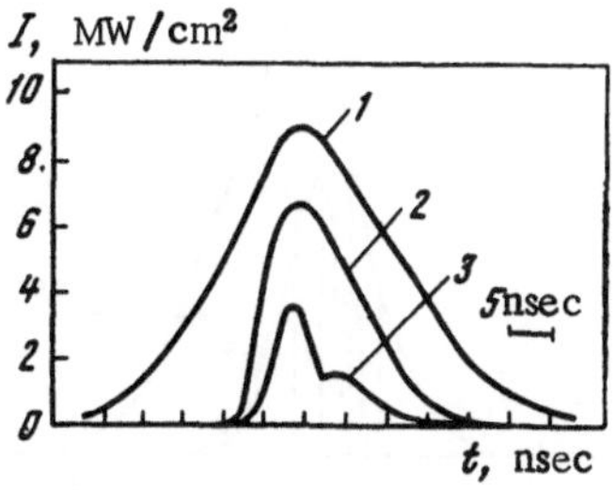

Fig. 20. Parallel amplification of a Stokes pulse under saturation conditions (amplified signal repeats the shape of the pump pulse): 1) pump; 2) amplified signal; 3) input signal (×100).

focal spots was obtained using a steplike attenuator, which was a wedge formed by two semitransparent mirrors [98].

Parallel Amplifier

The results of an investigation of the dynamics of a parallel amplifier are presented in Figs. 18-20. Oscillograms were obtained for different intensities of the input signal and pump wave so that both linear and saturated amplification conditions were studied. Figure 19 shows oscillograms whose scale was selected in accordance with the calorimetric measurements under linear (unsaturated) amplification conditions. This amplification did not change the shape of the pulses (Fig 19). The process was described by Eq. (2.25) in which we could ignore the saturation factor:

$$Y_0^+ \exp [(1 + Y_0^+) b\xi] \ll 1.$$

The saturated amplification regime is illustrated in Figs. 18 and 20. It is clear from Fig. 20 that, in this case, there was a considerable change in the shape of the signal pulse, which was governed by the pump radiation. This amplification regime was described by Eq. (2.25) in which the saturation factor played an important role. Substitution in Eq. (2.25) of the numerical values of $g = 1.6 \cdot 10^{-2}$ cm/MW [50], L = 50 cm, $I_S/I_p^0 = 4 \cdot 10^{-3}$, and the maximum pump intensity $I_p^0 = 9$ MW/cm^2 gave the saturation value of the gain K = 200. The experimental value was K = 185. The linear amplification corresponding to pump radiation of 9 MW/cm^2 intensity was 10^3.

Opposed Amplifier

In contrast to the preceding (parallel) case, transient processes which affected the front of the signal played an important role in the opposed amplification. Figure 21 shows oscillograms obtained in an experiment in which the input signal front was nearly vertical. The scale of the oscillograms was selected in accordance with the results of the energy measurements. These oscillograms revealed the characteristic features of the nonsteady (transient) opposed amplification of such a steplike signal: a single short pulse appeared at the front, in agreement with the results reported in [76, 88].

The results of an investigation of steady opposed and parallel amplification are presented in Fig. 22 in the form of dependences of the quantum efficiency on the linear gain increment. We found, in particular, that the steady-state formulas (2.25) and (2.33) described satisfactorily the processes occurring in parallel and opposed amplifiers. Under saturation conditions, the quantum efficiency of the conversion to the first Stokes component reached a value close to 100%. We also deduced from Fig. 22 that, for the same pump and signal intensities, the saturation effect occurred earlier in a parallel amplifier than in an opposed amplifier. This could be understood quite easily by going back to Figs. 11 and 13, showing the distributions of the pump intensity along the active medium under saturation conditions. It is clear from Fig. 13 that, in the opposed amplification case, the pump intensity fell strongly at the front of the active medium so that the signal traversed a large part of this medium practically without ampli-

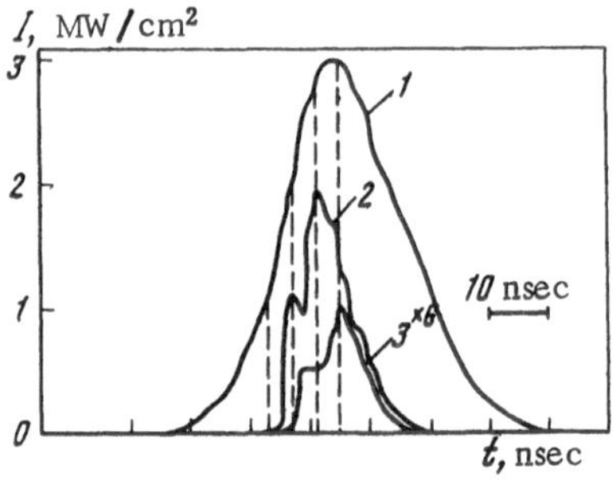

Fig. 21. Deformation of the input signal in opposed amplification: 1) pump; 2) amplified signal; 3) input signal.

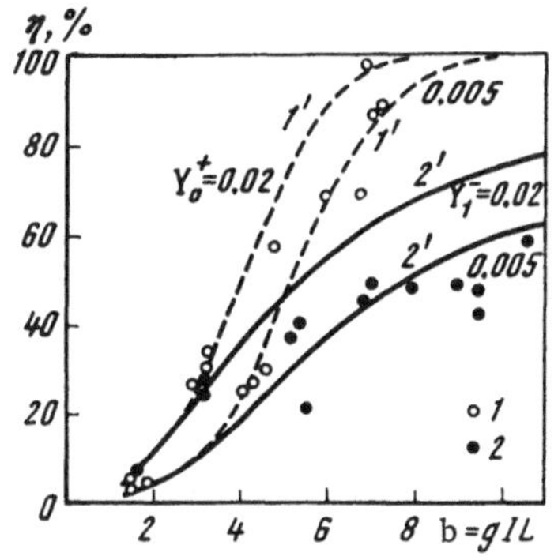

Fig. 22. Quantum efficiency of the conversion of pump radiation into the first Stokes component (calculated from the pulse-average photon flux density) plotted as a function of the gain increment b under steady-amplification conditions: 1) experimental results; 1') corresponding theory for steady parallel amplification; 2) experimental results; 2') corresponding theoretical results for steady opposed amplification; Y_0^+ and Y_1^- are the intensities of the input signal normalized to the maximum pump intensity in the case of parallel and opposed amplifiers, respectively.

fication. Conversely, in the parallel amplification case (Fig. 11), the signal reached that part of the medium where the pump radiation (and, consequently, the gain) had the maximum values. In the absence of saturation, the relative directions of propagation of the signal and pump radiations were unimportant.

The results of measurements of the angular distribution of a signal in a parallel amplifier are given in Fig. 23: it was found that the proportion of the signal energy contained within a given angle decreased with rising quantum efficiency.

§ 2. Influence of Pump Radiation Spectrum on Amplification Due to Stimulated Scattering

Physical Considerations

Let us consider a medium of length L whose gain is g and for which the spontaneous scattering line width is $\Delta\nu_l$. Let us assume that this medium is excited by a pump wave of

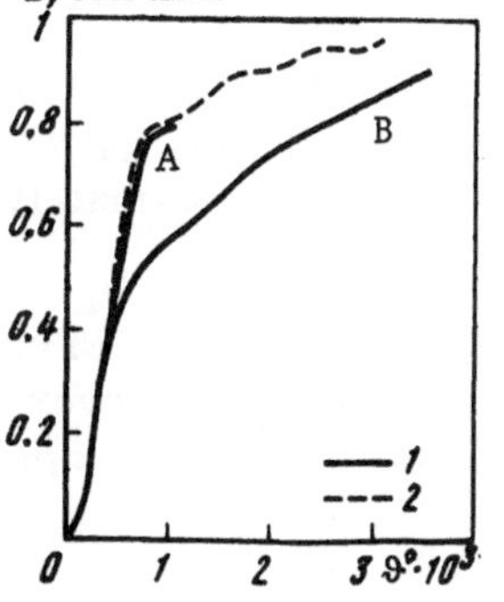

Fig. 23. Proportion of the energy E of an amplified signal contained within an angle ϑ plotted as a function of ϑ (parallel amplifier): 1) experimental results; 2) theoretical dependence for a plane wave transmitted through an aperture in a screen whose diameter is equal to the diameter of the amplified beam; A) b = 0.9, Y_0^+ = 0.04, η = 15%; B) b = 6.3, Y_0^+ = 0.04, η = 80%.

intensity I_p. The amplification of light at the frequency of the first Stokes component obeys the exponential law $\exp(gI_pL)$. The gain is of the form $g = A/\Delta\omega$ and, in the SRS case, we have $A \approx \lambda_S^2 \sigma N/4$, where λ_S is the wavelength of the first Stokes component; σ is the integrated scattering cross section; N is the concentration of molecules. In the SBS case, we have $A = \gamma^2 k^2/n_S^3 \rho v_{hs}$, where γ is the photoelastic constant of the medium; ρ is the density of the medium; n_S is the refractive index; $k = 2\pi/\lambda$ is the wave vector; v_{hs} is the velocity of hypersound. Allowing for the final pump line width $\Delta\nu_p$, we can show that

$$\Delta\omega \approx 2\pi(\Delta\nu_p + \Delta\nu_l). \quad (3.1)$$

Under wide-band pumping conditions, when $\Delta\nu_p \gg \Delta\nu_l$, the gain should always decrease with increasing $\Delta\nu_p$:

$$g \sim A/\Delta\nu_p. \quad (3.2)$$

However, the experiments carried out by the present author and others [80] demonstrated that this was not true. It was found that, when the pump intensity was sufficiently high, the efficiency of conversion of the pump radiation into the Stokes component, governed by the gain, could be independent of the spectral line width of the pump radiation for a wide range of this width from 10^{-3} to 40 cm^{-1}. Theoretical studies were later made [81-84] of the various features of the stimulated scattering in the field of wide-band pump radiation. According to the theory developed in [81-83], the ratio of the parallel gain in the wide-band pumping case g_w to the corresponding gain g_0 under monochromatic excitation conditions should be

$$\frac{g_w}{g_0} = \begin{cases} \frac{\Delta\nu_l}{\Delta\nu_p} & \text{for} \quad I_p < I_p^{th} = \frac{2\pi\Delta\nu_p(cm^{-1})}{g_0}|v'|, \\ 1 & \text{for} \quad I_p \gg I_p^{th}. \end{cases} \quad (3.3)$$

Here, I_p is the pump intensity, I_p^{th} its threshold value, and $v' = (v_S - v_p)c/v_p v_S$, where v_S and v_p are group velocities of the Stokes signal and pump wave. The expression (3.3) means that if I_p exceeds a certain threshold value, the amplification in the wide-band case is exactly the same as in the narrow-band case.

The physical meaning of Eq. (3.3) is as follows. Let us assume that a pump wave traveling across an active medium at a group velocity v_p is represented by a narrow-band random process with a specturm whose width $\Delta\omega_p = 2\pi\Delta\nu_p$ is governed by $\tau_{fl} \approx 2/\Delta\omega_p$, which is the time constant of the fastest fluctuation. We shall now consider the stimulated scattering process occurring during the fastest fluctuation time τ_{fl}. In this case, we may assume that the gain and pump intensity are constant. A signal traveling at the group velocity v_S passes through the medium in which the gain per unit length is gI_p (cm^{-1}). Amplification is possible if the difference between the transit times of the signal and pump radiations across a distance equal to the amplification length $l_a = 1/gI_p$ is less than the fluctuation time τ_{fl} (phase-matching conditions):

$$\frac{1}{gI_p}\left(\frac{1}{v_p} - \frac{1}{v_S}\right) \ll \tau_{fl} \approx (2\pi\Delta\nu_p)^{-1}.$$

Hence, we obtain the inequality of Eq. (3.3). The factor v' clearly decreases with decreasing difference between the group velocities of the signal and pump waves. In the opposed amplification case we have $|v'| \sim 2$ and the threshold excitation intensity is approximately two orders of magnitude higher than in the parallel amplification case.

These estimates explain the results obtained in [80]: when the excitation intensity is sufficiently high, the amplification is independent of the pump spectrum. Moreover, this analysis predicts that, in the wide-band pumping case, the dependence of the gain increment on the pump intensity should have a threshold and be strongly asymmetric relative to the forward–backward direction.

Experimental Investigation of the Influence of Pump Spectrum on SRS Amplification

1. Apparatus and Measurement Method. Experimental results indicated that the amplification was asymmetric and had a threshold. The present author and his colleagues investigated the dependence of the SRS gain increment on the pump intensity for different ratios of the spectral width of the pump line $\Delta\nu_p$ to the spectral width of the spontaneous Raman scattering line $\Delta\nu_l$ [85]. A study was made of the two principal variants of collinear amplification of an external signal, i.e., parallel and opposed amplifiers (in these cases, the pump and signal were either parallel or opposed).

A Q-switched phosphate neodymium glass laser [110] was used as the pump source. The spectrum of this laser could be varied from $\Delta\nu_p \leq 3 \cdot 10^{-2}$ cm^{-1} (Q switching by a saturable filter) to $\Delta\nu_p = 0.3$ cm^{-1} (Q switching by a rotating prism and mode selection: two plane-parallel plates of different thickness were used as the exit mirror of the resonator). The active medium was liquid nitrogen [111, 112]. It is clear from Table 2 that in this case $\Delta\nu_l =$ 0.067 cm^{-1} [49] and the gain was g = 10^{-2} cm/MW [58]. Thus, exciting radiation with a spectrum of width $\Delta\nu_p = 0.3$ cm^{-1} was of the wide-band type for liquid nitrogen ($\Delta\nu_p > \Delta\nu_l$), whereas, in the $\Delta\nu_p = 3 \cdot 10^{-2}$ cm^{-1} case, the radiation was of the narrow-band type ($\Delta\nu_p < \Delta\nu_l$). According to Eq. (3.3), the threshold pump intensity should be $I_p^{th} = 1.8$ MW/cm^2. At this intensity there should be no undesirable secondary phenomena such as self-focusing, significant transformation into the Stokes component in the absence of an external signal (this was observed for $I_p \approx 6$-8 MW/cm^2), and so on. In the opposed amplification case, the corresponding threshold should be $I_p^{th} = 300$ MW/cm^2. Thus, wide-band pump radiation was chosen and, at the same time, a relatively low threshold I_p^{th} was used to avoid the influence of secondary nonlinear effects.

A block diagram of the apparatus employed is shown in Fig. 24. The radiation emitted by the neodymium laser was split by a mirror M_3 into two beams. One was focused by a lens L_2 into a 50-cm long cell filled with liquid nitrogen. The first SRS Stokes component was used as an external signal which was passed to an SRS amplifier located in the lower part of the same cell.* The energy of the light pulses was measured with calorimeters C_1-C_3, whereas the shape and duration of the pump pulses were recorded with a coaxial photocell and an oscillograph. The spectral composition of the radiation was determined with a Fabry–Perot interferometer.

Measurements were made of the value of $\ln K_g = \ln(E_S^{out}/E_S^{in})$ as a function of the intensity of the pump radiation which excited an active medium of constant length L = 50 cm (E_S^{in} and E_S^{out} were, respectively, the energies of the Stokes signal at the entry into and exit from the Raman amplifier). If the gain obeyed the exponential law, the gain increment should be proportional to the pump intensity I_p because $\ln K_g = gI_pL$. The deviation of the dependence of $\ln K_g$ on I_p from linearity indicated a change in the gain g.

2. Results of Experiments and Discussion. The experimental results are plotted in Fig. 25. Under narrow-band pumping conditions (dependences I and III), the "narrow-band" gain was practically constant ($g_0 \approx 10^{-2}$ cm/MW) and was independent of the relative directions of the pump and signal. Completely different dependences (II, IV) were observed in the wide-band case. As long as the pump intensity was below a certain threshold ($I_p^{th} \sim 3$ MW/cm^2), the

* The intensities of the external Stokes signal and of the pump radiation were selected to ensure that no saturation occurred (linear amplification).

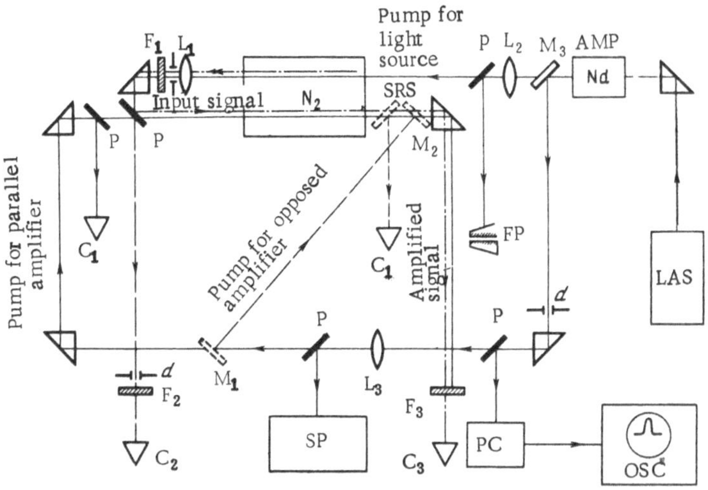

Fig. 24. Block diagram of the apparatus used in investigations of the influence of the pump radiation spectrum on the amplification based on stimulated scattering: C_1-C_3 are calorimeters; F_1-F_3 are selective filters transmitting the Stokes signals and suppressing the pump radiation; P are plane-parallel glass plates; SP is an STÉ-1 spectrograph; PC is a coaxial photocell; OSC is an I2-7 oscillograph; LAS is a Q-switched phosphate neodymium-glass laser; FP is a Fabry–Perot etalon (l = 0.5 cm); AMP is a neodymium-glass amplifier; the continuous arrows represent pump radiation and the chain arrows the SRS signal; the dashed line identifies that part of the system which is used in studies of opposed Stokes amplifiers.

wide-band gain g_w in the parallel amplification case was considerably lower than the narrow-band gain ($g_w \ll g_0$). When the threshold was exceeded the value of g_w rose rapidly, approaching g_0. In other words, the amplification became of the same kind as in the narrow-band case. In the opposed amplification case, the wide-band gain was practically zero ($g_w \approx 0$) throughout the investigated range of pump intensities.

The threshold nature of the dependence of the gain increment on the pump intensity was responsible for the failure of the first experiments [113], in which attempts were made to determine the dependence of the SRS gain on the width of the pump spectrum. In these experiments, the gain of nitrobenzene was estimated from the efficiency of conversion of the pump radiation into the first Stokes component. The pump source was a Q-switched ruby laser. This laser was operated either: 1) with mode selection, generating four axial modes (the spectral width of the emission line of the laser was then about $2 \cdot 10^{-2}$ cm^{-1}); 2) without mode selection (in this case, the width of the emission spectrum was 0.5 cm^{-1}). It was established that,

Fig. 25. Experimental dependences of ln K_g on the pump intensity I_p for variants I and II of parallel amplifiers and variants III and IV of opposed amplifiers: I, III) narrow-band pumping: $\Delta\nu_p \leqslant 0.03$ cm^{-1}; II, IV) wide-band pumping: $\Delta\nu_p = 0.3$ cm^{-1}.

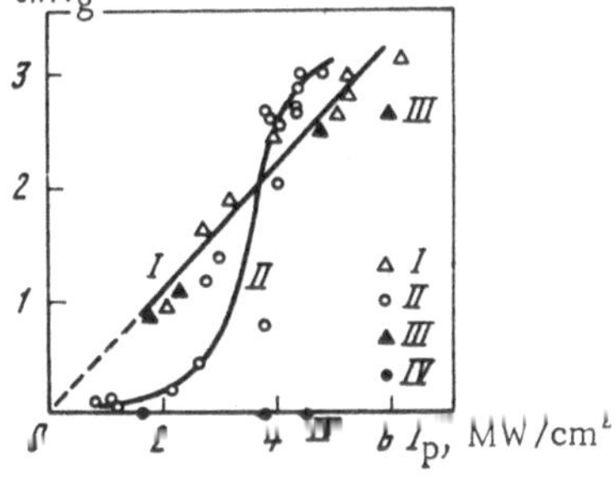

for a given pump intensity, $I_p = 19.5$ MW/cm^2, the gain was the same (~0.6%) irrespective of the regime in which the ruby laser was operated.

These results can easily be explained. In fact, according to Eq. (3.3), the threshold pump intensity for $\Delta\nu_p = 0.5$ cm^{-1} is $I_p^{th} = 1.5$ MW/cm^2. The pump radiation used in [113] was of $I_p = 19.5$ MW/cm^2 intensity, i.e., almost an order of magnitude higher than I_p^{th}. Consequently, amplification occurred in exactly the same way as in the narrow-band case. It should be pointed out that even when the pump intensity is less than the threshold, the reduction in the gain for $\Delta\omega_p = 0.5$ cm^{-1} should be slight because the pump radiation is not of the wide-band type: the spectrum of the pump radiation is still narrower than the spontaneous scattering line, i.e., $\Delta\nu_p < \Delta\nu_l \approx 1.5$ cm^{-1} (Table 2).

§ 3. Competition between Different Types of Stimulated Scattering

Conditions for Simultaneous Development and Competition between Different Types of Stimulated Scattering

Let us assume that a medium which is active in respect of two types of stimulated scattering is excited by a Gaussian pulse. According to Eq. (2.10), the ratio of the gain increments at a moment t is given by the expression

$$\frac{b_1(t)}{b_2(t)} = \begin{cases} 1)\ \dfrac{g_{01}}{g_{02}} & \text{for } t \gg 2bT_2^{(1,2)}, \\ 2)\ \left(\dfrac{g_{01}T_2^{(2)}}{g_{02}T_2^{(1)}}\right)^{1/2} & \text{for } t \ll bT_2^{(1,2)}, \\ 3)\ \dfrac{g_{01}IL}{(2g_{02}IL)^{1/2}}\left(\dfrac{T_2^{(2)}}{t}\right)^{1/2} & \text{for } T_2^{(1)} \ll t \ll T_2^{(2)}, \end{cases} \tag{3.4}$$

where $T_2^{(1)}$ and $T_2^{(2)}$ are, respectively, the transverse relaxation times of each of the two types

TABLE 4. Stimulated Brillouin Scattering (SBS) in Liquid Nitrogen and Oxygen*

Substance	T, °K	γ	n	ρ, g/cm^3	v · 10^{-4}, cm/sec [112]	α, cm^{-1} [111]	τ_{SBS}, nsec	t_{st}, nsec	g_{SBS} · 10^2, cm/MW
Nitrogen	77	0.32	1.18	0.8	8.69	954	12	210	1.5
Oxygen	90	0.36	1.2	1.14	9.11	774	14	248	1.4
	77	0.36	1.2	1.2	10.05	1054	9.4	164	0.85

*The quantities g, γ, and τ in this table are calculated using the formulas [13, 41]

$$g_{SBS} = \frac{\gamma^2 \tau_{SBS} k^2}{2cn^3\rho v},\quad \gamma = \frac{1}{3}(n^2-1)(n^2+1),\quad t_{st} = \frac{\tau g I L}{2}\left(1+\frac{2}{\sqrt{gIL}}\right),\quad g_{STS} = \frac{\tau_{STS} k\,(\partial\varepsilon/\partial T)}{n^2\rho C_p}\delta.$$

Here and later, the following notation is used: γ is the photoelastic constant; $\tau_{SBS} = 1/\alpha v$ and $\tau_{STS} = 1/\varkappa q^2$ are the SBS and STS relaxation times, respectively; α is the absorption coefficient of sound; v is the velocity of sound; χ is the thermal diffusivity; k is the wave vector for $\lambda = 0.6943$ μ; n is the refractive index; ρ is the density; t_{st} is the time taken to establish steady-state conditions; I is the intensity of the pump radiation; L is the length of the scattering volume (gIL = 25); ε is the permittivity of the medium employed; δ is the absorption coefficient of light; C_p is the specific heat at constant pressure; T is the temperature; λ_S is the wavelength of the first Stokes component; Δ is the width of the Raman scattering line. The values of t_{st} were calculated using the above formula and the appropriate values of τ and g.

TABLE 5. Stimulated Raman Scattering (SRS) in Liquid Nitrogen and Oxygen

Substance	T, °K	λ_S, μ	Δ, cm^{-1} [50]	τ_{SRS}, nsec	t_{st}, nsec	$g_{SRS} \cdot 10^2$(exp.), cm/MW [50]
Nitrogen	77	0.828	0.067	0.08	1.4	1.6
Oxygen	90	0.778	0.117	0.045	0.8	1.6

of scattering. It is clear from the above relationships (see Tables 4-6) that when subnanosecond and picosecond pump pulses are used, only SRS is possible [114-119]. In fact, typical values of the relaxation times for SRS and SBS (Tables 2 and 5) are, respectively, $T_2^{SRS} \sim 10^{-11}$ sec and $T_2^{SBS} \sim 10^{-9}$ sec. This corresponds approximately to case 3 in Eq. (3.4). We shall now assume that $t = 10^{-10}$ sec and that the pump intensity I and length of the active medium L are selected so that $b = gIL = 25$. Substituting the values of $T_2^{SRS, SBS}$ and $g_{SRS} \approx g_{SBS} \approx 10^{-2}$ cm/MW into Eq. (3.4), we find that the ratio of the gain increments is $b_{SRS}/b_{SBS} \approx 35$.

Simultaneous Excitation of SRS, SBS, and STS in Liquid Nitrogen and Liquid Oxygen

We investigated stimulated scattering in liquid nitrogen and liquid oxygen [96] and observed simultaneously three types of stimulated scattering: SRS, SBS, and STS (associated with absorption). It is clear from Tables 4-6 that the steady-state values of the gain for all three types of scattering are similar. Nevertheless, until the experiments reported in [96], only SRS had been observed in liquid nitrogen [50, 57, 58, 120-122] and in liquid oxygen [50]. This was due to the fact that in these experiments on liquid N_2 and O_2, the excitation sources were nanosecond light pulses. It is clear from Tables 4-6 that, in this case, the steady-state value of the gain is attained only for SRS. The gain for the stimulated Raman scattering is highest and this scattering is excited preferentially. In our experiments [96] we used 0.5 μsec pulses. Therefore, the duration of these pulses exceeded the time needed for the excitation of any one of these types of scattering so that they were all observed simultaneously.

The apparatus employed is shown schematically in Fig. 26. A master laser was operated in a special way so that single light pulses of 0.15-0.2 J energy and 0.5 μsec duration were emitted [70, 123]. A combination of a screen with an aperture and a saturable filter made it possible to reduce the width of the emission spectrum to a value below 0.03 cm^{-1}. A light pulse produced by this master laser passed through a Faraday isolator and reached the input of an amplifier, where its energy was increased to 5-8 J. This amplification was accompanied by some reduction in the pulse duration but the shape of the pulse was selected to minimize this effect. An additional isolator was used at the amplifier exit: this isolator was a Glan–Thompson prism in combination with a quarter-wave plate; the pump radiation passing through

TABLE 6. Stimulated Thermal Scattering (STS) Associated with Absorption in Liquid Nitrogen and Oxygen

Substance	T, °K	C_p, J/g · deg [112]	$\chi \cdot 10^4$ cm^2/sec [112]	τ_{STS}, nsec	$(\partial\varepsilon/\partial T)_p \cdot 10^3$	t_{st}, nsec	δ, 10^{-3} cm^{-1}	$g_{STS} \cdot 10^2$, cm/MW
Nitrogen	77	2.1	8.4	25	2	440	5	1
Oxygen	90	1.71	7.7	27	2	470	7	1.31
	77	1.67	8.3	25	2	440	7	1.1

Notes. 1) The values of n and ρ are the same as in Table 4. 2) The numerical value of $(\partial\varepsilon/\partial T)_p$ is unknown; the value given is typical of the majority of liquids.

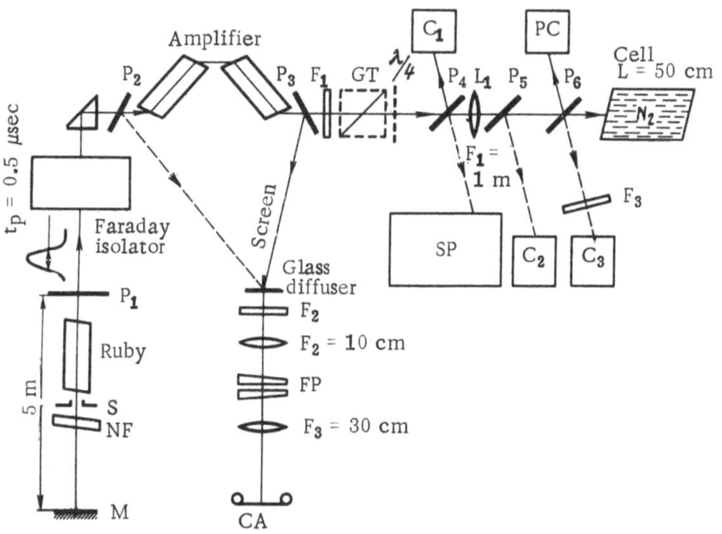

Fig. 26. Block diagram of the apparatus used for observation of SRS, SBS, and STS in liquid N_2 and O_2: P_1-P_6 are plane-parallel glass plates; C_1-C_3 are calorimeters; F_1 is an attenuating filter; F_2 is a selective filter transmitting $\lambda = 0.7\ \mu$ and absorbing $\lambda = 0.828\ \mu$; F_3 is a selective filter transmitting $\lambda = 0.828\ \mu$ and absorbing $\lambda = 0.7\ \mu$; GT is a Glan–Thompson polarizer (prism); M is a mirror; NF is a narrow-band filter; S is a screen with an aperture; CA is a photographic camera; PC is a coaxial photocell (FÉK); SP is an STÉ-1 spectrograph. The dashed lines represent the radiation reflected from the cell with liquid nitrogen or liquid oxygen.

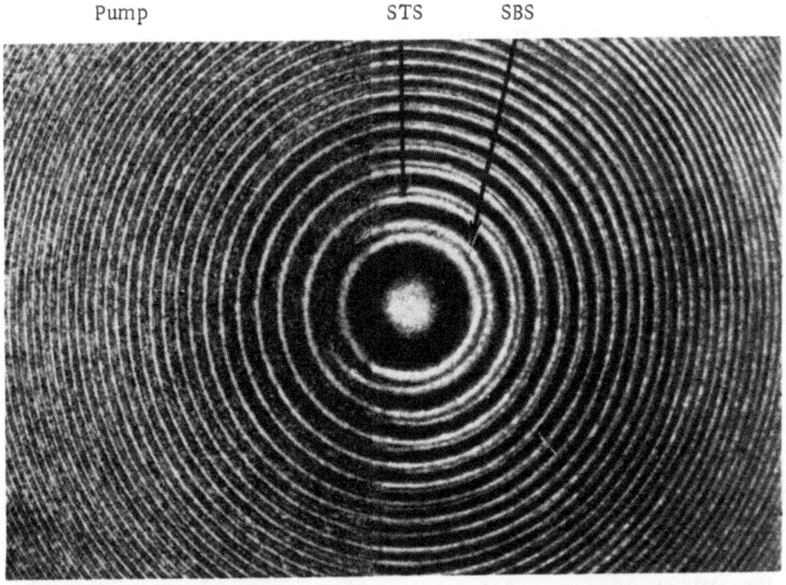

Fig. 27. Interferograms of pump radiation (left half) and back-scattered radiation (right half) obtained for liquid nitrogen at T = 77°K using a Fabry–Perot interferometer with mirrors 3 cm apart.

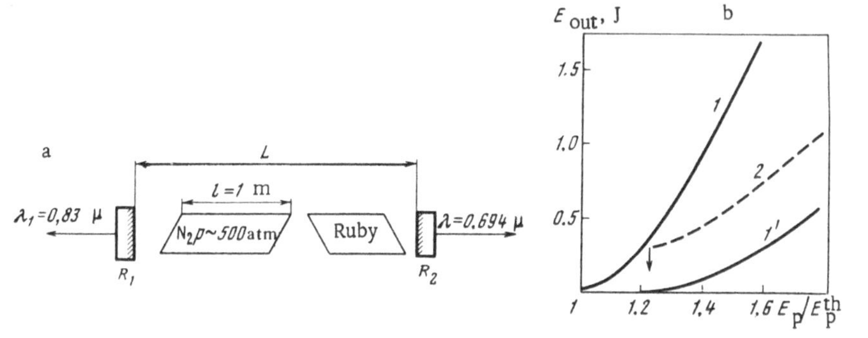

Fig. 28. Block diagram of the apparatus used in investigations of Q switching by SBS (a) and dependences of the output energy on the excess of the pump energy E_p over the threshold E_p^{th} under free-oscillation conditions (b): 1) p = 1 atm, L = 2.5 m; 1') p = 1 atm, L = 7.5 m; 2) giant pulses, p = 500 atm, L = 75 m. Here, R_1 = 99% for λ = 0.694 μ and 12% for λ = 0.83 μ; R_2 = 60% for λ = 0.694 μ and 97% for λ = 0.83 μ.

this prism reached a cell with liquid N_2 (or O_2) and the temperature in the cell was maintained at T = 77°K using a special cooling jacket. The spectra of the incident and backscattered light were recorded simultaneously with a Fabry–Perot etalon. These spectroscopic measurements of the backscattered radiation (representing up to 30% of the total energy) indicated the occurrence not only of SRS but also of thermal or Rayleigh scattering (STS), associated with the absorption and SBS. The thresholds of these effects were identical to within experimental error. A typical interferogram is shown in Fig. 27. A strong Rayleigh scattering line and a weaker SBS line were usually observed. In some experiments, we observed only SBS or only STS in combination with SRS.

§ 4. Formation of Light Pulses with the Aid of Stimulated Scattering

Stimulated scattering (Table 3) can be used to form picosecond [86, 88], subnanosecond [87], nanosecond [89, 90-92], and microsecond [94, 95] pulses. Light pulses of 20-30 psec duration were generated in [86, 88] as a result of opposed SRS amplification under saturation conditions. About 10% of the pump energy was transformed into these ultrashort pulses and the power of these pulses could be an order of magnitude greater than the power of the pump radiation.

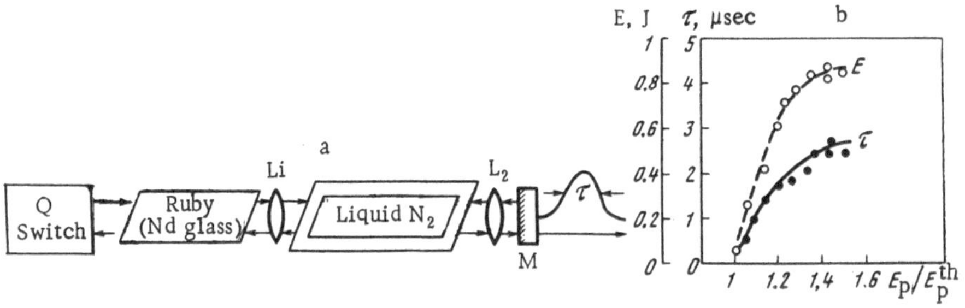

Fig. 29. Block diagram of the apparatus used for lengthening light pulses (a) and dependences of the pulse duration τ and pulse energy E on the excess of the pump energy over the threshold value E_p/E_p^{th} (b); L_1 and L_2 are confocal lenses.

The SBS [89, 92] and STS [93] effects were used successfully to generate giant nanosecond pulses in a Q-switched resonator. A cell containing the active medium was placed inside the resonator of a solid-state laser. The backward SBS Stokes component was amplified in the cell and in the active laser rod. This was equivalent to an increase in the reflection coefficient of one of the resonator mirrors. This "mirror" played a decisive role during the later stages and coherent radiation was emitted from the resonator in which the mirror was a cell with the active medium. This effect was achieved in the earliest investigations by the use of intense light fluxes. Light was focused either into the cell [89] or special preliminary Q switching was employed [90]. A different system (Fig. 28) was used in [91]: this system was free of focusing elements and no preliminary Q switching was necessary. It made it possible to avoid such undesirable phenomena as optical breakdown in the active medium, self-focusing, etc. The active medium was nitrogen gas under a pressure of 500 atm. The development of SBS generated pulses of ~ 0.3 J energy and ~ 4 nsec duration. The high intensity of the optical field in the resonator ensured efficient conversion into the first Stokes component of SRS ($\lambda_1 = 0.83\ \mu$). The generation of $\lambda = 0.694\ \mu$ pulses was accompanied by the emission of light at $\lambda_1 = 0.83\ \mu$ in the form of pulses of ~ 0.2 J energy and ~ 3 nsec duration. A further increase in the pumping rate produced a series of high-power pulses at wavelengths of 0.694 and 0.83 μ, and up to 70% of the free-oscillation energy was transformed into these pulses.

A method for lengthening pulses emitted by a Q-switched ruby laser was described in [94]: in this method, nonlinear losses due to SRS in liquid nitrogen were introduced into the resonator. These losses were proportional to the light intensity in the resonator. Their influence on the stimulated emission was similar to that of a variable feedback which controlled the output light flux.

This method was applied successfully in [95] to a neodymium–glass laser which was Q-switched by a rotating prism. A schematic diagram of the apparatus and the main results obtained are given in Fig. 29. This method made it possible to lengthen the output pulses from several tens of nanoseconds to several microseconds without significant energy losses.

LITERATURE CITED

1. E. J. Woodbury and W. K. Ng, Proc. IRE, 50:2367 (1962).
2. G. Placzek, The Rayleigh and Raman Scattering, US AEC Report UCRL-TRANS-526(L) (1962), 206 pp.
3. V. A. Zubov, M. M. Sushchinskii, and I. K. Shuvalov, Usp. Fiz. Nauk, 83:197 (1964).
4. N. Bloembergen, Am. J. Phys., 35:989 (1967).
5. V. N. Lugovoi, Introduction to the Theory of Stimulated Raman Scattering [in Russian], Nauka, Moscow (1968).
6. R. W. Minck, R. W. Terhune, and W. G. Rado, Appl. Phys. Lett., 3:181 (1963).
7. R. W. Minck, E. E. Hagenlocker, and W. G. Rado, Phys. Rev. Lett., 17:229 (1966).
8. N. Bloembergen, G. Bret, P. Lallemand, A. S. Pine, and P. Simova, IEEE J. Quantum Electron., QE-3:197 (1967).
9. F. M. Johnson, J. A. Duardo, and G. L. Clark, Appl. Phys. Lett., 10:157 (1967).
10. J. J. Barrett and A. Weber, J. Opt. Soc. Am., 60:70 (1970).
11. S. A. Akhmanov, B. V. Zhdanov, A. I. Kovrigin, and S. M. Pershin, ZhETF Pis'ma Red., 15:266 (1972).
12. R. Y. Chiao, C. H. Townes, and B. P. Stoicheff, Phys. Rev. Lett., 12:592 (1964).
13. I. L. Fabelinskii, Molecular Scattering of Light, Plenum Press, New York (1968).
14. V. S. Starunov and I. L. Fabelinskii, Usp. Fiz. Nauk, 98:441 (1969).

15. R. Y. Chiao, E. Garmire, and C. H. Townes, Proc. Intern. School of Physics "Enrico Fermi," Course 31, Quantum Electronics and Coherent Light, Ravenna, 1963, publ. by Academic Press, New York (1964), p. 326.
16. C. K. N. Patel, E. D. Shaw, and R. J. Kerl, Phys. Rev. Lett., 25:8 (1970).
17. C. K. N. Patel, IEEE J. Quantum Electron., QE-7:306 (1971).
18. C. K. N. Patel, Appl. Phys. Lett., 19:400 (1971).
19. C. K. N. Patel, Phys. Rev. Lett., 28:649 (1972).
20. C. K. N. Patel, Appl. Phys. Lett. 18:274 (1971).
21. P. A. Fleury and J. F. Scott, Phys. Rev. B, 3:1979 (1971).
22. R. L. Aggarwal, B. Lax, C. E. Chase, C. R. Pidgeon, and D. Limbert, Appl. Phys. Lett., 18:383 (1971).
23. R. G. Mellish, R. B. Dennis, and R. L. Allwood, Opt. Commun., 4:249 (1971).
24. S. R. J. Brueck and A. Mooradian, Appl. Phys. Lett., 18:229 (1971).
25. R. K. Allwood, R. B. Dennis, R. G. Mellish, S. D. Smith, B. S. Wherett, and R. A. Wood, J. Phys. C. 4:L126 (1971).
26. R. A. Wood, R. B. Dennis, and J. W. Smith, Opt. Commun., 4:383 (1972).
27. C. Irslinger, R. Grisar, H. Wachering, H. G. Hafele, and S. D. Smith, Phys, Status Solidi b, 48:797 (1971).
28. A. Mooradian, S. R. J. Brueck, E. J. Johnson, and J. A. Rossi, Appl. Phys. Lett., 21:482 (1972).
29. S. K. Kurtz and J. A. Giordmaine, Phys. Rev. Lett., 22:192 (1969).
30. J. Gelbwachs, R. H. Pantell, H. E. Puthoff, and J. M. Yarborough, Appl. Phys. Lett., 14:258 (1969).
31. V. I. Bespalov, A. M. Kubarev, and G. A. Pasmanik, Izv. Vyssh. Uchebn. Zaved. Radiofiz., 13:1433 (1970).
32. R. M. Herman and M. A. Gray, Phys. Rev. Lett., 19:824 (1967).
33. D. H. Rank, C. W. Cho, N. D. Foltz, and T. A. Wiggins, Phys. Rev. Lett., 19:828 (1967).
34. G. I. Zaitsev, Yu. I. Kyzylasov, V. S. Starunov, and I. L. Fabelinskii, ZhETF Pis'ma Red., 6:802 (1967).
35. Yu. I. Kyzylasov, V. S. Starunov, and I. L. Fabelinskii, ZhETF Pis'ma Red., 9:383 (1969).
36. D. V. Vlasov and V. S. Starunov, Zh. Eksp. Teor. Fiz., 61:1785 (1971).
37. I. M. Aref'ev and V. V. Morozov, ZhETF Pis'ma Red., 9:448 (1969).
38. W. H. Lowdermilk and N. Bloembergen, Phys. Rev. A, 5:1423 (1972).
39. A. P. Veduta and B. P. Kirsanov, Zh. Eksp. Tepr. Fiz., 56:1175 (1969).
40. J. F. Holzrichter and J. McMahon, Proc. Seventh Intern. Conf. on Quantum Electronics, Montreal, 1972, Digest of Technical Papers, Institute of Electrical and Electronic Engineers, New York (1972), Paper L1, p. 55.
41. G. Eckhardt, R. W. Hellwarth, F. J. McClung Jr, S. E. Schwarz, D. Weiner, and E. J. Woodbury, Phys. Rev. Lett., 9:455 (1962).
42. H. Takuma and D. A. Lennings, Appl. Phys. Lett., 4:110 (1964).
43. J. H. Dennis and P. E. Tannenwald, Appl. Phys. Lett., 5:58 (1964).
44. E. B. Aleksandrov, A. M. Bonch-Bruevich, N. N. Kostin, and V. A. Khodovoi, Opt. Spektrosk., 19:982 (1965).
45. D. P. Bortfeld and W. R. Sooy, Appl. Phys. Lett., 7:283 (1965).
46. G. Eckhardt, IEEE J. Quantum Electron., QE-2:1 (1966).
47. H. E. Puthoff, R. H. Pantell, and B. G. Huth, J. Appl. Phys., 37:860 (1966).
48. B. G. Huth, N. V. Karlov, R. H. Pantell, and H. E. Puthoff, IEEE J. Quantum Electron., QE-2:763 (1966).
49. W. R. L. Clements and B. P. Stoicheff, Appl. Phys. Lett., 12:246 (1968).
50. J. B. Grun, A. K. McQuillan, and B. P. Stoicheff, Phys. Rev., 180:61 (1969).
51. P. V. Avizonis, A. H. Guenther, T. A. Wiggins, R. V. Wick, and D. H. Rank, Appl. Phys. Lett., 9:309 (1966).

52. N. Bloembergen, G. Bret, P. Lallemand, A. S. Pine, and P. Simova, IEEE J. Quantum Electron., QE-3:197 (1967).
53. P. V. Avizonis, K. C. Jungling, A. H. Guenther, R. M. Heimlich, and A. J. Glass, J. Appl. Phys., 39:1752 (1968).
54. F. J. McClung Jr and D. H. Close, J. Appl. Phys., 40:3978 (1969).
55. A. J. Glass and J. M. McMahon, IEEE J. Quantum Electron., QE-5:1 (1969).
56. P. V. Avizonis and R. M. Heimlich, J. Appl. Phys., 40:3650 (1969).
57. A. Z. Grasyuk, V. F. Efimkov, I. G. Zubarev, V. I. Mishin, and V. G. Smirnov, ZhETF Pis'ma Red., 8:474 (1968).
58. V. V. Bocharov, M. G. Gangardt, A. Z. Grasyuk, I. G. Zubarev, and E. A. Yukov, Zh. Eksp. Teor. Fiz., 57:1585 (1969).
59. V. V. Bocharov, M. G. Gangardt, A. Z. Grasyuk, et al., Preprint No. 104 [in Russian], Lebedev Physics Institute, Academy of Sciences of the USSR, Moscow (1969).
60. A. K. McQuillan, W. R. L. Clements, and B. P. Stoicheff, Phys. Rev., 1:628 (1970).
61. E. P. Ippen, Appl. Phys. Lett., 16:303 (1970).
62. R. H. Stolen, E. P. Ippen, and A. R. Tynes, Appl. Phys. Lett., 20:62 (1972).
63. F. A. Korolev, V. I. Odintsov, and E. Yu. Sokolova, ZhETF Pis'ma Red., 13:112 (1971).
64. H. Takuma and D. A. Jennings, Appl. Phys. Lett., 5:239 (1964).
65. E. B. Aleksandrov, A. M. Bonch-Bruevich, N. N. Kostin, and V. A. Khodovoi, Zh. Eksp. Teor. Fiz., 49:1435 (1965).
66. A. S. Pine, Phys. Rev., 149:113 (1966).
67. M. Denariez and G. Bret, Phys. Rev., 171:160 (1968).
68. W. Röther, Z. Naturforsch. a, 25:1120 (1970).
69. V. I. Kovalev, V. I. Popovichev, V. V. Ragul'skii, and F. S. Faizullov, ZhETF Pis'ma Red., 14:503 (1971).
70. A. Z. Grasyuk, V. I. Popovichev, V. V. Ragul'skii, and F. S. Faizullov, Kvant. Elektron. (Mosc.), No. 1, 70 (1971).
71. V. A. Alekseev and I. I. Sobel'man, Zh. Eksp. Teor. Fiz., 54:1834 (1968).
72. M. Maier and G. Renner, Opt. Commun., 3:301 (1971).
73. A. Z. Grasyuk, I. G. Zubarev, V. I. Mishin, and V. G. Smirnov, Preprint No. 32 [in Russian], Lebedev Physics Institute, Academy of Sciences of the USSR, Moscow (1973); Kvant. Elektron. (Mosc.), No. 5(17), 27 (1973).
74. B. A. Akanaev, S. A. Akhmanov, and R. V. Khokhlov, ZhETF Pis'ma Red., 1(4):4 (1965).
75. D. Pohl, M. Maier, and W. Kaiser, Laser Unconv. Opt. J., No. 12, 12 (1968).
76. A. Z. Grasyuk, I. G. Zubarev, V. I. Mishin, and V. G. Smirnov, Preprint No. 14 [in Russian], Lebedev Physics Institute, Academy of Sciences of the USSR, Moscow (1973); Kvant. Elecktron. (Mosc.), No. 5(17), 27 (1973).
77. A. J. Glass, IEEE J. Quantum Electron., QE-3:516 (1967).
78. P. A. Wolff, IEEE J. Quantum Electron., QE-2:659 (1966).
79. M. L. Bhaumik, Am. J. Phys., 35:330 (1967).
80. V. V. Bocharov, A. Z. Grasyuk, I. G. Zubarev, and V. F. Mulikov, Zh. Eksp. Teor. Fiz., 56:430 (1969).
81. Yu. E. D'yakov, ZhETF Pis'ma Red., 11:362 (1970).
82. S. A. Akhmanov, Yu. E. D'yakov, and A. S. Chirkin, ZhETF Pis'ma Red., 13:724 (1971).
83. Yu. E. D'yakov, Kratk. Soobshch. Fiz., No. 7, 43 (1971).
84. S. A. Akhmanov and A. S. Chirkin, Statistical Phenomena in Nonlinear Optics [in Russian], Moscow State University (1971).
85. A. Z. Grasyuk, I. G. Zubarev, and N. V. Suyazov, ZhETF Pis'ma Red., 16:237 (1972).
86. M. Maier, W. Kaiser, and J. A. Giordmaine, Phys. Rev. Lett., 17:1275 (1966).
87. W. H. Culver, J. T. A. Vanderslice, and V. W. T. Townsend, Appl. Phys. Lett., 12:189 (1968).

88. M. Maier, W. Kaiser, and J. A. Giordmaine, Phys. Rev., 177:580 (1969).
89. D. Pohl, Phys. Lett. A, 24:239 (1967).
90. A. J. Alcock and C. De Michelis, Appl. Phys. Lett., 11:42 (1967); Appl. Phys. Lett., 11:185 (1967).
91. A. Z. Grasyuk, V. V. Ragul'skii, and F. S. Faizullov, ZhETF Pis'ma Red., 9:11 (1969).
92. V. I. Popovichev, V. V. Ragul'skii, and F. S. Faizullov, Kvant. Elektron. (Mosc.), No. 5(11), 126 (1972).
93. F. Gires and P. Soep, Proc. IEEE, 56:1613 (1968).
94. A. Z. Grasyuk (Grasiuk), V. F. Mulikov, and L. Csillag, Nuovo Cimento B, 64:300 (1969).
95. A. Z. Grasyuk, I. G. Zubarev, and V. F. Mulikov, Kratk. Soobshch, Fiz., No. 2, 27 (1971).
96. M. G. Gangardt, A. Z. Grasyuk, and I. G. Zubarev, Preprint No. 86 [in Russian], Lebedev Physics Institute, Academy of Sciences of the USSR, Moscow (1971); Kvant. Elektron. (Mosc.), No. 6, 118 (1971).
97. V. N. Lugovoi, Opt. Spektrosk., 27:649 (1969); Zh. Eksp. Teor. Fiz., 56:683 (1969).
98. V. V. Ragul'skii and F. S. Faizullov, Opt. Spektrosk., 27:707 (1969).
99. L. Mandel and E. Wolf, Rev. Mod. Phys., 37:231 (1965).
100. A. L. Mikaélyan, M. L. Ter-Mikaélyan, and Yu. G. Turkov, Solid-State Lasers [in Russian], Sovet-skoe Radio, Moscow (1967).
101. M. Maier and G. Renner, Phys. Lett. A, 34:299 (1971).
102. C. S. Wang, Phys. Rev., 182:482 (1969).
103. S. A. Akhmanov, K. N. Drabovich, A. P. Sukhorukov, and A. S. Chirkin, Zh. Eksp. Teor. Fiz., 59:485 (1970).
104. K. Dareé and W. Kaiser, Phys. Rev. Lett., 26:816 (1971).
105. J. Ducuing, C. Joffrin, and J. P. Coffinet, Opt. Commun., 2:245 (1970).
106. W. H. Culver and E. J. Seppi, J. Appl. Phys., 35:3421 (1964).
107. K. N. Drabovich, Zh. Prikl. Spektrosk., 12:411 (1970).
108. P. G. Kryukov and V. S. Letokhov, Usp. Fiz. Nauk, 99:169 (1969).
109. C. L. Tang, J. Appl. Phys., 37:2945 (1966).
110. A. Z. Grasyuk, I. G. Zubarev, and V. F. Mulikov, Zh. Prikl. Spektrosk., 15:806 (1971).
111. N. B. Vargaftik, Handbook on Thermophysical Properties of Gases and Liquids [in Russian], Fizmatgiz, Moscow (1963).
112. L. Bergmann, Ultrasonics, Bell, London (1938).
113. F. J. McClung Jr, W. G. Wagner, and D. Weiner, Phys. Rev. Lett., 15:96 (1965).
114. D. von der Linde, M. Maier, and W. Kaiser, Phys. Rev., 178:11 (1969).
115. E. E. Hagenlocker, R. W. Minck, and W. G. Rado, Phys. Rev., 154:226 (1967).
116. R. L. Carman, F. Shimizu, C. S. Wang, and N. Bloembergen, Phys. Rev. A, 2:60 (1970).
117. R. L. Carman, M. E. Mack, F. Shimizu, J. Reintjes, and N. Bloembergen, Bull. Am. Phys. Soc., 15:87 (1970); M. E. Mack, R. L. Carman, J. Reintjes, and N. Bloembergen, Appl. Phys. Lett., 16:209 (1970).
118. M. J. Colles, Opt. Commun., 1:169 (1969).
119. R. R. Alfano and S. L. Shapiro, Phys. Rev. Lett., 26:1247 (1971).
120. B. P. Stoicheff, Proc. Intern. School of Physics "Enrico Fermi," Course 31, Quantum Electronics and Coherent Light, Ravenna, 1963, publ. by Academic Press, New York (1964), p. 306.
121. N. G. Basov, A. Z. Grasyuk, V. F. Efimkov, and V. A. Katulin, Fiz. Tverd. Tela, 9:88 (1967).
122. V. V. Bocharov, M. G. Gangardt, A. Z. Grasyuk, I. G. Zubarev, and E. A. Yukov, Zh. Eksp. Teor. Fiz., 57:1585 (1969).
123. V. I. Popovichev, V. V. Ragul'skii, and F. S. Faizullov, Kvant. Elektron. (Mosc.), No. 1, 135 (1971).

THEORETICAL INVESTIGATION OF THE KINETICS OF CHEMICAL LASERS*

V. I. Igoshin

This paper is concerned with some aspects of the theory of chemical lasers. Mathematical models are developed for processes occurring in $H_2 + F_2$ and $D_2 + F_2 + CO_2$ mixtures. A detailed analysis is made of the influence of various factors on the population inversion.

§ Introduction

A chemical laser is a system in which a population inversion and coherent emission are due to the transformation of the free energy of the reagents released in elementary interactions of atoms and or molecules. The transformation is understood to represent the conversion of the chemical energy into the energy of electronic, vibrational, and rotational excitations of molecules or of electronic excitations of atoms without dissipation as heat. Thus, the energy needed for the operation of a chemical laser is supplied by the reaction itself in the course of modification of chemical bonds. Chemical compounds are characterized by a high concentration of energy per unit volume of matter. This can be illustrated by a simple estimate. In the cases of interest to us the energy yield of a reaction is of the order of 10^2 kcal/mole. Consequently, when the reagents are at a pressure of 1 atm, the density of the stored chemical energy is 10 J/cm^3.

The construction of a chemical laser based on the explosive reaction between hydrogen and chlorine, initiated by flash photolysis [1], was followed by the discovery of many chemical reactions which result in laser emission. The progress made up to 1971 is described in [2, 3]. Recently it has been demonstrated that high-power pulse and cw chemical lasers can be built. The most interesting results have been obtained for the $H_2 + F_2$ and $D_2 + F_2 + CO_2$ systems. The first report of the stimulated emission from an $H_2 + F_2$ mixture was given in [4]. The idea of achieving a population inversion by transferring the excitation energy from the "hot" product of a chain reaction (DF) to a "cold" gas (CO_2) was first put in practice in [5]. A mixture of D_2, F_2, and CO_2 was used in [5]. The method of introducing CO_2 into a mixture of D_2 and F_2 was found to be very promising. The $D_2 + F_2 + CO_2$ system [5] was found to have an energy yield an order of magnitude higher than that of the H_2 (D_2) + F_2 system. Stimulated emission was reported in [6] for a mixture of D_2, F_2, and CO_2 at pressures of the order of the atmospheric

---

* This paper is based on the thesis submitted to obtain the degree of Candidate of Physicomathematical Sciences. The work was carried out under the direction of Doctor of Physicomathematical Sciences Professor A. N. Oraevskii.

value. The use of high pressures in the working mixture made it possible to raise the output power of pulse lasers to a value of the order of 1 MW.

The need to master chemical pumping methods has made studies of the kinetic processes and mechanisms of population inversion in chemical lasers one of the most topical problems in quantum electronics and high-energy chemistry. The main purpose of mathematical modeling of the kinetics of chemical lasers is to give a qualitative and quantitative interpretation of the processes occurring in the active medium of a laser and to identify the main factors governing the efficiency of the conversion of chemical energy into coherent radiation.

Since the processes occurring in chemical lasers represent kinetically complex energy-conversion stages, such an interpretation can be provided by developing detailed models of chemical laser systems bearing in mind the complex interrelationship between physicochemical phenomena.

In an earlier communication [7] the present author and Oraevskii used the example of the $H_2 + Cl_2$ reaction in developing a kinetic model of a chemical laser based on a chain reaction resulting in a vibrational excitation of the products. This model allows for the main interactions in a chemical laser including the most important elementary chemical events, exchange of energy, spontaneous and stimulated emission of radiation. A stimultaneous computer solution of nonlinear equations of describing chemical kinetics, vibrational relaxation, and generation of laser radiation shows that a population inversion exists only during the initial stage of the reaction when the consumption of the reagents is still very small (of the order of 1%). The efficiency of such a laser does not exceed 10^{-3}-10^{-4}%. The present paper reports the results of a theoretical analysis of the kinetics of chemical lasers utilizing $H_2 + F_2$ and $D_2 + F_2 + CO_2$ mixtures. The main advantage of the approach employed is that it utilizes the minimum number of approximations and allows for the influence of the time envelope of the initiating pulse, heating of a mixture in the course of the reaction, temperature dependences of the rate constants of elementary processes, and distribution of the energy in elementary chemical events. Since the main processes are allowed for, it is possible to determine their relative importance in the excitation and deexcitation of laser levels. These advantages of detailed modeling of the kinetics of the processes involved can be utilized fully only if comprehensive information is available on the rates of elementary processes.

In spite of the fact that extensive investigations of elementary processes in chemical lasers have been carried out recently, the required information was not available at the time the calculations given below were carried out. In describing several processes it was necessary to make a priori calculations of the rate constants or to extrapolate the available experimental data over a wide range of temperatures using basic theories. A considerable experimental and theoretical effort in studies of elementary events may still be needed. Therefore, at this stage one should guard against overestimating the ability to predict exactly the behavior of kinetically complex systems. However, even if the predictions are proved to be incorrect, the work done in such predictions will help to reveal the key factors and force a more careful analysis of the alternative mechanisms of the overall process. In the light of this discussion some of the results of calculations of the laser characteristics given below should be regarded simply as estimates. The system $D_2 + F_2 + CO_2$ is used to illustrate the influence of changes in some of the kinetic (rate) constants on the results of calculations. In the final section we shall summarize the conclusions which seem to be less critically dependent on the values of little-known kinetic (rate) constants.

§ 1. Characteristics of the Kinetics of Pulse Chemical Lasers

In the present section we shall use the simplest kinetic model to reveal some of the principal effects of relaxation in chemical pumping and we shall formulate the general requirements that a chemical process must satisfy for stimulated emission to occur and for a chemical laser to operate efficiently. The kinetic approach makes it possible to divide chemical lasers into two groups: 1) systems with direct population inversion in an elementary "active" chemical reaction; 2) systems with population inversion of vibrational levels which is the result of energy transfer from "hot" molecules excited by a reaction to "cold" molecules forming the laser medium. We shall consider separately the kinetics of these systems.

Systems with Direct Population Inversion in Elementary Reaction

In this case we shall consider the following basic processes: chemical pumping of active levels, relaxation, and stimulated emission. For the sake of simplicity we shall discuss a two-level system but this choice does not alter the essence of the conclusions drawn. We shall express the rate equations in the form

$$\left.\begin{aligned} \frac{dx^*}{dt} &= \alpha^* w_0 - w_1 x^* - \sigma c\rho\,(x^* - x), \\ \frac{dx}{dt} &= \alpha w_0 + w_1 x^* + \sigma c\rho\,(x^* - x), \\ \frac{d\rho}{dt} &= \sigma c\rho\,(x^* - x) - \frac{\rho}{\tau}, \end{aligned}\right\} \tag{1}$$

where x^* and x are the population densities in the upper and lower active levels, respectively; α^* and α are the probabilities of formation of the reaction product in the upper and lower laser levels, respectively; ρ is the photon density in the resonator; w_0 is the chemical reaction rate; w_1 is the relaxation probability; σ is the stimulated transition cross section; c is the velocity of light; τ is the photon lifetime in the resonator.

In the quasisteady approximation $[d\rho/dt = 0,\ (d/dt)(x^* - x) = 0]$ we find that the radiation power density $P_L = \hbar\omega_L \rho/\tau$, in which $\hbar\omega_L$ is the energy of a laser (stimulated) photon, is given by

$$P_L(t) = \frac{1}{2}\hbar\omega_L\left[(\alpha^* - \alpha)\,w_0 - w_1\left(\int_0^t w_0(t')\,dt' + \Delta\right)\right], \tag{2}$$

where $\Delta = 1/\sigma c\tau$ is the threshold inversion density; $\frac{1}{2}\left(\int_0^t w_0(t')\,dt' + \Delta\right)$ is the population density of the upper level during stimulated emission obtained on the assumption that $x + x^* = \int_0^t w_0(t')dt'$ and $x^* = x + \Delta$.

The radiation energy emitted from a unit volume of the medium $\mathcal{E}_L$ during a laser pulse is

$$\mathcal{E}_L = \int_{t_0}^{t_1} P_L(t)\,dt = I_1 - I_2, \tag{3}$$

where t_0 and t_1 are found by assuming that $P_L(t) = 0$ at $t = t_0$ and $t = t_1$, and that $P_L(t) > 0$ at $t_0 < t < t_1$,

$$I_1 = \frac{1}{2}\hbar\omega_L(\alpha^* - \alpha)\int_{t_0}^{t_1} w_0(t)\,dt, \tag{4}$$

$$I_2 = \frac{1}{2}\hbar\omega_L\int_{t_0}^{t_1} w_1\left(\int_0^t w_0\,dt' + \Delta\right)dt. \tag{5}$$

Two important conclusions on the kinetics of systems of this kind follow from Eqs. (2)-(5): a) in the absence of relaxation processes ($w_1 = 0$) the radiation energy is independent of the reaction rate w_0 and is governed only by the mass of reagents which remains and by the energy distribution in an elementary event ($\mathcal{E}_L = I_1$) ; b) if the relaxation processes are allowed for, the time evolution of a chemical reaction is important. Relaxation gives rise to a threshold reaction rate below which stimulated emission does not occur (the threshold demands that $w_0 > w_1\Delta$) and to a dependence of the radiation energy on the duration of a chemical reaction. The quantity I_2 represents the energy losses due to relaxation processes. Since w_1 is finite, the value of I_2 tends to zero in the limit $t_1 - t_0 \to 0$. The radiation energy then rises approaching a limiting value given by Eq. (4). An analysis of Eqs. (4) and (5) shows that $I_2 \ll I_1$ for $t_1 - t_0 \ll 1/w_1$.

Systems with Energy Transfer

In this case allowance is made for the chemical pumping of the vibrational degrees of freedom of the reaction product, energy transfer, relaxation, and stimulated emission. The rate equations are now

$$\left.\begin{aligned}\frac{dq}{dt} &= w_0 - (w_1 + w_3)\, q,\\ \frac{dx^*}{dt} &= w_3 q - w_2 x^* - \sigma c\rho\,(x^* - x),\\ \frac{d\rho}{dt} &= \sigma c\rho\,(x^* - x) - \frac{\rho}{\tau},\end{aligned}\right\} \tag{6}$$

where q is the density of the vibrational quanta excited by the reaction in question at a rate w_0; w_1 is the probability of relaxation of the vibrational quanta; w_2 is the probability of relaxation of the upper laser level; w_3 is the probability of transfer of a vibrational quantum from the reaction product to the "active" molecules.

The equation for the rate of population of the lower laser level is not specified. Applying the same quasisteady approximation, we find that the density of the output radiation power is

$$P_L = \hbar\omega_L\left(\frac{w_3}{w_1 + w_3}\, w_0 - w_2\,(x + \Delta) - \frac{dx}{dt}\right), \tag{7}$$

where allowance is made for the fact that during stimulated emission we have $x^* = x + \Delta$. It follows from Eq. (7) that stimulated emission occurs only if

$$\frac{w_3}{w_1 + w_3}\, w_0 > w_2\,(x + \Delta), \tag{8}$$

where x is the equilibrium population of the lower laser level. This condition governs the threshold reaction rate w_0. Equation (8) is obtained ignoring the term dx/dt in Eq. (7). Moreover, it follows from Eq. (7) that an energy-transfer laser operates efficiently if

$$w_3 \gg w_1. \tag{9}$$

This condition imposes restrictions on the composition of the working mixture.

The condition (8) gives only the lower limit of the reaction rate. In principle, the rate of reaction in laser systems with energy transfer may have an upper limit because the time for the transfer of the vibrational excitation to the "active" molecules is finite. This effect will be considered in greater detail for the $D_2 + F_2 + CO_2$ system.

§ 2. Kinetic Models of $H_2 + F_2$ and $D_2 + F_2 + CO_2$ Systems And Calculation Methods

The interpretation (model) of nonequilibrium kinetics of the systems under consideration is based on the following elementary processes which include the elementary stages of the

H_2 (D_2) + F_2 reaction known from the literature, vibration-vibrational and vibration-translational exchange of energy in a mixture, and relaxation of the active levels of the CO_2 molecule:

(0) $F_2 + h\nu \xrightarrow{k_0(t)} 2F$, photoinitiation,

(1) $F + H_2(D_2) \underset{k_{-1}^{(n)}}{\overset{k_1^{(n)}}{\rightleftarrows}} HF(DF)(v=n) + H(D)$,

(2) $H(D) + F_2 \xrightarrow{k_2^{(n)}} HF(DF)(v=n) + F$,

(3) $HF(DF)(v=n) + F_2 \xrightarrow{k_3^{(n)}} HF(DF)(v=0) + 2F$,

(4) $H_2(D_2)(v=n) + F_2 \xrightarrow{k_4^{(n)}} HF(DF)(v=0) + H(D) + F$,

(5) $HO_2(DO_2) + F_2 \xrightarrow{k_5} HF(DF)(v=0) + O_2 + F$,

(6) $H(D) + O_2 + M \xrightarrow{k_6} HO_2(DO_2) + M$,

(7) $F + F + M \underset{k_{-7}}{\overset{k_7}{\rightleftarrows}} F_2 + M$,

(8) $H(D) + H(D) + M \underset{k_{-8}^{(n)}}{\overset{k_8^{(n)}}{\rightleftarrows}} H_2(D_2)(v=n) + M$,

(9) $H(D) + F + M \underset{k_{-9}^{(n)}}{\overset{k_9^{(n)}}{\rightleftarrows}} HF(DF)(v=n) + M$,

(10) $HF(DF)(v=n) + M \underset{P_{n-1\,n}}{\overset{P_{n\,n-1}}{\rightleftarrows}} HF(DF)(v=n-1) + M$,

(11) $H_2(D_2)(v=n) + M \underset{R_{n-1\,n}}{\overset{R_{n\,n-1}}{\rightleftarrows}} H_2(D_2)(v=n-1) + M$,

(12) $HF(DF)(v=n) \xrightarrow{A_{n\,n-1}} HF(DF)(v=n-1)$,

(13) $HF(DF)(v=n) + HF(DF)(v=m) \underset{P_{n+1\,n}^{m-1\,m}}{\overset{P_{n\,n+1}^{m\,m-1}}{\rightleftarrows}}$
$\rightleftarrows HF(DF)(v=n+1) + HF(DF)(v=m-1)$,

(14) $H_2(D_2)(v=n) + H_2(D_2)(v=m) \underset{R_{n+1\,n}^{m-1\,m}}{\overset{R_{n\,n+1}^{m\,m-1}}{\rightleftarrows}}$
$\rightleftarrows H_2(D_2)(v=n+1) + H_2(D_2)(v=m-1)$,

(15) $H_2(D_2)(v=n) + HF(DF)(v=m) \underset{Q_{n+1\,n}^{m-1\,m}}{\overset{Q_{n\,n+1}^{m\,m-1}}{\rightleftarrows}}$
$\rightleftarrows H_2(D_2)(v=n+1) + HF(DF)(v=m-1)$,

(16) $DF(v=n) + CO_2(v_1 v_2^l v_3) \underset{k_{-16}}{\overset{k_{16}}{\rightleftarrows}} DF(v=n-1) + CO_2(v_1 v_2^l v_3+1)$,

relaxation of the upper laser level 00^01 due to the transfer of energy from asymmetric vibrations of CO_2 to deformation and symmetric vibrations:

(17) $CO_2(00^01) + M \underset{k_{-17}}{\overset{k_{17}}{\rightleftarrows}} CO_2(v_1 v_2^l 0) + M$,

and relaxation of the lower laser level 10^00 due to the thermalization of the energy of asymmetric and deformation vibrations:

(18) $CO_2(01^10) + M \underset{k_{-18}}{\overset{k_{18}}{\rightleftarrows}} CO_2(00^00) + M$.

The rate constants of the elementary chemical reactions are listed in Table 1. The kinetic model and the rate constants of the elementary processes assumed in the calculations are analyzed below.

Processes (1)-(2)

Both chain-propagation reactions (1) and (2) produce vibrationally excited HF (DF) molecules. The rate constants k_1 and k_2 are selected on the basis of the results reported in [8], ob-

TABLE 1. Rate Constants of Elementary Chemical Processes in $H_2 (D_2) + F_2$ and $D_2 + F_2 + CO_2$ Systems

Reaction No.	Rate constant, $cm^3 \cdot mole^{-1} \cdot sec^{-1}$	Value of n	
		H	D
1	$k_1^{(n)} = \alpha_n \cdot 1.2 \cdot 10^{14} \exp(-1000/RT)$	$n \geqslant 0$	$n \geqslant 0$
	$k_{-1}^{(n)} = 10^{13} \exp(-35\,000/RT) \exp(E_n/RT)$	$n \leqslant 3$	$n \leqslant 4$
	$k_{-1}^{(n)} = 10^{13}$	$n \geqslant 4$	$n \geqslant 5$
2	$k_2^{(n)} = \beta_n \cdot 1.2 \cdot 10^{14} \exp(-2400/RT)$	$n \geqslant 0$	$n \geqslant 0$
3	$k_3^{(n)} = 1.2 \cdot 10^6 \exp(-1200/RT) \exp(-\Delta E_n/RT)$	$n \leqslant 3$	$n \leqslant 4$
	$k_3^{(n)} = 1.2 \cdot 10^6 \exp(-1200/RT)$	$n \geqslant 4$	$n \geqslant 5$
4	$k_4^{(n)} = 10^6 \exp(-9000/RT) \exp(E_n/RT)$	$n = 0$	$n \leqslant 1$
	$k_4^{(n)} = 10^6$	$n \geqslant 1$	$n \geqslant 2$
5	$k_5 = 6 \cdot 10^{12} \exp(-12000/RT)$		
6	$k_6 = 5.75 \cdot 10^{14} \exp(4350/RT)$		
7	$k_7 = 1.26 \cdot 10^{14} T^{1/2}$		
	$k_{-7} = 1.38 \cdot 10^{13} \exp(-31120/RT)$		
8	$k_8^{(0)} = 10^{16}$		
	$k_8^{(n)} = 0$	$n \geqslant 1$	$n \geqslant 1$
	$k_{-8}^{(n)} = 8.1 \cdot 10^{16} T^{-1/2} \exp(-103\,240/RT) \exp(E_n/RT)$	$n \geqslant 0$	$n \geqslant 0$
9	$k_9^{(0)} = 10^{16}$		
	$k_9^{(n)} = 0$	$n \geqslant 1$	$n \geqslant 1$
	$k_{-9}^{(n)} = 1.13 \cdot 10^{19} T^{-1} \exp(-134\,000/RT) \exp(E_n/RT)$	$n \geqslant 0$	$n \geqslant 0$

tained by measuring the rates of reactions $F + H_2$ and $H + F_2$. The influence of the isotopic effect has not yet been investigated experimentally and, therefore, in the analysis of the $D_2 + F_2 + CO_2$ system it is desirable to vary the values of k_1 and k_2 within certain ranges. The distributions of the HF (DF) molecules between the vibrational levels after the elementary reactions (1) and (2) are described by the sets of numbers α_n and β_n, respectively. The individual population constants of the vibrational levels are given by the relationships $k_1^{(n)} = \alpha_n k_1$ and $k_2^{(n)} = \beta_n k_2$, where $\sum_n \alpha_n = \sum_n \beta_n = 1$. The numbers α_n and β_n govern the fraction of the energy of the $H_2 + F_2 \rightarrow 2HF$ reaction (130 kcal mole) evolved as the vibrational energy of the HF (DF) molecules. The energy diagrams of the $F + H_2$ (D_2) and H (D) + F_2 reactions are given in Fig. 1. Since there have been no direct measurements of the energy distribution in the elementary stages (1) and (2) in the reaction $F_2 + D_2$, it is interesting to vary α_n and β_n as described below. The constant $k_{-1}^{(0)}$ for the D + DF reaction is assumed to be the same as for the H + HF reaction measured in [9]. It is also assumed that the constants $k_{-1}^{(n)}$ ($n \geq 1$) have the same preexponential factor as $k_{-1}^{(0)}$ and the activation energy of the process H (D) + HF (DF) (v = n) is less than the

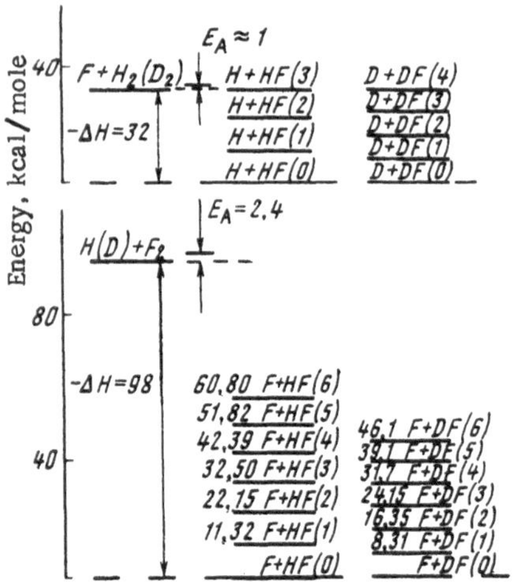

Fig. 1. Energy diagram of the $F + H_2$ (D_2) and H (D) + F_2 reactions.

activation energy of the process H (D) + HF (DF) (v = 0), and the difference is equal to the energy of the vibrational excitation of HF (DF) (v = n). It is assumed that the processes D + DF ($v \geq 5$) and H + HF ($v \geq 4$) occur without an activation energy because the energy of the vibrations of DF (v = 5) is 39 kcal/mole and that of HF (v = 4) is equal to 42 kcal/mole, both of which exceed the activation of the reaction H (D) + HF (DF) (v = 0). According to [10], the reaction F + HF (DF) occurs at a negligible rate at T < 6000°K.

Processes (3)-(5)

Possibility of branching in chain chemical reactions of excited molecules, formed in exothermal elementary events of chain propagation, was suggested in [11]. The hypothesis that the reaction between fluorine and hydrogen is of the branched-chain type, including reactions of excited particles, was confirmed experimentally in [12]. Kapralova et al. [12] related the branching to the process (3). Kinetically, this process is equivalent to the dissociation of fluorine at the expense of the vibrational energy of the excited HF (DF) molecules. A branching mechanism involving vibrationally excited hydrogen (deuterium) molecules is proposed in [13]. Vibrational excitation of H_2 (D_2) may occur as a result of the transfer of energy from HF (DF). Both branching mechanisms are considered in the present paper [processes (3) and (4)]. The rate constants of the processes (3) and (4) are selected on the basis of estimates of the rate of branching in the reaction of fluorine with hydrogen [8]. The reaction HF (DF) (v = n) + $F_2 \rightarrow$ HF (DF) (v = 0) + 2F is energetically possible if $v \geq 5$ (in the case of DF) and $v \geq 4$ (in the case of HF). The process is endothermal for the lower vibrational levels and a contribution for the translational motion energy is needed for the reaction. This is allowed for by introducing into the rate constant of the reaction (3) the factor $\exp(-\Delta E_n/RT)$, where ΔE_n is the energy defect depending on the number of the level.

Active chain centers are nucleated in the elementary event H_2 (D_2) (v = 0) + $F_2 \rightarrow$ H (D) + HF (DF) + F. This process is insignificant if the reaction is initiated optically or by an electric discharge but it is very important in the case of purely thermal initiation. The rate constant of this reaction is not known. The lower limit of the activation energy for the endothermal reaction is 8 kcal/mole. In the calculations reported below this energy is increased to 9 kcal/mole. In this case the branching in accordance with the reaction (4) requires excitation of D_2 to the second vibrational level and of H_2 to the first level or higher. The preexponential factor of the constants $k_4^{(n)}$ is selected in such a way that the branching rate is in agreement with observations.

The rate of the process (5) is estimated in [14], where it is shown also that when the reaction of fluorine with hydrogen is inhibited by oxygen there were not only the first and second ignition points but also a third chain limit resulting from the competition between the elementary process (5) and the loss of the HO_2 (DO_2) radicals on the walls of the reaction chamber. It should be pointed out that the branching of the chain in accordance with the reaction (5) is characterized by a relatively low rate and plays a secondary role in the kinetics of the system in the investigated range of pressures ($\lesssim$1 atm).

Processes (6)-(9)

The initial reagents H_2 (D_2) and F_2 can be used at considerable pressures only in the presence of an inhibitor. The addition of oxygen sets an upper limit to the pressures at which self-ignition is observed. The existence of the upper pressure limit is due to the trimolecular termination of the chain (6). As shown in [6], under certain conditions a mixture of D_2, F_2, and CO_2 does not ignite (this is probably due to the quenching of the excited DF molecules, responsible for the branching, by the vibration-vibrational exchange of energy between DF and CO_2) and the reagents can be mixed even in the absence of oxygen. However, the initial reagents usually contain a small amount of oxygen (up to 5%) as an impurity. Therefore, the loss of the active centers in accordance with the reaction (6) is allowed for in the calculations. The value of the quantity M which occurs in the reactions (6)-(9) is a function of the composition of the mixture. However, it is assumed in the calculations below that different components have the same efficiency in the processes (6)-(9). Since the recombination of atoms and the dissociation of molecules are slow processes compared with the excitation and relaxation of the laser levels, small errors in the determination of the rate constants of these processes have little effect on the final results of machine calculations. The rate constants of the reactions (6)-(9) are taken from [15]. An additional factor (E_n/RT) introduced into the dissociation rate constants allows for the increase in the rate of dissociation of the molecules H_2 (D_2) and HF (DF) in the excited vibrational levels.

Processes (10)-(11)

The temperature dependence of the rate of vibration-translational relaxation of the HF (DF) and H_2 (D_2) molecules can be calculated using the Swartz-Slawsky-Herzfeld method [16] and can be written in the form

$$P(T) = AT^{1/3}\exp\left[\left(\frac{1}{2}\Delta E + \varepsilon\right)/kT\right]\exp\left[\beta T^{-1/3}\right],$$

where ΔE is the energy defect of the process; ε is the parameter of the intermolecular interaction potential; A and β are factors which depend on the serial number of the level but not on T.

Numerical values of the preexponential factor in the constants P_{nn-1} and R_{nn-1} are assumed to be such that the rate of relaxation at some temperature agrees with the experimental data. The vibrational relaxation of hydrogen and deuterium was investigated in [17] in the temperature range 1100-3000°K. The value of R_{10} was identified with the experimental value of the rate of deactivation of H_2 (D_2) as a result of interaction with H_2 (D_2) at 1600°K $\{p\tau_{D_2-D_2} = (2.7 \pm 0.3) \cdot 10^{-10}\exp[(110 \pm 1.5)/T^{1/3}]$ sec · atm, $p\tau_{H_2-H_2} = (3.9 \pm 0.8) \cdot 10^{-10}\exp[(100 \pm 2)/T^{1/3}]$ sec · atm$\}$. The rate of the vibration-translational relaxation of DF was not measured. The numerical value of the probability of quenching of DF by D_2 at 470°K was assumed to be the same as the probability of quenching of HF by H_2 [18] ($p\tau = 3.3 \cdot 10^{-5}$ sec · atm, deactivation probability $2 \cdot 10^{-6}$). The efficiency of the other components of the mixture in the process of deactivation of H_2 (D_2) and HF (DF) was assumed to be equal to the efficiency of H_2 (D_2). This assumption probably introduces major errors into our calculations.

Process (12)

This process describes the spontaneous emission of radiation from the HF (DF) molecules. The contribution of the spontaneous emission to the rate of dissipation of the vibration al energy is important only at very low pressures (below 10 mm Hg). The Einstein coefficients A_{nn-1} for the first three vibrational transitions are taken from [18]. The Einstein coefficients for the vibrational quantum numbers $n \geq 3$ are calculated in the harmonic oscillator approximation using the experimental value of A_{32} ($A_{nn-1} = (n/3)A_{32}$, $n \geq 3$). Thus, the following values of A_{nn-1} (sec^{-1}) are assumed in the calculations: $A_{10} = 150$, $A_{21} = 260$, $A_{32} = 340$, $A_{43} = 450$, $A_{54} = 570$, $A_{65} = 680$.

Processes (13)-(15)

The vibrational relaxation model of the molecules HF (DF) and H_2 (D_2) allows for the interaction between a finite number of vibrational levels, namely seven levels of the HF (DF) molecules (v = 0, 1, 2, ..., 6) and three levels of the H_2 (D_2) molecules (v = 0, 1, 2). A quantitative description of the vibration-vibrational exchange of energy can be based on the Swartz-Slawsky-Herzfeld theory [16]. It should be noted that the a priori calculation of the rate of the vibration-vibrational exchange of energy would not be sufficiently reliable in the present stage of the development of the theory. Therefore, we shall analyze the influence of the variation of the rates of the processes (13)-(15) on the characteristics of a $D_2 + F_2 + CO_2$ laser. The results of an analysis of this influence are given below.

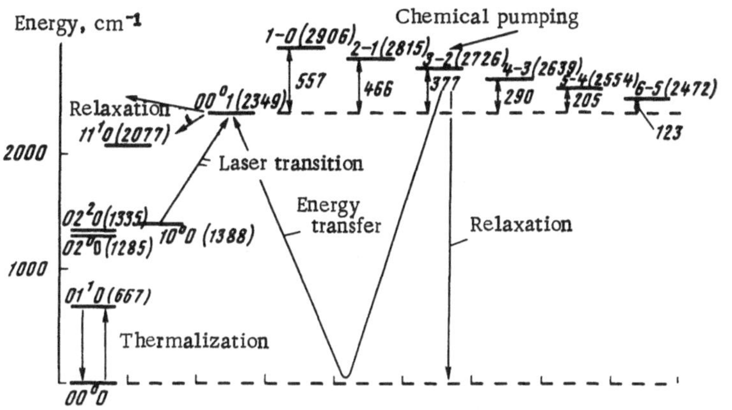

Fig. 2. Energy level diagram of the CO_2 molecule and frequencies of single-quantum vibrational transitions ($v \to v-1$) in DF. The main processes in a DF + CO_2 laser are identified.

Process (16)

Figure 2 gives the energy diagram of quantum transitions in the DF + CO_2 system. The 00^01 level of the CO_2 molecule (upper laser level) is excited by the vibration-vibrational exchange of energy between DF and CO_2. Since this process is responsible for the population inversion of the laser levels of CO_2 (the levels 00^01 and 10^00), the results of calculations depend critically on the selected rate of this process. An estimate of the rate constant of the energy transfer from DF to CO_2 is obtained in [19, 20]. In the kinetic calculations the constant k_{16} is assumed to be in the range 10^{-13}-10^{-12} cm^3/sec. The influence of variation within this range will be discussed later.

Processes (17)-(18)

An analysis of the relaxation processes in the CO_2 molecule will be made here on the assumption that a local thermodynamic equilibrium is established for each normal mode. Then, the distribution function of the populations depends on time only via the "internal" temperatures T_1, T_2, and T_3:

$$n_{v_1v_2v_3} = n_{CO_2}\left[e^{-v_1\frac{\vartheta_1}{T_1}}\left(1-e^{-\frac{\vartheta_1}{T_1}}\right)\right]\left[(v_2+1)\,e^{-v_2\frac{\vartheta_2}{T_2}}\left(1-e^{-\frac{\vartheta_2}{T_2}}\right)^2\right]\left[e^{-v_3\frac{\vartheta_3}{T_3}}\left(1-e^{-\frac{\vartheta_3}{T_3}}\right)\right],$$

where T_j is the temperature of the j-th mode; n_{CO_2} is the density of the CO_2 molecules; $n_{v_1v_2v_3}$ is the density of the CO_2 molecules which are in the vibrational state (v_1, v_2, v_3). The temperature T_j is governed by the energy E_j stored in the j-th mode:

$$T_j = \vartheta_j \ln^{-1}(1 + n_{CO_2} R\vartheta_j / E_j).$$

Thus, under quasiequilibrium conditions of the kind described above the vibrational kinetics can be described by just three rate equations each of which governs the rate of change of energy in the respective mode. The validity of this approach is justified by the fact that the resonance vibration-vibrational exchange of energy establishes a thermal equilibrium within each mode quite rapidly ($\sim 10^{-8}$ sec at 1 atm and 300°K) compared with the time scale of the establishment of equilibria between various modes and with the time scale of the vibration-translational relaxation. Moreover, if we introduce some additional simplifications which follow from the particular problem in question, we obtain a much simpler but still realistic relaxation model for the CO_2 molecule. Since the vibrational quantum of a deformation mode is approximately half the vibrational quantum of a symmetric mode, the states with identical values of $2v_1 + v_2$, v_3, l (for example, the states 10^00 and 02^00) are in Fermi resonance. Therefore, the modes ν_1 and ν_2 are strongly coupled by the vibration-vibrational exchange and are characterized by a single relaxation time. The rate constant of the interchange of the quanta $h\nu_1 \rightleftharpoons 2h\nu_2$ estimated in [2] is 10^6 sec^{-1} · Torr^{-1}. Since the rate of this process is very high, the temperatures of the symmetric and deformation modes are similar. We are quite justified in assuming that $T_1 = T_2$.

The simultaneous relaxation of the levels E_1 and E_2 is due to the vibration-translational transfer of energy to the levels $(0, v_2, 0)$. This process limits the rate of relaxation of the lower laser level 10^00. We can estimate the thermalization time of the levels E_1 and E_2 using the rate constant k_{18} (vibration-translational relaxation of the ν_2 mode) given in [22]: $k_{18}(CO_2-CO_2) = 0.485 \cdot 10^2 T^{10/3}$ cm^3 · mole^{-1} · sec^{-1}, $k_{18}(CO_2-HE) = 0.41 \cdot 10^9 T$ cm^3 · mole^{-1} · sec^{-1}. The derivation of these relationships (and of the rate constants of the exchange of energy between the ν_3 mode, on the one hand, and the ν_1 and ν_2 modes, on the other) is made in [22] on the basis of the results reported in [23-25], and additional calculations are made of the probabilities in the temperature range 1000-2500°K. These dependences describe, to within a factor of ~2, the observed rates of relaxation at room and higher temperatures. The relaxation time of the E_1 and E_2 levels at 1000°K and for densities corresponding to pressures of 200 Torr CO_2, 400 Torr He (the pressures are reduced to 300°K) is 10^{-7} sec. This time is short compared with the chemical reaction time and the duration of stimulated emission (~10^{-5} sec at the working-mixture pressures under consideration). The fast relaxation makes the vibrational temperatures T_1 and T_2 approximately equal to the gas temperature T. In the calculation of the populations of the laser levels it is assumed that $T_1 = T_2 = T$. In this approximation the vibrational kinetics of CO_2 is described by one rate equation which governs the rate of change of the energy of the asymmetric mode (00^0v).

In the derivation of the above expressions allowance is made for the exchange of quanta between DF and CO_2 (rate constant k_{16}) and for the transfer of the energy of the asymmetric mode via the 00^01 level to the deformation and symmetric modes [the rate constant is $k_{17}(CO_2-M)$, which depends on the nature of the collisional partner M]. The values and temperature dependences of the constants $k_{17}(CO_2-M)$ are given by the formulas: $k_{17}(CO_2-CO_2) = 0.43 \cdot 10^4 T^{5/2}$ cm^3 · mole^{-1} · sec^{-1} [22], $k_{17}(CO_2-He) = 0.95 \cdot 10^3 T^{5/2}$ cm^3 · mole^{-1} · sec^{-1} [22], $k_{17}(CO_2-D_2) = 4.6 \cdot 10^9$ cm^3 · mole^{-1} · sec^{-1} [25], and $k_{17}(CO_2-Ar) = 0.9 \cdot 10^9$ cm^3 · mole^{-1} · sec^{-1} [26].

System of Rate Equations

A. The following notation will be used in the reaction kinetics equations:

ρ is the initial concentration of a mixture (mole/cm^{-3}),

τ is the dimensionless time related to normal time t (sec) by $\tau = at$,

a is the dimensional scale coefficient (sec^{-1}),

$$x_n = [HF(DF)(v=n)]/\rho,$$
$$v_n = [H_2(D_2)(v=n)]/\rho,$$
$$u = [F_2]/\rho,$$
$$y = [F]/\rho,$$
$$z = [H(D)]/\rho,$$
$$w = [O_2]/\rho,$$
$$f = [HO_2(DO_2)]/\rho,$$
$$g = [He]/\rho,$$
$$h = [CO_2]/\rho,$$
$$n_{00^01} = [CO_2(00^01)]/\rho,$$
$$n_{10^00} = [CO_2(10^00)]/\rho,$$
$$s = \sum_n x_n + \sum_n v_n + y + z + w + f + g + h,$$

$$s_1 = \mu_1\left(\sum_n x_n\right) + \mu_2\left(\sum_n v_n\right) + \mu_3 y + \mu_4 z + \mu_5 w + \mu_6 f + \mu_7 g + \mu_8 h,$$

μ_i is the efficiency of component i in quenching HF (DF).

B. The rate equations for the concentrations of the components of a mixture are:

$$\frac{a}{\rho}\frac{dx_n}{d\tau} = \sum_m L_n^m, \qquad n = 0, \ldots, 6, \quad m = 1, \ldots, 11,$$

where

$$L_n^1 = \alpha_n k_1 y v_0, \qquad n = 0, \ldots, 6,$$
$$L_n^2 = \beta_n k_2 z u, \qquad n = 0, \ldots, 6,$$
$$L_n^3 = -k_{-1}^{(n)} x_n z, \qquad n = 0, \ldots, 6,$$
$$L_n^4 = (P_{n+1\,n} x_{n+1} - P_{n\,n+1} x_n + P_{n-1\,n} x_{n-1} - P_{n\,n-1} x_n)\, s_1, \qquad n = 0, \ldots, 6,$$
$$L_n^5 = \left(\sum_m P_{n+1\,n}^{m-1\,m} x_{m-1}\right) x_{n+1} - \left(\sum_m P_{n\,n+1}^{m\,m-1} x_m\right) x_n + \left(\sum_m P_{n-1\,n}^{m+1\,m} x_{m+1}\right) x_{n-1} - \left(\sum_m P_{n\,n-1}^{m\,m+1} x_m\right) x_n, \qquad m, n = 0, \ldots, 6,$$
$$L_n^6 = \left(\sum_m Q_{m-1\,m}^{n+1\,n} v_{m-1}\right) x_{n+1} - \left(\sum_m Q_{m\,m-1}^{n\,n+1} v_m\right) x_n + \left(\sum_m Q_{m+1\,m}^{n-1\,n} v_{m+1}\right) x_{n-1} - \left(\sum_m Q_{m\,m+1}^{n\,n-1} v_m\right) x_n, \quad m = 0, 1, 2, \quad n = 0, \ldots, 6,$$
$$L_n^7 = -\frac{A_{n\,n-1}}{\rho} x_n + \frac{A_{n+1\,n}}{\rho} x_{n+1}, \qquad n = 0, \ldots, 6,$$
$$L_0^8 = \left(\sum_n k_4^{(n)} v_n\right) u + \left(\sum_n k_3^{(n)} x_n\right) u + k_5 f u,$$
$$L_n^8 = 0, \qquad n = 1, \ldots, 6,$$
$$L_n^9 = -k_{-9}^{(n)} x_n s, \qquad n = 0, \ldots, 6,$$
$$L_0^{10} = k_9^{(0)} \rho s y z,$$
$$L_n^{10} = -k_3^{(n)} x_n s, \qquad n = 1, \ldots, 6,$$
$$L_n^{11} = h k_{16}\,[x_{n+1} - (1 + \varphi) x_n + \varphi x_{n-1}], \qquad n = 0, \ldots, 6,$$
$$\varphi = \exp(-3380/T_3 - 804/T),$$
$$\frac{a}{\rho}\frac{dv_n}{d\tau} = \sum_m N_n^m, \qquad n = 0, 1, 2, \quad m = 1, 2, 3, 4,$$

where

$$N_n^1 = -k_4^{(n)} v_n u - k_{-8}^{(n)} v_n s, \qquad n = 1, 2,$$
$$N_0^1 = -k_4^{(0)} v_0 u - k_{-8}^{(0)} v_0 s - k_1 y v_0 + k_8^{(0)} \rho s z^2 + \left(\sum_n k_{-1}^{(n)} x_n\right) z,$$
$$N_n^2 = \left(\sum_m Q_{n+1\,n}^{m-1\,m} x_{m-1}\right) v_{n+1} - \left(\sum_m Q_{n\,n+1}^{m\,m-1} x_m\right) v_n + \left(\sum_m Q_{n-1\,n}^{m+1\,m} x_{m+1}\right) v_{n-1} - \left(\sum_m Q_{n\,n-1}^{m\,m+1} x_m\right) v_n, \qquad n = 0, 1, 2,$$
$$N_n^3 = (R_{n+1\,n} v_{n+1} - R_{n\,n+1} v_n + R_{n-1\,n} v_{n-1} - R_{n\,n-1} v_n)\, s, \qquad n = 0, 1, 2,$$
$$N_n^4 = \left(\sum_m R_{n+1\,n}^{m-1\,m} v_{m-1}\right) v_{m+1} - \left(\sum_m R_{n\,n+1}^{m\,m-1} v_m\right) v_n + \left(\sum_m R_{n-1\,n}^{m+1\,m} v_{m+1}\right) v_{n-1} - \left(\sum_m R_{n\,n-1}^{m\,m+1} v_m\right) v_n, \qquad n = 0, 1, 2,$$
$$\frac{a}{\rho}\frac{du}{d\tau} = -\left(\sum_n k_4^{(n)} v_n\right) u - \left(\sum_n k_3^{(n)} x_n\right) u - k_2 z u + k_7 \rho s y^2 - k_5 f u - k_{-7} s u - k_0 u/\rho,$$
$$\frac{a}{\rho}\frac{dy}{d\tau} = 2k_{-7} s u + \left(\sum_n k_4^{(n)} v_n\right) u - k_1 y v_0 + k_2 z u + 2\left(\sum_n k_3^{(n)} x_n\right) u + k_5 f u - 2k_7 \rho s y^2 + \left(\sum_n k_{-9}^{(n)} x_n\right) s - k_9^{(0)} \rho s y z + \left(\sum_n k_{-1}^{(n)} x_n\right) z + 2k_0 u/\rho,$$
$$\frac{a}{\rho}\frac{dz}{d\tau} = 2\left(\sum_n k_{-8}^{(n)} v_n\right) s + \left(\sum_n k_4^{(n)} v_n\right) u - \left(\sum_n k_{-1}^{(n)} x_n\right) z + \left(\sum_n k_{-9}^{(n)} x_n\right) s + k_1 y v_0 - k_2 z u - k_6 \rho s z w - 2k_8^{(0)} \rho s z^2 - k_9^{(0)} \rho s y z,$$
$$\frac{a}{\rho}\frac{dw}{d\tau} = -k_6 \rho s z w + k_5 f u,$$
$$\frac{a}{\rho}\frac{df}{d\tau} = k_6 \rho s z w - k_5 f u.$$

C. The rate equation for the energy of the CO_2 mode ν_3 is

$$\frac{a}{\rho^2}\frac{dE_3}{d\tau}=w_1 h\left(\sum_{n\geq 1} x_n-\varphi\sum_{n\geq 0} x_n\right)-h\left(\sum_i w_2^{(i)}\xi_i\right)\left(1-e^{-\frac{\vartheta_1}{T}}\right)\times$$
$$\times\left(e^{-\frac{\vartheta_3}{T_3}}-e^{-\frac{\vartheta_3}{T}}\right)\left(1-e^{-\frac{\vartheta_2}{T}}\right)^2\left(1-e^{\frac{\vartheta_1}{T}}\right),$$

where

$$w_1=R\vartheta_3 k_{16},\quad w_2^{(i)}=R\vartheta_3 k_{17}(CO_2-M_i),$$
$$\vartheta_1=1920^\circ K,\quad \vartheta_2=960^\circ K,\quad \vartheta_3=3380^\circ K,$$

ξ_i is the relative concentration of the i-th component of the mixture, and

$$\left(\sum_i \xi_i=1\right).$$

D. The thermal balance equation is obtained by considering the evolution and absorption of heat as a result of chemical reactions and because of the vibrational relaxation. The thermal balance equation can be expressed in the form

$$C\frac{a}{\rho}\frac{dT}{d\tau}=\sum_i Q_i W_i-\sum_i E_n(x)\left(\frac{a}{\rho}\frac{dx_n}{d\tau}\right)_C-\sum_n E_n(v)\left(\frac{a}{\rho}\frac{dv_n}{d\tau}\right)_C-\frac{a}{\rho^2}\frac{dE_3}{d\tau},$$

where

$$\left(\frac{a}{\rho}\frac{dv_n}{d\tau}\right)_C=N_n^2+N_n^3+N_n^4,\quad \left(\frac{a}{\rho}\frac{dx_n}{d\tau}\right)_C=L_n^4+L_n^5+L_n^6+L_n^{11},$$

Q_i and W_i are the thermal effect and the rate of the i-th elementary chemical process; E(x) and $E_n(v)$ are the energies of the n-th vibrational levels of the $HF_n(DF)$ and H_2 (D_2) molecules, respectively; C is the specific heat of the mixture at a constant volume.

In calculations of the specific heat the rotational degrees of freedom are assumed to be fully excited. The contribution of the vibrational degrees of freedom of the F_2 and O_2 molecules to the specific heat and the contributions of the ν_1 and ν_2 modes of CO_2 are described by the Planck–Einstein function, and the specific heat of DO_2 is assumed to have the classical value. The system given above consists of seventeen first-order nonlinear differential equations. Numerical integration of these equations was performed by the Runge-Kutta method with an automatic selection of the calculation steps; this was done using BÉSM-4 and M-220 computers.

E. The efficiency of a $D_2 + F_2 + CO_2$ laser was calculated as follows. In the quasisteady approximation the radiation power density P_L is (MW/cm^3 for $a = 10^6$ sec^{-1})

$$P_L=4{,}18\frac{h\nu_L}{h\nu_3}\frac{\rho^2}{a}\left[w_1 h\left(\sum_{n\geq 1} x_n-\varphi\sum_{n\geq 0} x_n\right)-\left(\sum_i w_2^{(i)}\xi_i\right)(n_{10^00}+\Delta)-h\nu_3\frac{a}{\rho}\frac{dn_{10^00}}{d\tau}\right],$$

where $h\nu_L$ is the energy of a laser photon; $h\nu_3$ is the energy of the asymmetric vibrational mode (cal/mole); Δ is the threshold inversion density.

The solution of the rate equations gives P_L as a function of time. The specific energy of the useful radiation E_L is described by the integral

$$E_L\ (J/cm^3)=\int_{\tau_1}^{\tau_2} P_L(\tau)\,d\tau,$$

where the integration interval is defined by the conditions $P_L(\tau) > 0$, $(n_{00^01} - n_{10^00}) > 0$ when $\tau_1 < \tau < \tau_2$. The laser efficiency is $\eta = E_L/Q$, where Q is the density of the stored chemical energy. In these calculations it was assumed that $\Delta = 0$. The quantum efficiency of stimulated emission was defined as the ratio of the number of quanta emitted by a unit volume of the medium to the number of pump quanta absorbed by the same volume.

§ 3. $H_2 + F_2$ System

Experimental Results

Among several tens of the investigated reactions resulting in stimulated emission the results have been promising for the hydrogen fluorination reaction $H_2 + F_2$. Stimulated emission from an $H_2 + F_2$ mixture in a gas discharge was reported in [4]. N. G. Basov et al. [27] achieved stimulated emission due to vibration-rotational transitions in HF molecules formed from a mixture of H_2 and F_2 as a result of flash photolysis, and they investigated the spectral composition of the emitted radiation. Further studies of chemical lasers utilizing hydrogen and fluorine mixtures were reported in [19, 28-30]. It is clear from the results reported in [19, 27] that the emission spectrum of the HF molecules is rich and that it includes transitions in the P branch of the vibrational bands from 1-0 to 6-5.

The delay of a stimulated emission pulse relative to the initiating pulse was found to vary with the transition. An analysis of the emission spectrum indicated that within each vibration-rotational band the maximum of the emission energy corresponded to transitions with similar values of J: P(6)-P(8) [19].

When each value of the vibrational quantum number v was attributed to an emission energy E(v) in the band $v \rightarrow v-1$, measurements of the relative energy yield from an HF laser could be expressed in the form E(1) : E(2) : E(3) : E(4) : E(5) : E(6) = 0.0455 : 1 : 0.244 : 0.138 : 0.149 : 0.0454. This distribution reflected to some extent the ratio of the rates of population of specific vibrational levels in the course of the chemical reaction. The function E(v) was found to have two maxima. The first was located at v = 2, corresponding to a vibrational excitation energy of ~22 kcal/mole. The second, less pronounced, maximum was found at v = 5, corresponding to ~52 kcal/mole. This form of the E(v) curve reflected the complex mechanism of the chemical reaction in which the excited product HF* was formed in two different elementary reactions with different energy yields:

$$F + H_2 \rightarrow HF^* + H, \quad \Delta H = -32 \text{ kcal/mole},$$
$$H + F_2 \rightarrow HF^* + F, \quad \Delta H = -98 \text{ kcal/mole}.$$

Investigations of the stimulated emission from $H_2 + F_2$ [30] demonstrated that this system could have a high quantum efficiency of the generation of coherent radiation amounting to about 10^2. The influence of an inert diluent (He) on the operation of a laser based on the $H_2 + F_2$ reaction was investigated in [31]. The results indicated that a considerably higher energy and longer pulses could be obtained from an HF laser utilizing H_2, F_2, and diluent than from nonchain systems or from the $H_2 + F_2$ system without a diluent.

Simplified Kinetic Scheme of the Reaction between Hydrogen and Fluorine

The $H_2 + F_2$ reaction occurs in accordance with the scheme [11, 12]

$$(1) \quad F + H_2 \xrightarrow{k_1} HF + H,$$
$$(2) \quad H + F_2 \xrightarrow{k_2} HF^* + F,$$

(3) $HF^* + F_2 \xrightarrow{k_3} HF + 2F$, (7) $H \xrightarrow{k_7}$ wall,

(4) $HF^* + M \xrightarrow{k_4} HF + M$, (8) $F \xrightarrow{k_8}$ wall,

(5) $H + O_2 + M \xrightarrow{k_5} HO_2 + M$, (9) $HF^* \xrightarrow{k_9}$ wall,

(6) $HO_2 + F_2 \xrightarrow{k_6} HF + O_2 + F$, (10) $HO_2 \xrightarrow{k_{10}}$ wall.

A possible energy-branching reaction in this scheme is (3). The process (4) describes the vibration-translational relaxation of the excited molecules. The chain is propagated by the reactions (1) and (2). The process (5) is the trimolecular chain-termination reaction, associated with the presence of oxygen. The processes (7)-(10) describe the loss of radicals at the chamber walls. The process (6) is the chain-branching reaction. An alternative mechanism of energy branching is described by a scheme [13] which includes vibrationally excited hydrogen molecules:

(3a) $HF^* + H_2 \rightarrow HF + H_2^*$,
(3b) $H_2^* + F_2 \rightarrow H + F + HF$,
(4c) $H_2^* + M \rightarrow H_2 + M$.

The processes (1)-(10) in a mixture explain the observed ignition limits. Competition between the processes (3) and (7)-(9) gives rise to the lower ignition limit. The upper limit is due to the competition between the processes (3) and (5). Inclusion of the process (6) gives rise to the third limit of chain ignition and this limit is due to the competition between the processes (6) and (10). The ignition limits depend on the composition of the mixture and the size of the reaction chamber.

Calculation of the Branching Factor

Inside the ignition region of the pressure-temperature (p-T) diagram the rate of excitation of the upper laser level is proportional to $W_0 e^{st}$ [32], where W_0 is a quantity which depends on the energy of the initiating pulses; s is the chain branching factor. If the condition $st_0 \gg 1$ is satisfied, where t_0 is the duration of the laser output pulses, the threshold reaction rate resulting in stimulated emission can be achieved for a small energy input. This circumstance makes it attractive to use branched chain reactions in chemical lasers and promises to provide a basis for a laser which would need practically no external energy. Therefore, it is interesting to study the behavior of the factor s in the p-T plane for a specific $H_2 + F_2$ mixture.

Using the scheme (1)-(10), we can write the rate equations for the concentrations of the active centers H, F, HF^*, and HO_2:

$$\frac{dn_i}{dt} = \sum_{j=1}^{4} \sigma_{ij} n_j, \tag{10}$$

where $n_1 = [F]$, $n_2 = [H]$, $n_3 = [HF^*]$, and $n_4 = [HO_2]$.

The elements of the matrix σ_{ij} are as follows:

$$\sigma_{11} = -(k_1[H_2] + k_8), \qquad \sigma_{33} = -(k_3[F_2] + k_4[M] + k_9),$$
$$\sigma_{12} = k_2[F_2], \qquad \sigma_{34} = \sigma_{41} = 0,$$
$$\sigma_{13} = 2k_3[F_2], \qquad \sigma_{42} = k_5[O_2][M],$$
$$\sigma_{14} = k_6[F_2], \qquad \sigma_{32} = k_2[F_2],$$
$$\sigma_{21} = k_1[H_2], \qquad \sigma_{43} = 0,$$
$$\sigma_{22} = -(k_2[F_2] + k_5[O_2][M] + k_7), \qquad \sigma_{44} = -(k_6[F_2] + k_{10}).$$
$$\sigma_{23} = \sigma_{24} = \sigma_{31} = 0,$$

The general solution of the system (10) is of the form

$$n_i = \sum_j N_{ij} e^{s_j t},$$

where s_j are the roots of the secular equation

$$s^4 + p_1 s^3 + p_2 s^2 + p_3 s + p_4 = 0, \tag{11}$$

where $p_k = (-1)^k a_k$, a_k is the sum of all the k-th order minors of the determinant $|\sigma_{ij}|$, obtained by crossing out 4-k rows with the numbers $\alpha_1, \ldots, \alpha_k$ and columns with the same numbers.

Thus, the coefficients p_k are of the form

$$p_1 = -(\sigma_{11} + \sigma_{22} + \sigma_{33} + \sigma_{44}),$$
$$p_2 = \sigma_{33}\sigma_{44} + \sigma_{22}\sigma_{44} + \sigma_{22}\sigma_{33} + \sigma_{11}\sigma_{44} + \sigma_{11}\sigma_{33} + \sigma_{11}\sigma_{22} - \sigma_{12}\sigma_{21},$$
$$p_3 = -(\sigma_{22}\sigma_{33}\sigma_{44} + \sigma_{11}\sigma_{33}\sigma_{44} + \sigma_{11}\sigma_{22}\sigma_{44} + \sigma_{14}\sigma_{21}\sigma_{42} - \sigma_{12}\sigma_{21}\sigma_{44}$$
$$+ \sigma_{11}\sigma_{22}\sigma_{33} + \sigma_{13}\sigma_{21}\sigma_{32} - \sigma_{12}\sigma_{21}\sigma_{33}),$$
$$p_4 = \sigma_{11}\sigma_{22}\sigma_{33}\sigma_{44} - \sigma_{21}\sigma_{12}\sigma_{33}\sigma_{44} + \sigma_{21}\sigma_{14}\sigma_{33}\sigma_{42} + \sigma_{21}\sigma_{13}\sigma_{32}\sigma_{44}.$$

Equation (11) was solved numerically by the Newton method on a BÉSM-4 computer and the same method was used to determine the largest root of the secular equation as a function of p and T.

The following values of the rate constants were assumed on the basis of the published results: $k_1 = 1.2 \cdot 10^{14} \exp(-1000/RT)$, $k_2 = 1.2 \cdot 10^{14} \exp(-2400/RT)$, $k_3 = 1.2 \cdot 10^{6} \exp(-1200/RT)$, $k_4 = 1.2 \cdot 10^{9}$, $k_5 = 3.62 \cdot 10^{14} \exp(1600/RT)$, $k_6 = 6 \cdot 10^{12} \exp(-1200/RT)$. All these constants were in units of $cm^3 \cdot mole^{-1} \cdot sec^{-1}$. The rate constants of the heterogeneous loss of radicals k_7-k_{10} were calculated from the formula $k = (r^2/8D + 2r/\varepsilon\bar{u})^{-1}$, where $D = {}^1/_3 u\lambda$ is the diffusion coefficient, $\bar{u}$ is the average velocity of particle; λ is the mean free path; r is the radius of the reaction chamber; ε is the probability of loss of a radical on the wall. The value of ε was varied within the range 10^{-5}-10^{-2}, which had an important influence on the position and shape of the ignition "peninsula." The results of calculations are presented in Figs. 3-5.

It should be pointed out that variation of the oxygen concentration in the mixture within the range 2-10% and of the value of ε within the range 10^{-2}-10^{-5} had relatively little influence on the branching factor s for $\rho > 2 \cdot 10^{-6}$ mole/cm³ and T > 600°K, i.e., well inside the ignition region. The calculations demonstrated a difficulty encountered in the utilization of the advantages of the reaction between fluorine and hydrogen as a branched chain process. The pres-

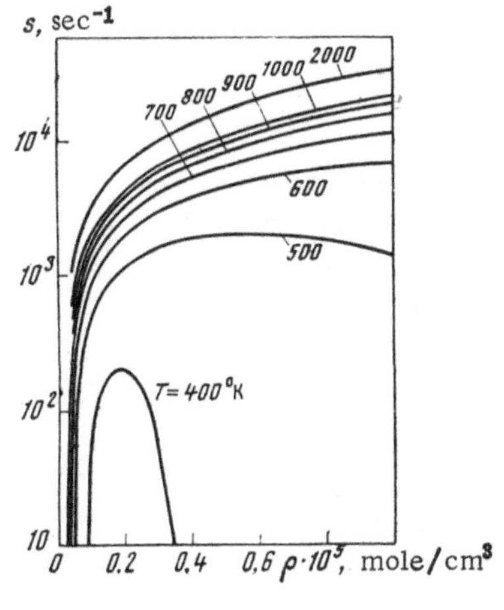

Fig. 3. Branching factor s for the H_2:F_2:O_2 = 0.45:0.45:0.1 mixture plotted as a function of the total concentration ρ at various temperatures. The probability of the loss of radicals on the chamber walls was 10^{-2} and the radius of the chamber was 1 cm.

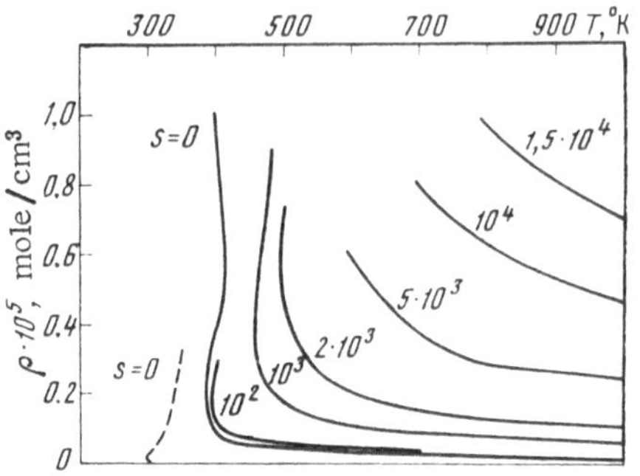

Fig. 4. Curves representing the branching factor s for the $H_2:F_2:O_2 = 0.45:0.45:0.1$ mixture ($\varepsilon = 10^{-2}$). The dashed curve s = 0 represents the ignition limit for $\varepsilon = 10^{-5}$.

sure in the working mixture in [4, 27] did not exceed 10 Torr, and the duration of the laser emission pulses was $\sim 10^{-5}$ sec. In this range of pressures the branching was a slow process compared with the duration of laser emission. It is clear from Fig. 3 that the experimentally determined branching factor s is of the order of 10^2-10^3. Consequently, $st_0 \ll 1$ and the branching does not make any significant contribution to the efficiency of the system. However, the high values of the rate constants of the elementary stages of the chain propagation, the energy yield of the stimulated emission, and the efficiency of conversion of the chemical energy into laser radiation in $H_2 + F_2$ mixtures are several orders of magnitude greater than the corresponding parameters of the chemical lasers investigated earlier. We shall now give the results of a numerical calculation of the kinetics of an $H_2 + F_2$ laser and we shall employ a model of the system described in § 2.

Analysis of Mechanisms of Excitation and Stimulated-Emission

Quenching

The computer program was designed to give information on the rates of the processes resulting in the excitation and deexcitation of the vibrational levels of HF.

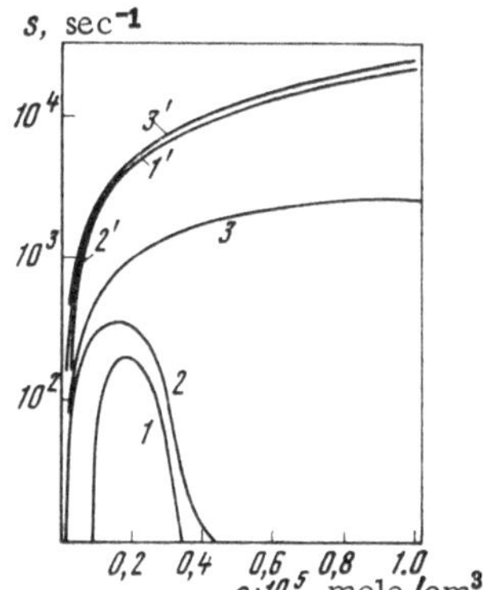

Fig. 5. Influence of the value of ε (revealed by curves 1 and 2 at T = 400°K and curves 1' and 2' at T = 100°K) and of the oxygen content in the mixture (revealed by curves 2 and 3 at 400°K and curves 2' and 3' at T = 1000°K) on the behavior of the factor s. Curves 1 and 1' represent $H_2:F_2:O_2 = 0.45:0.45:0.1$ and $\varepsilon = 10^{-2}$; curves 2 and 2' represent $H_2:F_2:O_2 = 0.45:0.45:0.1$ and $\varepsilon = 10^{-5}$; curves 3 and 3' represent $H_2:F_2:O_2 = 0.48:0.5:0.02$ and $\varepsilon = 10^{-5}$.

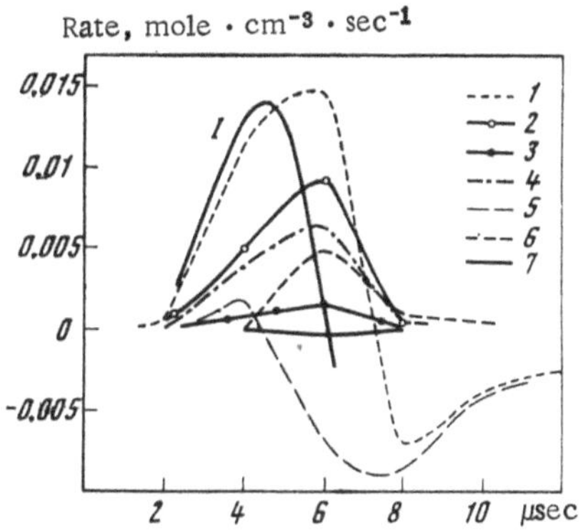

Fig. 6. Dynamics of the v = 1 level of the HF molecule in the $H_2:F_2:O_2$ = 4.5:4.5:1 Torr mixture: 1) d/dt[HF(v = 1)]; 2) F + H_2; 3) H + F_2; 4) VV(HF-H_2); 5) VV(HF-HF); 6) VT; 7) H + HF(v = 1). Here, I is the density of the inversion Δn in the $P_{2-1}(8)$ transition; Δn_{max} = $0.25 \cdot 10^{-5}$ mole/cm³.

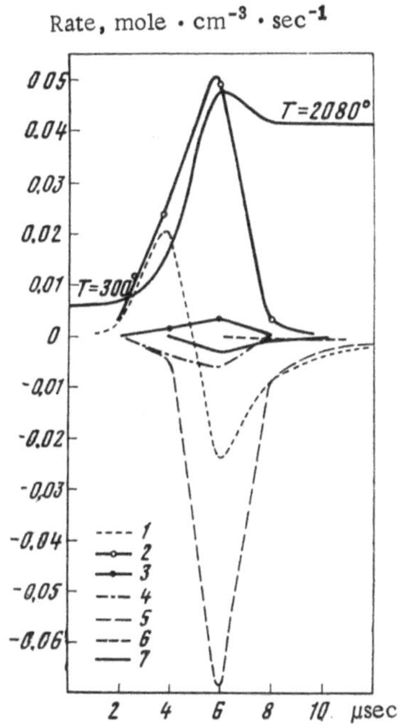

Fig. 7. Dynamics of the v = 2 level of the HF molecule for the $H_2:F_2:O_2$ = 4.5:4.5:1 Torr mixture: 1) d/dt[HF(v = 2)]; 2-6) same meaning as in Fig. 6; 7) H + HF(v = 2).

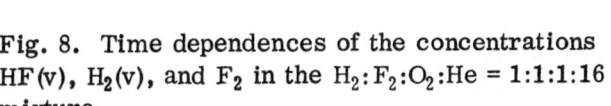

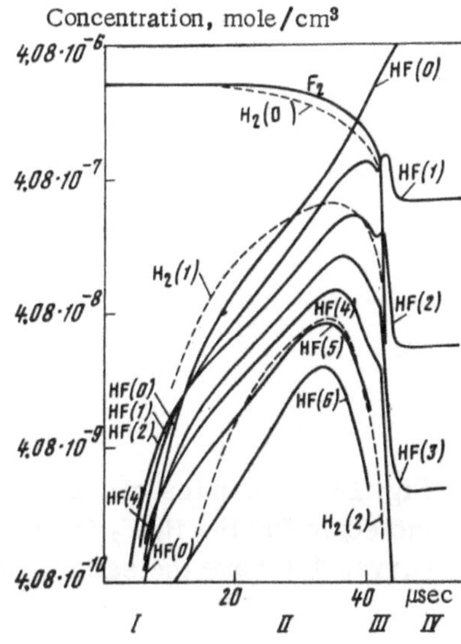

Fig. 8. Time dependences of the concentrations HF(v), H_2(v), and F_2 in the $H_2:F_2:O_2:He$ = 1:1:1:16 mixture.

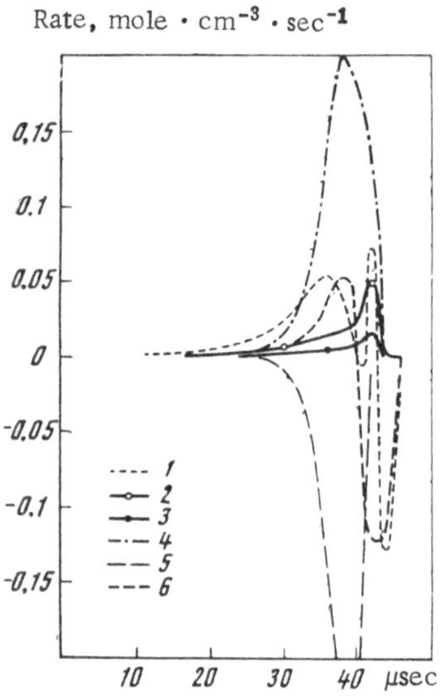

Fig. 9. Dynamics of the v = 1 level of the HF molecule in the H_2:F:O_2:He = 1:1:1:16 mixture; curves 1-6 have the same meaning as in Fig. 6.

The results of calculations of the kinetics of the v = 1, 2 levels are presented in Figs. 6-10. It follows from the kinetic curves that, in general, different vibrational levels of HF are populated and emptied by processes of different nature and the levels are independent of one another. Moreover, the contribution of a specific process to the formation of HF(v) depends on the composition and pressure of a mixture. At low pressures in a working mixture (~10 Torr) the v = 2 level of the HF molecule is emptied mainly by the VV exchange of energy in HF-HF collisions (Fig. 7). The contribution of the VT processes to the rate of deactivation of HF(v = 2) is slight. The VV exchange between HF and H_2 deactivates the v = 2 level and populates the v = 1 level of HF. A considerable role in the formation of HF(v = 1) is played by the VT processes. The main contribution to the excitation of the v = 1, 2 levels is made by the F + H_2 reaction and a small contribution by the H + F_2 process. The effects of chemical pumping are manifested simultaneously and they cannot be separated in time. At high pressures in a mixture (~760 Torr) and in the presence of a diluent the role of the VT relaxation becomes considerably greater (Figs. 9 and 10).

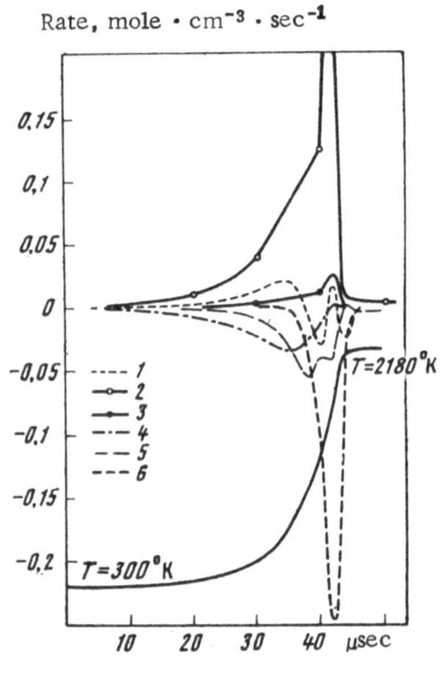

Fig. 10. Dynamics of the v = 2 level of the HF molecule for the H_2:F_2:O_2:He = 1:1:1:16 mixture; curves 1-6 have the same meaning as in Fig. 7.

It should be pointed out that neither the VT nor the VV processes reach a state of equilibrium at any stage of the reaction. This is due to a perturbation of the distribution of the HF molecules over the vibrational levels by the chemical reactions $H + F_2$ and $F + H_2$.

Population Inversion of Vibration-Rotational Levels of HF in the $H_2 + F_2$ System

In calculations of the population inversion we have to know the probabilities α_n and β_n of the excitation of vibrational levels of HF in the elementary reactions $F + H_2$ and $H + F_2$, respectively. The values of α_n and β_n were selected on the basis of the experimental data reported in [33, 34]: $\alpha_0 = \alpha_3 = \alpha_4 = \alpha_5 = \alpha_6 = 0$, $\alpha_1 = 0.154$, $\alpha_2 = 0.846$; $\beta_0 = 10^{-3}$, $\beta_1 = 0.055$, $\beta_2 = 0.083$, $\beta_3 = 0.138$, $\beta_4 = 0.185$, $\beta_5 = 0.276$, $\beta_6 = 0.262$. The calculated time dependences of the populations of vibrational levels of HF and of the population inversion of vibration-rotational levels as a result of the $v = 2 \to v = 1$ transition are plotted in Figs. 8 and 11. The population inversion was calculated on the assumption that the rotational temperature was equal to the gas temperature. The solutions obtained made it possible to identify four main stages of the reaction, which differed in the degree of disequilibrium of the distribution function of the vibration-rotational system of the levels of HF (Fig. 8).

I. During this stage a complete population inversion is achieved for some pairs of the vibrational levels, i.e., the transition temperature is negative for certain values of v:

$$T_{v,\,v-1} < 0.$$

In the adopted model of the $H_2 + F_2$ reaction a population inversion is due to the transitions $1 \to 0$, $2 \to 1$, $4 \to 3$, and $5 \to 4$. The strongest inversion is achieved as a result of the transitions $1 \to 0$ and $2 \to 1$.

II. During this stage a partial inversion occurs in vibration-rotational transitions, i.e.,

$$(T_{v,\,v-1}/T_r) > (h\nu_{v,\,v-1}/h\nu_r),$$

where $h\nu_{v,v-1}$ is the energy of a vibrational quantum; $h\nu_r$ is the energy of a rotational quantum involved in these transitions ($f < 20$); T_r is the rotational temperature.

III. During this stage the vibrational temperature falls rapidly and the rotational temperature rises so that the following condition is satisfied:

$$(T_{v,\,v-1}/T_r) < (h\nu_{v,\,v-1}/h\nu_r).$$

This results in a decay of the partial inversion and the stimulated emission is quenched.

IV. During the final stage the state of thermodynamic equilibrium is reached:

$$T_{v,\,v-1} = T_r = T.$$

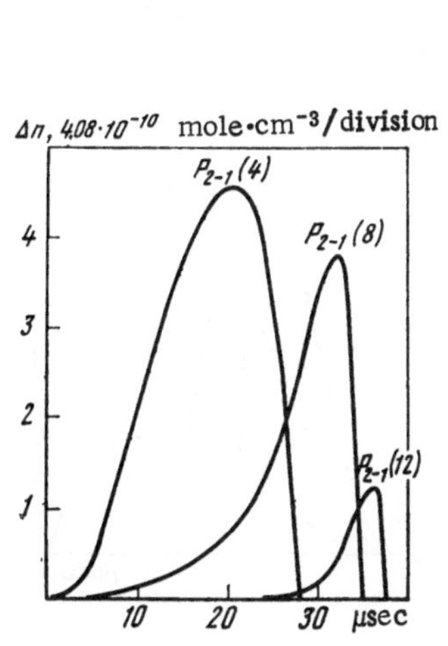

Fig. 11. Time dependences of the population inversion of vibration-rotational levels of HF in the H_2:F_2:O_2:He = 1:1:1:16 mixture.

At the end of stage I the consumption of the reagents is ~1%, whereas at the end of the stage II is is ~60%. Calculations showed that the absolute population inversion in individual vibration-rotational lines achieved during stage II within a given band is considerably greater than the total inversion during stage I (Figs. 8 and 11). Consequently, the main contribution to the energy of a laser pulse is made by stage II. Thus, a high energy of rotational quanta of HF ensures a longer lifetime of the partial inversion and is one of the decisive factors which makes it possible to use the advantages of the chain reaction in the generation of coherent radiation.

Estimate of Laser Efficiency

Considerable (and even insuperable) difficulties are encountered in the exact calculation of the energy of laser pulses because of a large number of stimulated transitions. An approximate estimate of the efficiency of the system can be obtained as follows. According to the calculations, the average inversion density for one vibration-rotational transition ($H_2 : F_2 : O_2 :$ He = 1 : 1 : 1 : 16 mixture) is $2 \cdot 10^{-9}$ mole/cm^3. The total number of transitions contributing to stimulated emission is 10-20. In this situation the density of the coherent radiaton energy is $(2\text{-}4) \cdot 10^{-4}$ cal/cm^3. The density of the stored chemical energy is 0.3 cal/cm^3. Consequently, the laser efficiency is of the order of 10^{-3}. This value is in good agreement with the experimental data [31] ($H_2 : F_2 :$ He = 4.5 : 1.5 : 45 Torr, density of stored chemical energy 10^{-2} cal/cm^3, density of radiation energy ~10^{-5} cal/cm^3).

Influence of Initiating Pulse Energy and Duration on Laser Characteristics

The time dependence of the initiation rate constant can be expressed in the form $k_0(t) = \gamma_1 t e^{-\gamma_2 t}$. A maximum of $k_0(t)$ occurs at a moment $t_{max} = \gamma_2^{-1}$. The energy of an initiation pulse is, apart from a dimensional constant, γ_1/γ_2^2. This value governs the degree of decomposition of F_2 as a result of photodissociation: $([F_2]_0 - [F_2]_\infty)/[F_2]_0 = 1 - e^{-\gamma_1/\gamma_2^2}$. In the integration variant presented in Figs. 6 and 7, we have $\gamma_2^{-1} = 5$ μsec and $\gamma_1/\gamma_2^2 = 0.1$. The curves in Figs. 8-11 are obtained for the following parameters of initiation pulses: $\gamma_2^{-1} = 30$ μsec and $\gamma_1/\gamma_2^2 = 0.0125 \cdot 10^{-1}$, which correspond to the formation of $3.2 \cdot 10^{16}$ fluorine atoms in 1 cm^3 as a result of photodissociation.

Figures 12 and 13 illustrate the influence of the energy and duration of initiation pulses on the time dependence of the population inversion. Both an increase in the energy of initiation pulses of constant duration and a reduction in the duration of pulses of constant energy enhance the maximum population inversion and shorten the duration of inversion pulses. These results are in qualitative agreement with observations [28]. It is clear from Fig. 13 that the use of short initiation pulses increases considerably the laser efficiency. When the value of t_{max} is reduced from 100 to 1 μsec (keeping the energy of initiation pulses constant), the maximum inversion increases by a factor of 50. Estimates based on the solutions obtained indicate that the efficiency of laser action initiated by short pulses (~1 μsec) may reach 0.01.

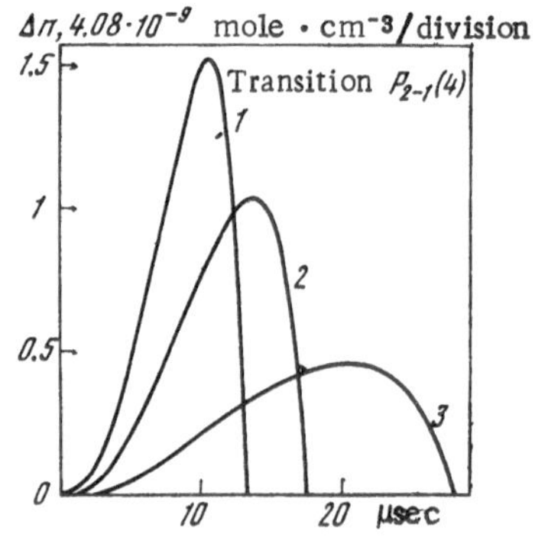

Fig. 12. Influence of the energy of initiation pulses of constant duration on the time dependences of the population inversion in the $H_2 : F_2 O_2 :$ He = 1:1:1:16 mixture. The initiation parameters were $\gamma_2^{-1} = 30$ μsec; $\gamma_1/\gamma_2^2 = 0.0125$ (curve 3), 0.0375 (curve 2), 0.075 (curve 1).

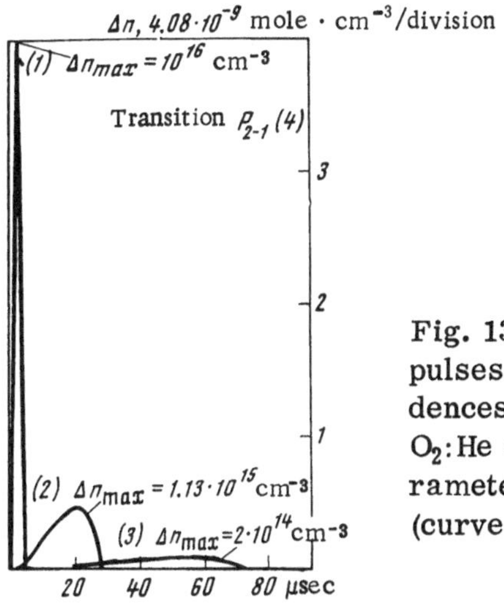

Fig. 13. Influence of the duration of initiation pulses of constant energy on the time dependences of the population inversion in the H_2:F_2:O_2:He = 1:1:1:16 mixture. The initiation parameters were γ_2^{-1} = 1 μsec (curve 1), 30 μsec (curve 2), 100 μsec (curve 3); γ_1/γ_2^2 = 0.0125.

Concluding Comments on Calculations

Reliable information on the rates of elementary processes over a wide range of temperatures is needed in predicting the operation of a chemical laser and in optimizing its parameters (composition of the mixture, temperature, initiation). Such information is not yet available. Therefore, in describing some of the processes it was necessary to use a priori calculations of the rate constants or to extrapolate the available experimental data over a wide range of temperature on the basis of the available fundamental theories. Least information was available on the VV energy exchange processes. The rate constants of these processes were calculated on the basis of the Swartz-Slawsky-Herzfeld (SSH) theory [16].

In the description of the VT relaxation in the adopted kinetic model, use was made of the information reported in [18] on the rate of quenching of HF by H_2 at T = 470°K. The efficiency of the other components of the mixture in this quenching process was assumed to be equal to the efficiency of H_2. The temperature dependences of the VT rates were described using the formulas given in [16]. The SSH theory and the theory of relaxation of hydrogen halides developed in [35] both predicted that the light molecule H_2 should be most efficient in the acquisition of the vibrational energy from HF. Therefore, one would expect the values of the VT rates used in the calculations of the kinetics of an HF laser to overestimate the contribution of the VT processes to the rate of decay of the population inversion. However, in recent years new experimental data on the vibrational relaxation of HF have become available and these could not be interpreted on the basis of the fundamental SSH theory or the theory given in [35]. It was found in [36] that the VT process HF(v = 1) + HF(v = 0) → 2HF(v = 0) occurred at a very high rate at 350°K ($p\tau$ = 1.5 · 10^{-8} sec · atm). According to the experimental results given in [35], the deactivation of HF by HF was slower in the temperature range 1500-4000°K [$p\tau$ = (1-3) · 10^{-7} sec · atm]. The observed temperature dependences of the VT rates (HF-HF) did not agree with the predictions of the theories under consideration. Shin [37] calculated the rates of the VT relaxation in HF-HF and DF-DF collisions making allowance for the dipole-dipole attraction forces between the molecules in the range 400-4000°K. His theory predicted the existence of a minimum rate of relaxation of HF interacting with HF at 800°K. Shin's calculations were in agreement with the experimental results at low temperatures (400°K). At high temperatures (1500-4000°K) the experimental values of the rates of the VT relaxation rates. were approximately an order of magnitude smaller than the calculated rates.

The rates of the VT relaxation in the HF-HF interaction, used in the HF laser model, were in satisfactory agreement with the experimental results at high temperatures (T ~ 1000°K) but were underestimated at low temperatures (T ~ 300°K). Nevertheless, the calculated laser efficiencies were in good agreement with the measurements. Further studies of the kinetic processes in this laser and the refinement of the model by comparing the theoretical predictions with the experimental results should give new information on the relaxation rates.

§ 4. $D_2 + F_2 + CO_2$ System

We shall report a calculation of the laser efficiency for a specific $D_2 + F_2 + CO_2$ mixture and we shall describe machine experiments in which several important kinetic parameters of this laser system were varied.

Influence of Initiation Rate on Time Dependences of Population Inversion and Laser Efficiency

The reaction kinetics equations were integrated for a homogeneous mixture with the parameters $v_0(0) = 0.0592$, $u(0) = 0.0592$, $h = 0.246$

$$g = 1 - \left(\sum_n v_n(0) + u(0) + h + w(0) + y(0) + z(0) + \sum_n x_n(0)\right),$$

$w(0) = 0.00132$, $v_1(0) = v_2(0) = z(0) = x_n(0) = f(0) = 0$ and this was done for an initial temperature $T(0) = 300°K$ and concentration $\rho = 4.08 \cdot 10^{-5}$ mole/cm^3. Four rates of initiation of the reaction were considered: $y(0) = 0.396 \cdot 10^{-4}$, $y(0) = 0.396 \cdot 10^{-3}$, $y(0) = 0.132 \cdot 10^{-2}$, and $y(0) = 0.396 \cdot 10^{-2}$. Numerical calculations of the time dependences of the population inversion of the laser levels gave results plotted in Fig. 14. The time dependences of the populations of the laser levels of CO_2 and of the difference between these populations are compared in Fig. 15 with an oscillogram of a stimulated emission pulse reported in [6] for identical conditions and the same total pressure in the mixture. The calculations agree with the experimental results in

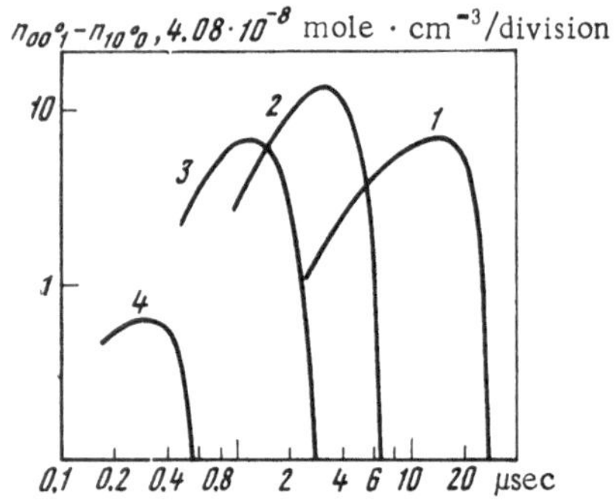

Fig. 14. Time dependences of the population inversion of the $D_2:F_2:CO_2:He = 1:1:4:11$ mixture obtained for different initiation levels (rates): 1) $y(0) = 0.396 \cdot 10^{-4}$; 2) $y(0) = 0.396 \cdot 10^{-3}$; 3) $y(0) = 0.132 \cdot 10^{-2}$; 4) $y(0) = 0.396 \cdot 10^{-2}$.

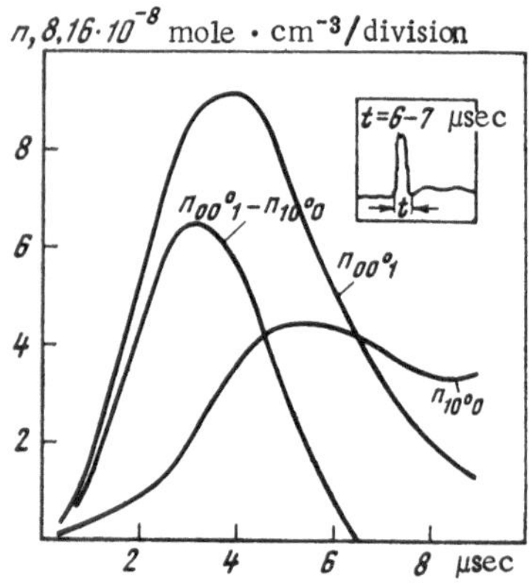

Fig. 15. Time dependences of the populations and difference between the populations of laser levels of the $D_2:F_2:CO_2:He = 1:1:4:11$ mixture for $y(0) = 296 \cdot 10^{-3}$. The inset shows an oscillogram of a laser pulse emitted by the same mixture.

TABLE 2

$y(0)$	t_0, μsec	$T(t_0)$, °K	$T_3(t_0)$, °K	$E_3(t_0)$, cal/cm^3	t_{max} μsec	$T(t_{max})$, °K	$T_3(t_{max})$, °K	$E_3(t_{max})$, cal/cm^3
$0.396 \cdot 10^{-4}$	27.5	730	1192	$0.423 \cdot 10^{-2}$	25	648	1201	$0.433 \cdot 10^{-2}$
$0.396 \cdot 10^{-3}$	6.5	1430	2505	$0.237 \cdot 10^{-1}$	6	1285	2550	$0,245 \cdot 10^{-1}$
$0,132 \cdot 10^{-2}$	2.5	1480	2711	$0.274 \cdot 10^{-1}$	3	1774	2845	$0.298 \cdot 10^{-1}$
$0.396 \cdot 10^{-2}$	0.6	856	1469	$0.754 \cdot 10^{-2}$	1.6	1902	2665	$0.266 \cdot 10^{-1}$

respect of the duration and energy of the output pulses. The results of calculation of the laser efficiency η as a function of y(0) are as follows:

$y(0)$	$0.396 \cdot 10^{-4}$	$0.396 \cdot 10^{-3}$	$0.132 \cdot 10^{-2}$	$0.396 \cdot 10^{-2}$
η	$0.6 \cdot 10^{-2}$	$0.31 \cdot 10^{-1}$	$0.35 \cdot 10^{-1}$	10^{-2}

Hence, we can see that the laser efficiency rises first with the initiation level, passes through a maximum, and then falls. The optimal value of y(0) for the mixture under investigation lies within the range $(0.396-1.32) \cdot 10^{-3}$.

The existence of an optimal initiation level, i.e., an optimal chemical reaction rate, can be explained quite simply. It is illuminating to compare the calculated maximum values of the vibrational temperature of the ν_3 mode of CO_2 achieved at the moment t_{max} with the value of the same temperature at the moment of disappearance of the population inversion of the laser levels t_0, and to compare the values of t_{max} and t_0 for different levels of initiation of the reaction. The necessary information is given in Table 2, which includes also the values of the vibrational energy of the ν_3 mode of CO_2 and of the gas temperature at the relevant moment.

It is clear from these results that when the reaction rate is high, we have $t_0 < t_{max}$. In this case the "bottleneck" of the chemical pumping process is the stage of transfer of excitation from DF to CO_2. Rapid heating of the working mixture results in a high population of the lower laser level and it stops stimulated emission. The major part of the vibrational energy of the ν_3 mode of CO_2 is lost uselessly. On the other hand, when the reaction rate is low, we have $t_{max} < t_0$. However, in this case the vibrational temperature of the ν_3 mode of CO_2 is low, since it is affected by the relaxation of the vibrational energy evolved during the reaction. The competition between the processes of excitation of the lower and upper laser levels gives rise to an optimal reaction rate. It is clear that this effect should disappear when the specific heat of the mixture is increased. The calculated optimal value of y(0) depends strongly on the rate constant of the energy exchange between DF and CO_2.

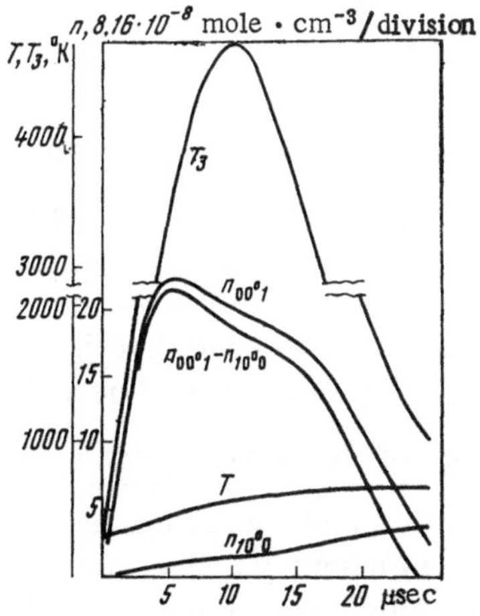

Fig. 16. Time dependences of populations, difference between the populations of laser levels, translational temperature of the medium, and vibrational temperature of the ν_3 mode of CO_2 in the $D_2:F_2:CO_2:M$ = 1:1:4:11 mixture (M is an inert gas with a specific heat of 30 kcal · mole^{-1} · deg^{-1}); $y(0) = 0.132 \cdot 10^{-2}$.

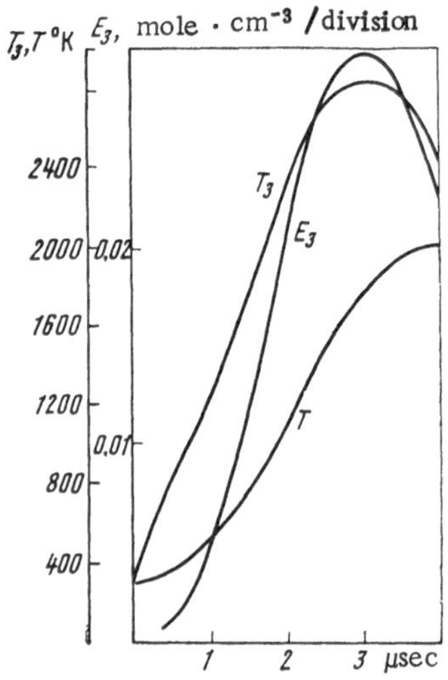

Fig. 17. Time dependences of the translational temperature of the D_2:F_2:CO_2:He = 1:1:4:11 mixture, and of the vibrational temperature and specific energy of the ν_3 mode of CO_2; y(0) = 0.132 · 10^{-2}.

Influence of Specific Heat of a Mixture on Laser Parameters

We carried out model calculations which revealed the role played by the working mixture temperature in the population inversion dynamics. The chemical kinetics equations were integrated for the same parameters of the mixture as in the variant represented by Fig. 14, with the exception of the specific heat of the inert gas. Figure 16 shows the time dependences of the temperature of the active medium, vibrational temperature of the ν_3 mode of CO_2, populations of the laser levels, and of the difference between the populations of the laser levels when the specific heat of the inert gas 30 cal · mole^{-1} · deg^{-1}, i.e., when this heat was increased by a factor of 10 compared with the specific heat of He. Under these conditions the vibrational temperature of the ν_3 mode differed very greatly from the translational temperature. The value of T_3 at the maximum was 4700°K, whereas the value of T did not exceed 700°K. The laser efficiency rose to 14%. The calculations demonstrated that the heating of a mixture which reached 2000°K under real conditions in chemical lasers (Fig. 17) was one of the major factors which limited the efficiency of conversion of chemical energy into light. The chemical efficiency of a laser utilizing a mixture with a sufficiently high specific heat was close to the maximum value found from the law of conservation of energy

$$\text{Efficiency} < \frac{h\nu_L}{Q}(\bar{v}_1 + \bar{v}_2),$$

where $\bar{v}_1 = \sum_v \alpha_v v$ and $\bar{v}_2 = \sum_v \beta_v v$ are the average numbers of the vibrational quanta excited in the elementary reactions F + D_2 and D + F_2, respectively; Q is the exothermal effect of the D_2 + F_2 = 2DF reaction, equal to 130 kcal/mole; $h\nu_L$ is the energy of a laser photon.

An investigation of the stimulated emission spectrum of a chemical laser utilizing a D_2 + F_2 mixture [19] demonstrated that the elementary reactions F + D_2 and D + F_2 were most likely to populate the v = 3 and v = 6–9 vibrational levels, respectively, of the DF molecule. Therefore, we assumed that $\bar{v}_1 + \bar{v}_2 = 9$. In this case the maximum theoretical efficiency of a D_2 + F_2 + CO_2 laser should be 19%.

Variation of Constants k_1 and k_2

In this and subsequent subsections we shall analyze the influence of the rate constants of several elementary processes on the calculated laser characteristics. In the initial integration variant, which was compared with the results of variation of the rate constants, we used the

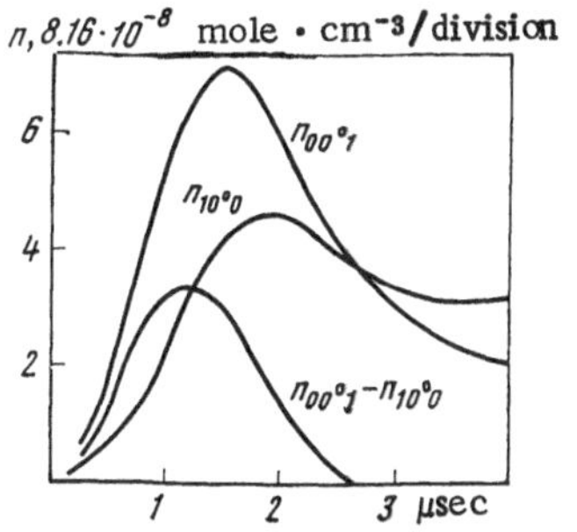

Fig. 18. Time dependences of the populations and difference between the populations of laser levels for the D_2:F_2:CO_2:He = 1:1:4:11 mixture; $y(0) = 0.132 \cdot 10^{-2}$.

following initial parameters: $v_0(0) = 0.0592$, $u(0) = 0.0592$, $h = 0.246$, $g = 0.63$, $w(0) = 0.00132$, $y(0) = 0.132 \cdot 10^{-2}$, $T(0) = 300$°K. The rate constants of the elementary processes F + D_2 and D + F_2 were reduced by a factor of 2. The maximum value of $n_{0001} - n_{1000}$ increased from $2.14 \cdot 10^{-7}$ mole/cm^3 in the initial variant, illustrated in Fig. 18, to $4.5 \cdot 10^{-7}$ mole/cm^3 and the duration of the inversion pulses increased from 2.7 to 5 μsec. The laser efficiency was 3.5% for the initial variant and it was practically unaffected. This machine experiment indicated that the selection of the constants k_1 and k_2 had a strong influence on the calculated gain of the medium and on the time characteristics of stimulated emission. The laser efficiency calculated in the range of optimal values of y(0) was affected less by the values of k_1 and k_2.

Variation of Constants α_n and β_n

We assumed the following values of the probabilities of population of the vibrational levels of DF in elementary processes F + D_2 and D + F_2: $\alpha_0 = \alpha_1 = \alpha_2 = \alpha_3 = 1/4$, $\alpha_4 = \alpha_5 = \alpha_6 = 0$, $\beta_0 = \beta_1 = \beta_2 = \beta_3 = \beta_4 = \beta_5 = \beta_6 = 1/7$. For these values of α_n and β_n the vibrational energy of the DF molecules formed in these reactions represented 28% of the total energy evolution per one "link" in the chain reaction (in the initial variant this quantity was 54%). Integration of the rate equations demonstrated that a population inversion of the laser levels of CO_2 should not take place because of the faster rise of the gas temperature. This serious disagreement between the calculated and experimental results indicated that the energy yield of the reaction transferred to the vibrations of DF exceeded greatly 28%.

Variation of Constants $P_{n\,n+1}^{m\,m-1}$ and $Q_{n\,n+1}^{m\,m-1}$

A tenfold reduction in the rate constants of the vibration-vibrational exchange of energy in the DF-DF and DF-D_2 collisions increased the duration of the population inversion pulses from 2.7 to 3 μsec and raised the maximum inversion from $2.74 \cdot 10^{-7}$ to $2.8 \cdot 10^{-7}$ mole/cm^3. On the other hand, a tenfold increase in the same rate constants reduced the duration of the population pulses from 2.7 to 2.35 μsec and reduced the maximum inversion from $2.74 \cdot 10^{-7}$ to $2.16 \cdot 10^{-7}$ mole/cm^3. These machine experiments established that the results of calculations of the laser characteristics depended very weakly on the constants $Q_{n\,n+1}^{m\,m-1}$ and $P_{n\,n+1}^{m\,m-1}$.

Variation of Constant k_{16}

In all the above calculations the rate constant of the vibration-vibrational exchange of energy between DF and CO_2 was assumed to be $3 \cdot 10^{-13}$ cm^3/sec. Integration of the rate equations for $k_{16} = 10^{-13}$ cm^3/sec demonstrated that in this case a population inversion of the laser levels did not take place. However, an increase of this constant to 10^{-12} cm^3/sec raised the laser efficiency from 3.5 to 9%. These machine experiments (calculations) revealed a strong dependence of the calculated characteristics of the laser radiation on the constant k_{16}. A comparison of the results of calculation of the efficiency with the results of measurements demonstrated that the value of k_{16} was close to $3 \cdot 10^{-13}$ cm^3/sec. The value of k_{16} was estimated in [20] to be (1-50) $\cdot 10^{-13}$ cm^3/sec. Thus, an analysis of the laser kinetics at high pressures gave a value of k_{16} which was close to the lower limit obtained in [20]. A study of the charac-

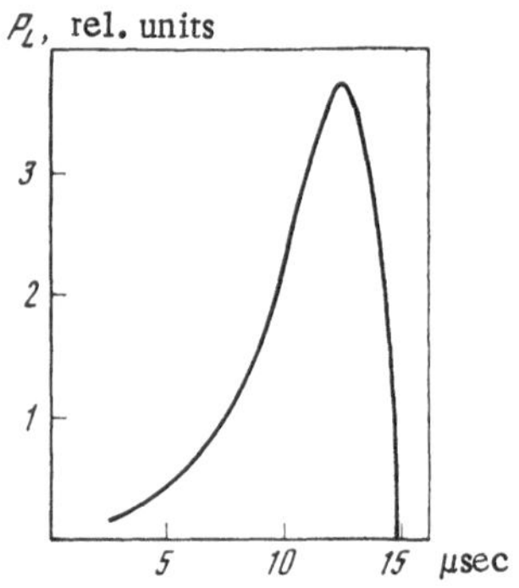

Fig. 19. Time dependences of the laser output power in relative units obtained from the D_2:F_2:CO_2:He:Ar = 1:3.15:10:10:75.85 mixture.

teristics of a chemical using a $D_2 + F_2 + CO_2$ mixture at low pressures [19] gave $k_{16} \approx 1.5 \cdot 10^{-12}$ cm^3/sec.

Thus, at this state of our knowledge the values of the rate constant of the VV energy exchange between DF and CO_2 deduced by analyzing various experiments differ quite considerably. This is understandable because the interpretation of such experiments is based on kinetic models in which the important processes are not only the exchange of energy between DF and CO_2 but also several other processes whose rates are not known reliably. In the case of the model of a $D_2 + F_2 + CO_2$ laser this applies particularly to the VT relaxation of DF interacting with DF. We can obtain a calculated value of the laser efficiency which agrees with the experimental results if we use a low value of k_{16} and of the DF−DF quenching rate constant. The present author is not aware of any experimental data on the rate of the latter process. A calculation reported by Shin [37] shows that in the range of temperatures of interest to us the VT relaxation due to the DF−DF interaction is slower (by about an order of magnitude) than the VT relaxation due to the HF−HF interaction. The maximum relaxation time occurs at to 600°K ($p\tau \approx 10^{-7}$ sec·atm). The DF−DF quenching rate constant assumed in our laser model is two orders of magnitude smaller than that calculated by Shin. Since the relative concentration of DF in a mixture does not exceed 10%, this may result in the rate of quenching of DF being an order of magnitude slower than the true value. If we use the D-F quenching rate given in [37] we find that the calculated and experimental values of the energy output agree if we increase by about an order of magnitude the rate constant* of energy transfer from DF to CO_2 (i.e., to values in the range 10^{-12}-10^{-11} cm^3/sec). This should increase the optimal value of the rate of initiation of the reaction y(0).

More detailed investigations of the laser kinetics and of elementary processes are needed before the points discussed above are settled finally.

Calculation of Laser Efficiency in the Case of Strong Dilution of $D_2 + F_2 + CO_2$ Mixtures with Inert Gas

According to the calculations reported above, the laser efficiency approaches its maximum value if the temperature of the mixture under consideration does not exceed 500-600°K in the course of the reaction. The gas temperature can be lowered by using a monatomic inert gas (for example, Ar) as a buffer. Since the specific heat of a monatomic gas is 3 kcal/mole, a high degree of dilution of the reagents is needed. We calculated the laser efficiency for the $D_2 : F_2 : CO_2$: He : Ar = 0.01 : 0.0315 : 0.1 : 0.1 : 0.7585 mixture at T(0) = 300°K for $\rho = 0.588 \cdot 10^{-4}$ mole/cm^3. In these calculations allowance was made for the time dependence $k_0(t) = \gamma_1 t e^{-\gamma_2 t}$, where $\gamma_2^{-1} = 15$ μsec. The parameter γ_1 was selected in such a way that the total amount of atomic fluorine produced during an initiation pulse was

* According to the recent measurements $k_{16} = 4 \cdot 10^{-12}$ cm^3/sec for DF(v = 1) and it rises to $1.6 \cdot 10^{-11}$ cm^3/sec for DF(v = 3) [38].

$1.6 \cdot 10^{-7}$ mole/cm^3 irrespective of the value of ρ. The constant k_{16} was assumed to be 10^{-12} cm^3/sec and allowance was made for the relaxation of the 00^01 level because of the interaction with D_2, CO_2, He, and DF. The constant $k_{17}(CO_2-DF)$ was taken from [39]. The time dependence of the laser output power P_L is plotted in Fig. 19. Under the stated conditions the calculated laser efficiency is 8.5%. The efficiency of a laser utilizing this mixture was calculated as a function of ρ. When ρ was increased from $0.588 \cdot 10^{-4}$ to $1.764 \cdot 10^{-3}$ mole/cm^3, the laser efficiency fell from 8.5 to 3.3%. The main factor which limited the laser efficiency at high values of ρ was the rise of the rate of recombination of the active centers in the chain reaction.

Conclusions

Detailed models of the $H_2 + F_2$ and $D_2 + F_2 + CO_2$ systems were developed and machine (computer) calculations were made of the kinetics of these systems.

A study was made of the kinetics of the processes occurring in a laser utilizing $H_2 + F_2$ mixtures. The $H_2 + F_2$ reaction was of the branched-chain type. A calculation of the branching velocity as a function of the temperature and pressure in the mixture revealed difficulties in the utilization of the advantages of this branched-chain reaction.

A kinetic scheme of the $H_2 + F_2$ reaction, which included over 100 elementary processes, was used in a calculation of the time dependences of the populations of the vibrational levels of the HF molecule, and of the complete and partial population inversion. An analysis was made of the excitation and quenching mechanisms.

The following important points were noted:

1. A complete inversion of the vibrational levels occurs only during the initial stage when only about 1% of the initial substances have reacted. A partial inversion is maintained for much longer so that when this inversion disappears the consumption of the reagents is up to 60%. The maximum population inversion occurs for the 1-0 and 2-1 transitions.

2. The main contribution to the output power is due to the $F + H_2$ reaction, whereas the contribution of $H + F_2$ is less since the molecules formed as a result of the latter reaction have a fairly wide distribution over the vibrational levels. This limits the laser efficiency.

3. The laser efficiency depends strongly on the rate of initiation when the energy is constant. In the case of brief illumination (~ 1 μsec) the chemical efficiency of a laser may reach 1% for a quantum efficiency of the order of 10.

A quantitative analysis was made of the kinetic processes in a chemical laser utilizing a mixture of D_2, F_2, and CO_2 and operating at high pressures.

According to the calculations one of the main factors limiting the chemical efficiency of such lasers is the heating of the gas medium. Dilution of this medium with a buffer gas of high heat capacity may increase the laser efficiency to 10-15% and the quantum efficiency to about 400.

The results of calculations were in good agreement with the experimental values of the laser parameters.

Literature Cited

1. J. V. V. Kasper and G. C. Pimentel, Phys. Rev. Lett., 14:352 (1965).
2. M. S. Dzhidzhoev, V. T. Platonenko, and R. V. Khokhlov, Usp. Fiz. Nauk, 100:641 (1970).
3. N. G. Basov, V. I. Igoshin, E. P. Markin, and A. N. Oraevskii, Kvant. Elektron. (Mosc.), No. 2, 3 (1971).

4. O. M. Batovskii, G. K. Vasil'ev, E. F. Makarov, and V. L. Tal'roze, ZhETF Pis'ma Red., 9:341 (1969).
5. N. G. Basov, V. T. Galochkin, L. V. Kulakov, et al., Kratk. Soobshch. Fiz., No. 8, 10 (1970).
6. N. G. Basov, S. I. Zavorotnyi, E. P. Markin, A. I. Nikitin, and A. N. Oraevskii, ZhETF Pis'ma Red., 15:135 (1972).
7. V. I. Igoshin and A. N. Oraevskii, Preprint No. 162 [in Russian], Lebedev Physics Institute, Academy of Sciences of the USSR, Moscow (1969); Khim. Vys. Energ., 5:397 (1971).
8. A. F. Dodonov, E. B. Gordon, and G. K. Lavrovskaya, et al., Mass-Spectrometric and Maser Determinations of Rate Constants of Elementary Reactions Typical of Hydrogen Halide Lasers (Preprint) [in Russian], Institute of Physical Chemistry, Academy of Sciences of the USSR, Moscow (1969).
9. T. A. Jacobs, R. R. Giedt, and N. Cohen, J. Chem. Phys., 43:3688 (1965).
10. J. A. Blauer, J. Chem. Phys., 72:79 (1968).
11. N. N. Semenov and A. E. Shilov, Kinet. Katal., 6:3 (1965).
12. G. A. Kapralova, E. M. Trofimova, and A. E. Shilov, Kinet. Katal., 6:977 (1965).
13. G. A. Kapralova, E. M. Margolina, and A. M. Chaikin, Kinet. Katal., 10:32 (1969).
14. V. L. Tal'roze, Existence of Third Self-ignition Limit of a Mixture of Fluorine and Hydrogen Inhibited by Oxygen (Preprint) [in Russian], Institute of Chemical Physics, of Sciences of the USSR, Moscow (1969).
15. V. N. Kondrat'ev, Rate Constants of Gaseous Reactions [in Russian], Nauka, Moscow (1970).
16. K. F. Herzfeld and T. A. Litovitz, Absorption and Dispersion of Ultrasonic Waves, Academic Press, New York (1959).
17. J. H. Kiefer and R. W. Lutz, J. Chem. Phys., 44:668 (1966).
18. G. K. Vasil'ev, E. F. Makarov, V. G. Papin, and V. L. Tal'roze, Spectroscopic Investigation of Radiative and Nonradiative Relaxation of Excited HF Molecules (Preprint) [in Russian], Institute of Chemical Physics, Academy of Sciences of the USSR, Moscow (1969).
19. N. G. Basov, V. T. Galochkin, V. I. Igoshin, L. V. Kulakov, E. P. Markin, A. I. Nikitin, and A. N. Oraevskii (Oraevsky), Appl. Opt., 10:1814 (1971).
20. G. K. Vasil'ev, E. F. Makarov, V. G. Papin, and V. L. Tal'roze, Zh. Eksp. Teor. Fiz., 61:97 (1971).
21. R. D. Sharma, J. Chem. Phys., 49:5195 (1968).
22. A. S. Biryukov, B. F. Gordiets, and L. A. Shelepin, Preprint No. 41 [in Russian], Lebedev Physics Institute, Academy of Sciences of the USSR, Moscow (1969).
23. B. F. Gordiets, N. N. Sobolev, and L. A. Shelepin, Zh. Eksp. Teor. Fiz., 53:1822 (1967).
24. W. A. Rosser Jr, A. D. Wood, and E. T. Gerry, J. Chem. Phys., 50:4996 (1969).
25. C. B. Moore, R. E. Wood, B.-L. Hu, and J. T. Yardley, J. Chem. Phys., 46:4222 (1967).
26. J. T. Yardley and C. B. Moore, J. Chem. Phys., 46:4491 (1967).
27. N. G. Basov, L. V. Kulakov, E. P. Markin, A. I. Nikitin, and A. N. Oraevskii, ZhETF Pis'ma Red., 9:613 (1969).
28. G. G. Dolgov-Savel'ev, V. A. Polyakov, and G. M. Chumak, Zh. Eksp. Teor. Fiz., 58:1197 (1970).
29. N. G. Basov, E. P. Markin, A. I. Nikitin, and A. N. Oraevskii (Oraevsky), IEEE J. Quantum Electron., QE-6:183 (1970).
30. G. G. Dolgov-Savel'ev, V. F. Zharov, Yu. S. Neganov, and G. M. Chumak, Zh. Eksp. Teor. Fiz., 61:64 (1971).
31. L. D. Hess, Appl. Phys. Lett., 19:1 (1971).
32. A. N. Oraevskii, Zh. Eksp. Teor. Fiz., 55:1423 (1968).
33. N. Jonatan, C. M. Melliar-Smith, and D. H. Slater, J. Chem. Phys., 53:4396 (1970).

34. K. L. Kompa, J. H. Parker, and G. C. Pimentel, J. Chem. Phys., 49:4257 (1968).
35. G. A. Kapralova, E. E. Nikitin, and A. M. Chaikin, Nonempirical Calculations of Probabilities of Vibrational Transitions in Hydrogen Halides (Preprint) [in Russian], Institute of Chemical Physics, Academy of Sciences of the USSR, Moscow (1969).
36. J. R. Airey and S. F. Fried, Chem. Phys. Letters, 8:23 (1971).
37. H. K. Shin, Chem. Phys. Letters, 10:81 (1971).
38. J. R. Airey and I. W. M. Smith, J. Chem. Phys., 57:1669 (1972).
39. R. S. Chang, R. A. McFarlanc, and G. J. Wolga, J. Chem. Phys., 56:667 (1972).

PLASMA HEATING AND NEUTRON GENERATION RESULTING FROM SPHERICAL IRRADIATION OF A TARGET WITH HIGH-POWER LASER RADIATION

N. G. Basov, O. N. Krokhin, G. V. Sklizkov, and S. I. Fedotov

A high-power laser system with a series-parallel amplifier configuration was used to heat plasma to high temperatures. An experimental investigation was made of the principal parameters of the output of this laser system, including the coherence and spectrum of the output radiation. The energy of the radiation was 400-1300 J and it was emitted in the form of pulses of 1-16 nsec duration; the contrast was $\sim 10^7$ and the brightness was $\sim 10^{17}\ W \cdot cm^{-2} \cdot sr^{-1}$. An analysis was made of the requirements which should be satisfied by a laser system capable of high-temperature heating of a plasma generated by spherical irradiation of a target and suitable for realization of physically beneficial thermonuclear fusion. An experimental study was made of the processes occurring during high-temperature heating of a plasma generated by spherical irradiation of an isolated solid target with a nine-beam laser. The x-ray emission of the plasma was investigated. It was found that up to 20% of the laser radiation energy absorbed by the plasma was converted into x rays. About 10^7 neutrons/pulse were obtained from a $(CD)_n$ plasma. Generation of neutrons in a spherically irradiated laser plasma was found to be a thermal process. The process of gas-dynamic expansion of the plasma was studied. The pressure on the target surface due to ablation reached $\sim 2 \cdot 10^8$ bar. At these pressures the plasma core was compressed strongly by a factor of ~ 40.

INTRODUCTION

It was reported in [1] that heating of a dense plasma by high-power laser radiation made it possible to reach thermonuclear temperatures and achieve controlled fusion. Calculations carried out in [1-4] demonstrated that the realization of physically beneficial thermonuclear fusion would require a laser with an output energy of at least 10^5 J in the form of light pulses of $\tau \leq 1$ nsec duration. Moreover, a sufficiently effective and symmetric supply of the energy of light pulses to a plasma would have to be provided. The dominant mechanism ensuring an efficient absorption of high-power laser radiation by a superdense plasma is not yet clear but the results reported in the present paper and elsewhere [5-6] suggest that under certain conditions a considerable proportion of the laser energy may be absorbed.

Serious technical difficulties are encountered in the construction of such lasers so that considerable attention is currently being paid to the possibility of reducing the threshold laser energy E_{th} needed for the initiation of thermonuclear fusion in laser plasmas. For example, an increase in the efficiency of the supply of energy to a laser plasma by optimization of the geometry, size, and composition of the target is considered analytically in [3, 7, 8]. This approach makes it possible to reduce the value of E_{th} but the reduction is one order of magnitude.

The most promising way of achieving controlled thermonuclear fusion is to increase the density of the heated target by cumulative compression in a spherical configuration [9]. Beneficial thermonuclear fusion requires that the Lawson criterion $n\tau > 10^{14}$ be satisfied; here, n is the density of particles in a plasma and τ is the plasma lifetime at a thermonuclear temperature of ~10 keV. The lifetime of a laser plasma is governed primarily by the inertial confinement time, which is given approximately by $\tau \sim d/V_s$, where d is the target size and V_s is the velocity of sound in the plasma. In principle, the plasma density can be increased by several orders of magnitude above the value found in a solid. For example, it is pointed out in [10] that compression of matter can be achieved by cumulation of a spherical converging shock wave. Waves of this kind can be generated, for example, by exploding a heavy spherical liner [8]. It is suggested in [11] that a special shaping of laser pulses may ensure isentropic compression of a homogeneous spherical target by a factor of $\sim 10^3$-10^4.

Cumulation processes which occur in the spherical irradiation of homogeneous targets by high-power lasers may be due to an intense pressure pulse which is produced by the expansion of the outer hot shell of the plasma or corona (this process is known as ablation). The required shape of the pressure pulse can be obtained by varying the shape of the laser pulse. One of the special cases of adiabatic compression is subjected to a numerical analysis in [11, 12], where it is shown that under certain conditions a compression may be $\sim 10^3$-10^4 and the density of the laser radiation flux at the end of a pulse may reach 10^{18}-10^{19} W/cm^2.

First experimental studies of the heating of plasma as a result of spherical irradiation of solid targets were reported in [9, 13]. Irradiation of a deuterated polyethylene $(CD_2)_n$ target of ~110 μ diameter with radiation produced by a high-power nine-channel laser with an output energy of ~214 J in the form of light pulses of ~2 nsec duration resulted in the emission of $\sim 10^7$ neutrons/pulse from the plasma [14]. This was considerably greater than the neutron yield resulting from the sharp focusing of radiation emitted by a single-beam laser [15-18].

The present paper describes a high-power laser system (HPL) designed for heating plasmas to high temperatures. The principal parameters of the output radiation of this laser are described and the possibility of development of such laser systems for the realization of controlled fusion are considered. The concluding chapter gives the results of an investigation of the efficiency of heating and principal parameters of a plasma resulting from the spherical irradiation of a solid with this high-power laser.

CHAPTER I

DESIRED PARAMETERS OF LASER SYSTEMS

The most important parameter of a high-power laser is the output energy for a given duration of light pulses. The most powerful currently existing laser is one with a series-parallel amplifier system [13], whose output energy is ~600 J in the form of light pulses of ~2 nsec duration. Further development of such lasers seems to be the most promising way of constructing laser systems for heating plasmas in order to achieve controlled fusion.

§ 1. Radiation Contrast

In the development of high-power lasers the most difficult problem to overcome is the achievement of a high contrast of the output radiation, i.e., of a high value of the ratio of the energy in a heating pulse to the energy of the radiation reaching the target before the arrival of this pulse. The latter energy is governed primarily by the superluminescence of the active elements and by the quality of the electrooptic switches. The contrast necessary for laser systems with an output energy of $\sim 10^3$ J is $\sim 10^6$. For energies in the range $\sim 10^5$-10^6 J it should exceed 10^8. The attainment of these high values of the contrast is a very urgent task which requires special study.

§ 2. Divergence of Radiation

Another equally important parameter of the radiation emitted by laser systems is the divergence angle α, whose value is governed primarily by the target size and the geometry of the focusing of the radiation on the target surface. The minimum divergence that a beam can have is known as the diffraction limit and it amounts to $\sim 2 \cdot 10^{-5}$ rad for the majority of modern lasers; however, if allowance is made for the distortions which occur in active elements, a realistic divergence (one which does not involve a serious loss of the energy) for neodymium lasers is $\sim 10^{-4}$ rad. We shall now consider in greater detail the problem of the required divergence in a many-beam laser.

Let us assume that the radiation emitted by a multibeam laser is focused on a spherical target of radius r_{00} using objectives and that each beam is focused separately by an objective with a focal length f and a back vertex intercept f_1. This focusing system is not essential to the discussion given below. When the objectives are packed as closely as possible on a sphere, a change in the focusing system alters only slightly the estimates of divergence given below.

We shall consider separately one focusing channel (Fig. 1). The following notation is used here: r_{00} is the radius of a target whose center coincides with the principal focal point of the objective; $2r_c$ is the diameter of the exit aperture of the beam emerging from the exit stage of the amplifier; f_3 is the distance from the exit surface of the objective from the focal plane; f_4 is the focal length of the entry component of the objective; $f_5 = f_4 - f_3$; 2r is the diameter of the focusing spot on the target surface; 2α is the divergence of the laser beam (it is assumed that when the angle is 2α all the laser energy is concentrated on the target); L is the distance from the entry component of the objective to the exit surface of the amplifier.

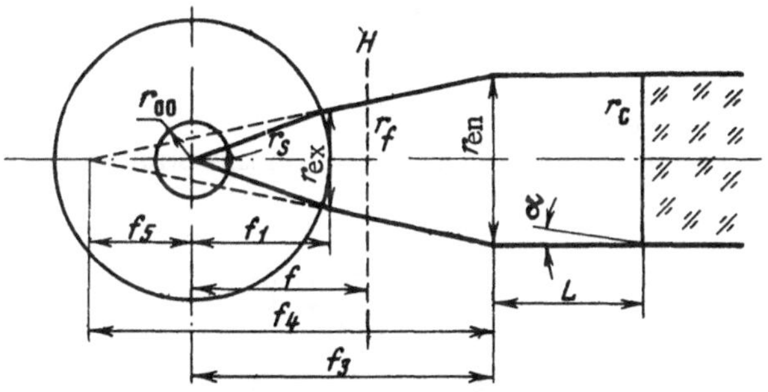

Fig. 1. Schematic diagram explaining a system for focusing radiation of a multibeam laser on a spherical target. The diagram shows one beam of the focusing system; all the other beams are focused in a similar manner. The diagram is not drawn to scale. It is assumed that $r_{en} \geq r_{ex}$, where r_{en} and r_{ex} are the radii of the entry and exit apertures of the objective. The packing of the elements on a sphere of radius f is governed by the dimensions of the exit component of the objective.

We shall now estimate the maximum value of the beam divergence which still ensures a uniform illumination of the target surface. This maximum value is the divergence for which the focusing spot on the target surface spreads by an amount equal to its diameter. This corresponds to

$$\alpha = r_s/(f - r_{00}), \tag{1}$$

where $r_s = r_f r_{00}/f$.

Bearing in mind that the principal plane of this system is located between the components of the objective, we find that

$$r_f = r_{en}\eta_5,$$

where

$$\eta_5 = (f + f_5)/f_4. \tag{2}$$

Assuming that the beam diameter at the entry to the objective is equal to the diameter of the aperture, we obtain

$$r_f = \eta_5 (r_0 + \alpha L). \tag{3}$$

Combining Eqs. (1)-(3), we obtain the following expression for the divergence

$$\alpha = \frac{r_{00} r_0 \eta_5}{f(f - r_{00}) - r_{00}\eta_5 L}. \tag{4}$$

The relationship between the required divergence, on the one hand, and the laser energy and temperature of the heated plasma, on the other, can be established by introducing the maximum ("breakdown") energy E_{br} which can be directed onto the exit component of the objective that has the smallest surface area. It is clear that the minimum focal length of the objective can be represented in the form

$$f_{min} = \left(\frac{E_l \eta_1}{4\pi E_{br}\eta_3}\right)^{1/2} (1 - \eta_4)^{-1}, \tag{5}$$

where $\eta_4 = 1 - (f_1/f)$; $\eta_1 = E_s/E_l$; E_s is the energy reaching the target surface; η_3 is the proportion of a sphere of radius f_1 filled by the exit components of the objectives.

On the other hand, the target radius, which governs the mass of the heated plasma for a given chemical composition of the target, depends on the plasma temperature at the end of the heating pulse:

$$r_{00} = \left(\frac{E_l \eta_1 \eta_2 \eta_D \eta_t}{2\pi n_D kT}\right)^{1/3}, \tag{6}$$

where $\eta_2 = E_n/E_s$; E_n is the energy absorbed in the target; n_D is the density of deuterium ions; $\eta_D = n_D/n$; n is the density of particles in the target; $\eta_t = E_t/E$; E_t is the thermal energy of the plasma.

Using Eqs. (5) and (6), we can rewrite Eq. (4) in the form

$$\alpha = \frac{r_0 \left(\frac{E_l \eta_1 \eta_2 \eta_D \eta_t}{2\pi n_D kT}\right)^{1/3} (1 - \eta_4)^2 \eta_5}{\frac{E_l \eta_1}{4\pi E_{br}\eta_3} - \left(\frac{E_l \eta_1 \eta_2 \eta_D \eta_t}{2\pi n_D kT}\right)^{1/3} (1 - \eta_4)\left[\left(\frac{E_l \eta_1}{4\pi E_{br}\eta_3}\right)^{1/2} + L(1 - \eta_4)\eta_5\right]}. \tag{7}$$

When the energy is increased, the distance from the laser to the focusing system rises proportionally to the linear dimensions of the exit stage so that L can be described by the following approximate formula:

$$L \approx 50\,(E_l)^{1/3}\ \text{cm}, \tag{8}$$

where E_l is the output energy of the laser in joules. This dependence corresponds approximately to the parameters of the existing systems and is of semiempirical nature.

Figures 2 and 3 give the values of $\alpha(E_l)$ for deuterium and polyethylene targets. Figure 2 illustrates the change in the radiation divergence of an HPL when the active elements in the system are of constant diameter, whereas Fig. 3 illustrates the same change for a constant number of channels with active elements of increasing cross sections. The plasma temperature is the parameter of the curves in Figs. 2 and 3. The figures show also the values of the diffraction divergence α_{diff}, corresponding to the diameters of the active elements used. It is clear from these curves that the maximum permissible divergence decreases strongly with rising energy and temperature. In the case corresponding to Fig. 2 the divergence reaches its maximum permissible value for energies of $\sim 10^5$ J. The values of α for $E_l = 10$ J are given by way of illustration and are of no practical value because lasers with output energies below 100 J are unsuitable for spherical irradiation: such energies are sufficient for heating a target of extremely small dimensions and a high-quality optical system would be needed for such a target.

The divergence requirements can be lowered at least several-fold if we assume the possibility of overlap of the focusing spots on the target surface. It is then necessary to carry out an exact calculation of the required intensity distribution in the far zone of the laser beam. However, in the plasma heating systems based on the use of cumulation phenomena in the targets as a result of spherical irradiation the effective radius of the target at the end of the heating period is much smaller than the initial value. Therefore, in systems of this kind the required divergence is close to the value predicted by Eq. (7). For example, if $E_l \sim 10^5$ J and

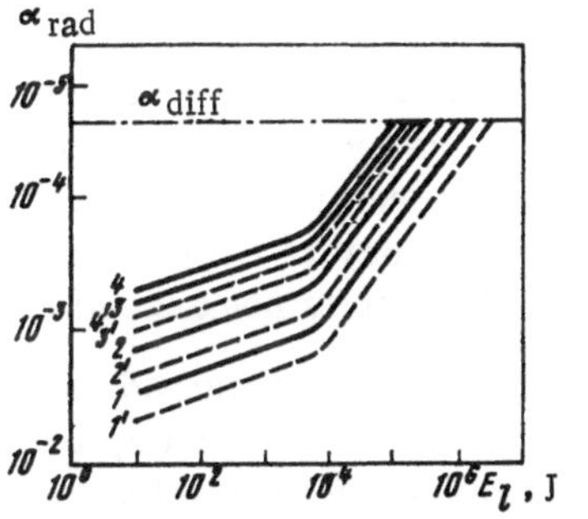

Fig. 2. Dependences $\alpha(E_l)$ for laser systems whose total output energy is increased by increasing the number of channels keeping constant the diameter of active elements. The curves plotted here correspond to Eq. (7) with the following parameters: $\eta_1 = 0.7$, $\eta_2 = 0.8$, $\eta_5 = 1$, $\eta_t = 0.5$, $\eta_D = 2/11$ for a CD_2 target and $\eta_D = 1/2$ for a D_2 target. In the energy range 10–10^3 J it is assumed that $r_f = r_{min} = 1/4$. Curves 1-4 represent a $(CD_2)_n$ target and curves 1'-4' a D_2 target with plasma temperatures of 10^2, 10^3, 10^4, and $2 \cdot 10^4$ eV, respectively.

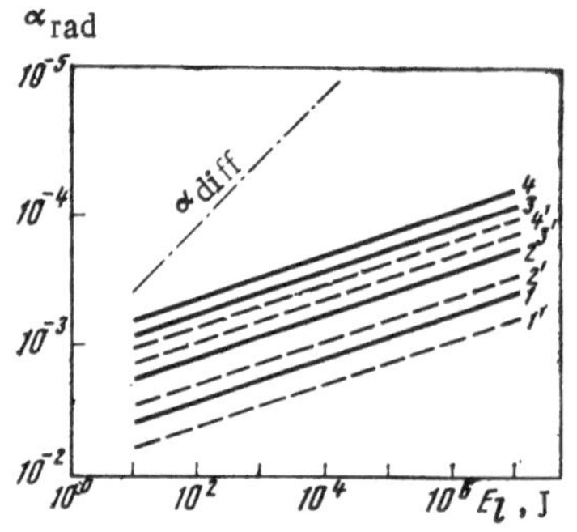

Fig. 3. Dependences $\alpha(E_l)$ for a high-power laser system in which the total output energy is increased by increasing the cross sections of the active elements keeping constant the number of the amplification channels. The values of the parameters η_1, η_2, η_5, η_t, and η_D and of the temperature are the same as in Fig. 2. The number of the exit beams is 12. No restrictions are imposed on the aperture of the focusing system.

the required temperature is $kT \sim 10^4$ eV, we find that the divergence required for a deuterium target compressed to a density $n_D \sim 10^{24}$ cm^{-3} is $\alpha \sim 10^{-5}$ rad. Thus, the need to achieve a small divergence of the beam may become the principal difficulty in the development of thermonuclear systems utilizing lasers characterized by high efficiencies of the conversion of electrical energy into light.

§ 3. Efficiency of Laser Systems

Another important parameter of lasers with output energies of $\sim 10^5$ J designed to heat plasmas to thermonuclear temperatures is the maximum efficiency η of the conversion of electrical energy into light. In the case of neodymium-glass lasers with series amplification stages the maximum conversion efficiency corresponds to input signals of high energy densities since in this case practically all the energy stored in the active element is transformed into radiation. In such cases the conversion efficiency is limited by the optical strength of the active elements and for currently used lasers it does not exceed ~0.3%.

It should be pointed out that this value is obtained on the assumption of relatively long exponentially decaying laser pulses with $\tau_l \sim$ 1-2 msec, and it can be increased considerably to a value of ~1-2% by ensuring that the shape of the laser pulses is optimal.

Systems with series-parallel high-power amplifiers (HPA) have a number of special features. These include particularly a beam splitter which not only contributes to optical losses but also alters considerably the energetics of the amplifier.

Formation of N beams at the exit from an HPA system with series-parallel amplifying stages can be achieved in various ways by altering the ratio of the number of series stages m and changing the division factor of light beams n. An analysis shows that the maximum value of the conversion efficiency is achieved when the division factor is equal to the gain in a single stage k and that this efficiency is equal to the value of η for a linear amplifier subjected to high-density input signals.

It should be pointed out that the division factor of a single stage is not constant but depends on the input signal energy, duration of light pulses, and pump beam energy. Under these

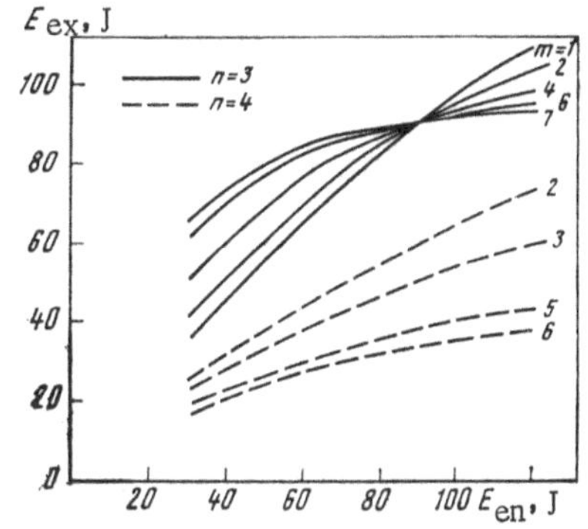

Fig. 4. Dependences of the exit energy of one beam on the energy entering a high-power amplifier. It is assumed that all the active elements of the amplifier are of the same diameter, which is 45 mm. The gain of each amplifier is assumed to vary linearly with the energy of the light pulses reaching its entry.

conditions the problem of the maximum value of η of a laser system is fairly difficult to solve analytically.

The results presented in the next chapter were used in analyzing the dependence of the energy in each beam at the exit from an HPA on the beam energy at the entry (for a fixed duration of the pulses). It was assumed that the amplification of strong signals was not greatly affected and the linear approximation was introduced for the function $k(E_{en})$.

Figure 4 shows the dependence of the exit energy of an HPA in each light beam on the entry energy of a light pulse. There are two series of curves corresponding to different values of the division factor n and different numbers of amplifying stages m. It is clear from these curves that in the optimal division case $k > n = 3$ for a given energy reaching the entry of an HPA each output beam is characterized by a higher energy than in the $k < n$ case. Moreover, in the former case a fairly high value of $\beta = (\Delta E_{ex}/\Delta E_{en})/(E_{ex}/E_{en})$, the relative stabilization coefficient of the exit energy, is obtained at high values of the latter. For example, if $n = 3$, $m = 5$, and $E_{en} = 90$ J, the value of β is ~5.7 whereas in the case of $n = 4$ and for the same values of m and E_{en} the value of β does not exceed ~1.6.

The efficiency of conversion of electrical energy into light for the whole HPA system in the case of optimal division is also considerably higher (Fig. 5). We can see that in the n = 3

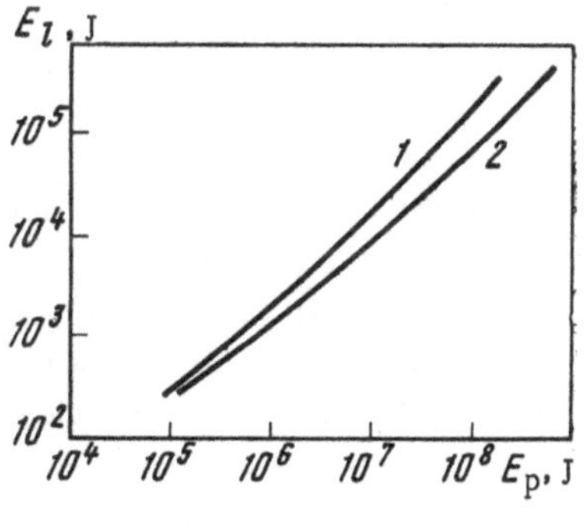

Fig. 5. Dependences of the total exit energy of a high-power amplifier on the energy of electrical storage devices used in the optical pumping of the active elements. Curves 1 and 2 correspond to two different values of the division factor of a beam (n = 3 and n = 4, respectively), subject to the following assumptions: $E_{en} = 90$ J, pump energy supplied to each amplifier 30 kJ, duration of pump pulses 1 msec, and diameter of active elements 45 mm.

case we can achieve an output energy of $\sim 10^5$ J if the pump energy stored electrically is ~ 45 MJ, whereas in the case of nonoptimal division the latter energy rises to ~ 200 MJ.

The reported calculations are based on the experimental information on the gain of four-lamp enclosures which were assumed to be the same in all the component cells of an HPA. The use of better enclosures and impovements in the circuit supplying optical pulse-discharge lamps, as well as the use of active elements with large end areas, should make it possible to increase η of such series-parallel laer systems to 2%.

Thus, the reported results show that lasers with series-parallel amplification stages can already provide (for reasonable stored energies) an output radiation of $\sim 10^5$ J energy which is needed for the realization of physically beneficial controlled fusion.

CHAPTER II

MASTER LASER AND SYSTEM OF PRELIMINARY AMPLIFYING STAGES

A master laser oscillator with preliminary amplifying stages, intended for use in high-power laser systems, should satisfy several specific requirements. Above all, switching of the Q factor of the master laser should be performed in a controllable manner because of the need for synchronization of a large number of diagnostic units characterized by high time resolution, possibility of simultaneous use of several lasers emitting at different wavelengths, and employment of systems for sudden introduction of a target into the laser focus and plasma heating. A laser pulse leaving the preliminary amplifying stages (PAS) should have the following parameters: the pulse duration should be continuously variable in the range 0.5-10 nsec, the pulse energy should be ~ 100 J, the beam divergence should not exceed 10^{-4} rad, and the energy contrast factor (ratio of the pulse energy to the total background) should be at least 10^5. The latter requirement arises from the fact that the background level at the exit from the system should not exceed the sublimation energy of the target.

Lasers with Q switching by an electrooptic Kerr modulator [19, 20] used so far have been unable to generate pulses with the required parameters. The situation is somewhat better in the case of lasers switched by Pockels cells. In this case the energy can be sufficiently high (~ 100 J) and the duration of the pulses can be $\tau \sim$ 3-4 nsec [5, 6, 17]. However, the beam divergence in such lasers is no better than $\sim 10^{-3}$ rad. The only exception is the laser system described in [16] capable of a divergence of $\sim 10^{-5}$ rad (for an energy of ~ 60 J), but this is achieved at the expense of a large optical path in the system, which is ~ 400 m. Laser systems employing a mode-locked master oscillator and a unit for selecting one beam can be used to produce sufficiently short pulses. However, these lasers are not controlled and their contrast factor does not exceed 10^3 [21].

We shall describe a single-mode laser, Q, switched by a Kerr modulator, which ensures – when used in conjunction with a system for controlling pulse duration and with preliminary amplifying stages – the parameters necessary for the injection of energy into a high-power amplifier (HPA) system.

§ 1. General Description of Apparatus

The optical part of the apparatus is shown in Fig. 6. A master laser oscillator produced pulses of ~ 35 nsec duration at midamplitude and the energy of these pulses was ~ 0.8 J. The beam divergence 2α was $\sim 2 \cdot 10^{-4}$ rad and the beam diameter was 4 mm. When the diameter of the aperture in a selecting screen in the resonator was increased, the output energy rose

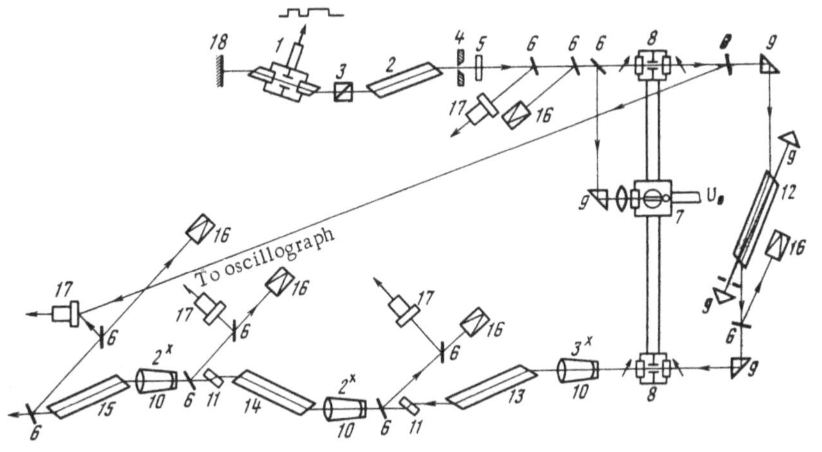

Fig. 6. Schematic diagram of the master laser oscillator and of the preliminary amplifying states (PAS) designed for parallel beams: 1) Q switch (modulator); 2) active element of the laser; 3) glass pile; 4) screen with aperture; 5) exit mirror; 6) beam splitters; 7) laser-triggered discharger; 8) shaping switches (modulators); 9) total-internal-reflection prisms; 10) telescopes; 11) saturable filters; 12-15) active elements in the amplifying stages; 16) calorimeters; 17) coaxial photocells; 18) dielectric mirror.

and the duration of the pulses became considerably shorter (~15 nsec). However, in view of the fact that the system for controlling the pulse duration included a fast-response electro-optic Kerr modulator [22] with an aperture of ~4 mm, the aperture in the screen was selected to be also 4 mm.

The system for controlling the pulse duration was a combination of a Kerr modulator and a laser-triggered discharger, which ensured accurate synchronization of the modulator with the beginning of a laser emission pulse [23-24]. This system made it possible to alter the duration of light pulses in discrete steps in the range 1-16 nsec. Typical oscillograms of the light pulses at the exit from this system were of the type shown in Fig. 7. A light pulse emerging from the duration control system reached the preliminary amplifying stages. These consisted of four stages: in the first stage an active element 300 mm long and of 20 mm diameter was used, whereas the other stages included neodymium rods of increasing diameter (20, 30, and 45 mm) with the pumped part of the rod ~600 mm long. The ends of all the rods were cut at the Brewster angle.

In the first two stages the beam was passed three times through each rod in order to increase the amplification efficiency. The total gain of each of these stages was ~30. The duration of the light pulses leaving these stages was ~2 nsec and the energy flux density was close to the safe limit for optical glass (~10 J/cm^2). When pulses of greater duration were ampli-

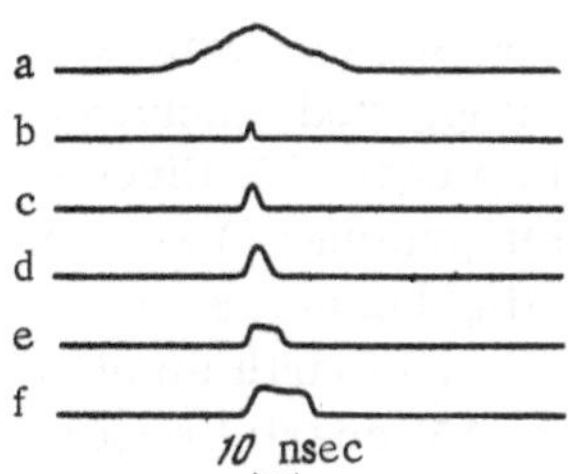

Fig. 7. Oscillograms of pulses: a) laser pulse; b-f) pulses of 1, 2, 4, 8, and 16 nsec duration at the exit from the shaping switch.

fied the gain was reduced by lowering the pump energy so that the energy at the exit did not exceed 1 J.

Further amplification of light pulses was carried out in two different ways using parallel and diverging beams. In the former case the flux density at the exit ends of the rods were matched to the gain and the directionality of the radiation was improved by placing Galilean optical collimators between the stages. The objectives of these collimators were lenses with minimal spherical aberrations. Three consecutive collimators with collimation factors of 3×, 2×, and 2× were used.

The presence of a large number of optical surfaces in the PAS system required special measures to prevent self-excitation. With this point in mind we placed an additional Kerr modulator between the first and second PAS stages; this modulator was controlled synchronously with the modulator in the system used to vary the duration of the light pulses. An electrical signal was applied to the Kerr modulator about 0.5 nsec ahead of the laser pulse and the duration of this signal exceeded the light pulse by 1 nsec. Additionally, saturable filters were located between the successive stages. These measures ensured that the contrast at the exit from the PAS system was at least 10^7.

The contrast was measured in control laser flashes with the discharger in the pulse duration system disconnected but using a 9-channel high-power amplifier (HPA) system [13]. The radiation emerging from the HPA system was focused on the surface of a spherical target of 100 μ diameter placed at the center of a vacuum chamber. Measurements indicated that the energy required to evaporate such a target was $\sim 10^{-3}$ J. Therefore, if the target was not damaged by control flashes, this indicated that the contrast at the exit of the PAS system was at least 10^6. This value should reach $\sim 10^7$ after amplification by the HPA system but this would require an experimental check.

The optical system of a laser used for amplification of diverging beams is shown in Fig. 8. The necessary matching of the laser beam diameter and the gain was ensured by a rapid increase of the cross section of the diverging beam. The required angle of divergence

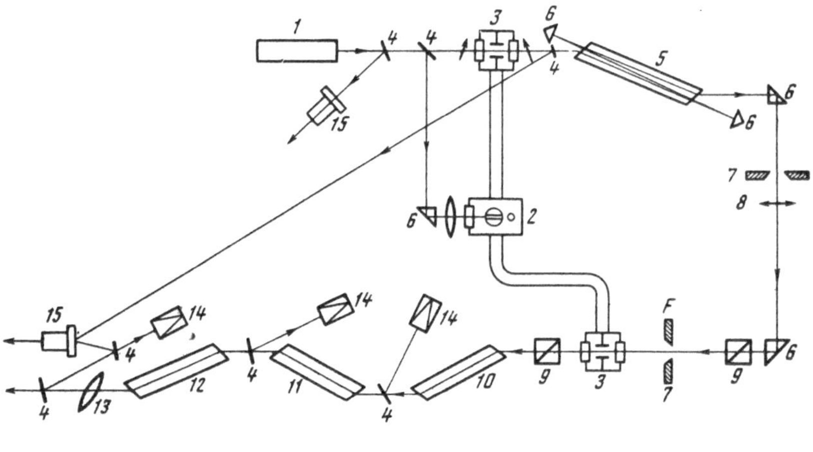

Fig. 8. Optical system of the laser and amplifying stages designed for diverging beams: 1) master laser; 2) laser-triggered discharger; 3) shaping Kerr switches (modulators); 4) beam splitters; 5) three-pass amplifying stage; 6) total-internal-reflection prisms; 7) screens with apertures; 8) positive lens producing diverging beams; 9) polarizers; 10-12) active elements of second, third, and fourth amplifying stages; 13) correcting lens; 14) calorimeters; 15) coaxial photocells.

was governed by the geometry of the active rods and by their distribution along the optic axis. When the total length of the system was $\sim 6 \cdot 10^2$ cm and the diameter of the exit rod was 4.5 cm, this angle was $2\alpha_0 \sim 7.5 \cdot 10^{-3}$ rad. Such a beam could be formed most readily by a negative lens with a focal length $f = d/2\alpha_0$, where d is the diameter of the beam reaching the lens. However, in view of the need to use a second modulator to ensure the necessary contrast of the radiation (the aperture of this modulator was about 4 mm) it was more convenient to use a positive lens. Moreover, it was possible to place a screen with a restricting aperture in the focal plane of this lens and this provided a very effective decoupling of the later stages from the three-pass amplifier and the master oscillator. In this case the second electrooptic modulator was placed behind the focus of the lens in the diverging beam. Glass piles were used as polarizers. The total absence of optical surfaces oriented at small angles with respect to the beam axis, which could participate in the self excitation of the system, made it possible to use this arrangement without additional decoupling (isolator) devices, so that the losses in the PAS system were reduced and the amplification efficiency was increased.

However, this arrangement had also important disadvantages Among the most important of these disadvantages were spherical aberrations, typical of wide beams, which occurred when the diverging beam passed through rods cut at an angle. The most serious among the aberrations was the astigmatism of the beam. The astigmatic segment at the focus of a lens with $f = 430$ mm was ~15 mm when the angle of divergence was $2\alpha_0 \sim 7 \cdot 10^{-3}$ rad. These considerable aberrations required special measures for their correction at the entry into the PAS system. This was done using complex lenses, which were combinations of spherical and cylindrical surfaces. A selection of the optimal radii of curvature of these lenses ensured that the divergence at the exit from the PAS system did not exceed $\alpha \sim 10^{-4}$ rad.

§ 2. Master Laser

We shall now consider the operation of the master laser oscillator (Fig. 6). The Q factor of the laser was switched by a Kerr modulator placed between the active element and a totally reflecting mirror. In contrast to the usual switching systems, the voltage applied to the modulator throughout the pumping time of the active element was turned off suddenly (this voltage was applied in the form of pulses because nitrobenzene withstood poorly static voltages). This switching system made it possible to reduce considerably the optical losses in the modulator caused by the inhomogeneity of the electric field in the Kerr cell. Using a special Kerr cell of a small optical thickness and with windows oriented at the Brewster angle, it was possible to reduce the total losses during a double pass in the resonator to a value not exceeding several percent.

The high optical quality of the resonator with such a modulator made it possible to use various methods for the formation of light pulses with given parameters. The transverse modes were selected by a screen placed inside the resonator. The aperture in the screen was selected by trial and error in such a way as to maximize the output energy of the laser and to minimize the divergence of the laser radiation; the restrictions imposed by the system for controlling the pulse duration were taken into account. The best results were obtained for an aperture of 4 mm in the screen. The angular distributions of the radiation emitted by the master laser were determined for different pump energies (Fig. 9).

The distributions obtained indicated that the total angle of divergence in which more than 90% of the beam energy was concentrated was $2\alpha \sim 1.06 \cdot 10^{-3}$ rad and the value of this angle did not change significantly for small changes in the pump energy. The diffraction divergence for an aperture of 4 mm diameter was $2\alpha_{diffr} = 0.65 \cdot 10^{-3}$ rad. Thus, the angular divergence of the master laser was ~1.6 times greater than the limit. However, since the output energy under these conditions was ~0.8 J, the brightness of the radiation was considerably higher than

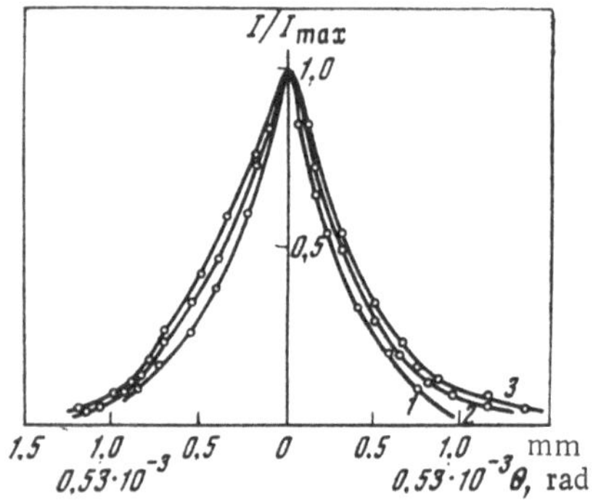

Fig. 9. Angular distributions of the laser radiation intensity obtained for pump energies of 14, 15, and 16 kJ (curves 1, 2, and 3, respectively).

the brightness obtained for single-mode lasers [25] and it amounted to $\sim 3 \cdot 10^{14}$ W $\cdot$ cm$^{-2} \cdot$ sr^{-1}. The number of transverse modes present in the laser radiation was estimated. The TEM_{mOq} mode was formed by plane waves travelling at an angle θ_m with respect to the resonator, where θ_m was found from the relationship $\sin\theta_m = m\lambda/2D$ (D is the transverse size of the resonator, which could be assumed to be equal to the diameter of the aperture in the screen for a plane-parallel resonator). The measured divergence angle of the beam was estimated to correspond to m = 4; thus, not more than 16 transverse modes were emitted by the laser.

§ 3. Investigation of Spatial-Temporal Coherence of Laser Radiation

The coherence of laser radiation is not related directly to the ability to concentrate energy as much as possible by focusing it on a target. However, the principal parameters of the radiation (its directionality and brightness) are functions of the coherence of the source. Therefore, studies of the coherence properties should make it possible to estimate correctly the possibility of using the radiation emitted by a given laser in studies of the interaction between light and matter.

The degree of spatial coherence of a laser beam, like its angular divergence, is related to the mode structure of the radiation. For example, in the presence of a single axial mode the radiation is coherent throughout the transverse cross section of a beam at any moment. The presence of several transverse modes impairs the spatial coherence.

The spatial coherence of the radiation emitted by the master laser was investigated in the usual way: the contrast γ of a pattern obtained with a Young interferometer was measured. The diameter of the interferometer apertures was $\sim 100\ \mu$ and the interferometer base d was varied from 0.23 to 2 mm. The spatial coherence was investigated with the interferometer located in the direct vicinity of the exit mirror of the laser and one of the interferometer apertures

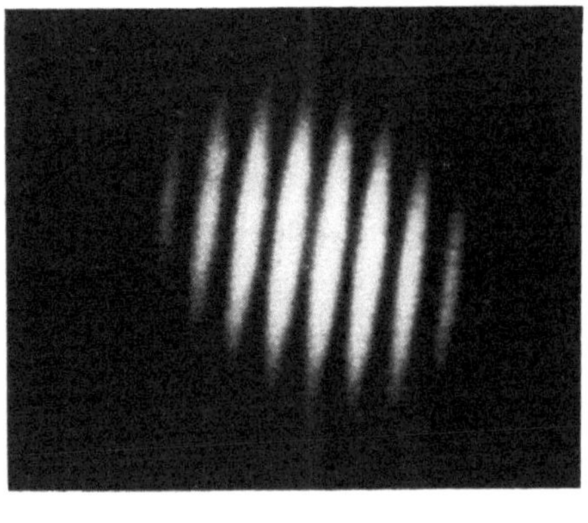

Fig. 10. Typical Young interferogram obtained for a distance of 1.38 mm between apertures.

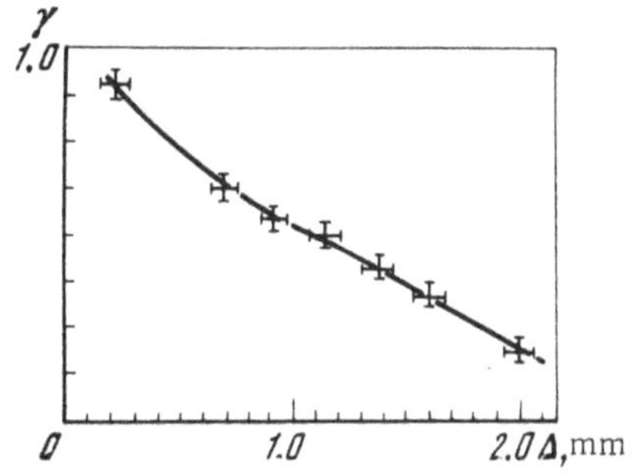

Fig. 11. Dependence of the interference fringe contrast on the distance between apertures on a Young interferometer (one aperture was located constantly on the beam axis).

was kept constantly on the laser beam axis. A typical interferogram obtained for $\Delta = 1.38$ mm was of the type shown in Fig. 10. The dependence of the interference pattern contrast on the distance between the investigated points in the beam was of the type shown in Fig. 11. It was found that near the beam axis ($\Delta \sim 300\ \mu$) the radiation was almost completely coherent in the spatial sense ($\gamma \sim 0.85$). The contrast fell at longer distances from the axis and it reached its minimum value (~0.35) at the beam edge. This relatively high spatial coherence over the beam cross section was evidence of a high directionality of the laser radiation and confirmed that only a limited number of transverse modes was present in the radiation.

The high spatial coherence of the output radiation enabled us to determine the structure of the wave front in one flash. In these measurements we used a Michelson interferometer one of whose arms was covered by a screen with an aperture of 100 μ diameter; this screen was placed in the plane of one of the mirrors. Thus, we recorded the pattern of the interference between the whole beam and a small part of the beam selected by the screen. A typical interferogram (Fig. 12) was composed of concentric rings. The radii of these rings were governed by the optical path difference between the interfering beams and by the phase of the radiation at the investigated points in the beam cross section.

The dependence of the ring radius ρ_m on the ring number (Fig. 13) was obtained by analyzing the interferogram in Fig. 12. This dependence was also calculated on the assumption of a plane wave front (curve 1 in Fig. 13). A comparison of the two curves in Fig. 13 indicated

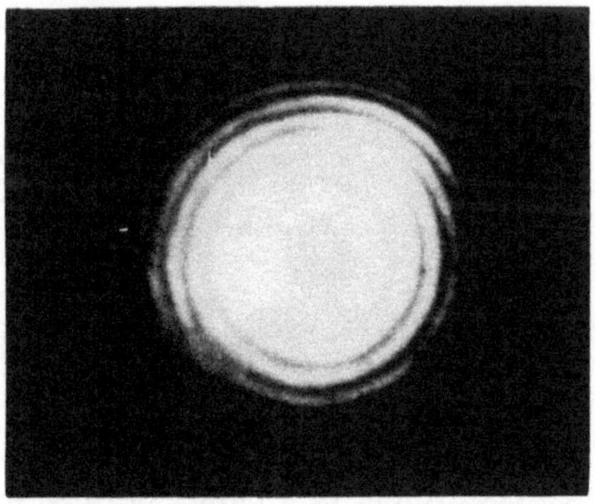

Fig. 12. Interference rings obtained for a beam restricted by an aperture of 100 μ diameter located on the axis of the beam.

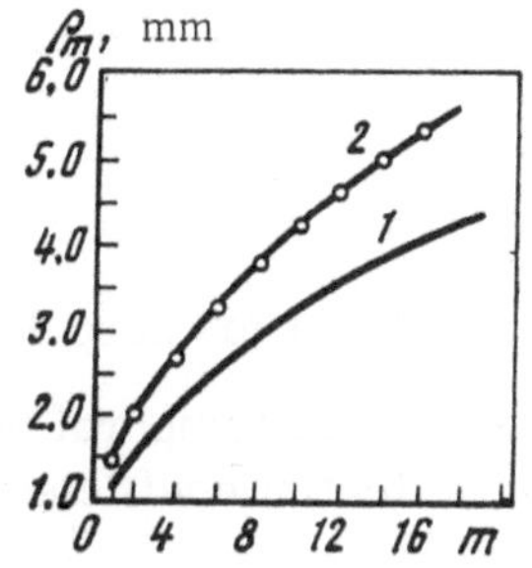

Fig. 13. Dependences of the ring radii ρ_m on the number m (m is defined by $\rho_m^2 = mB\lambda + \varphi_0/\pi$): 1) calculated curve for an ideal plane front; 2) experimental measurements.

that the wave front of the laser beam differed considerably from the plane form. The interferogram could be used to determine the divergence of the radiation by approximating the wave front with a large-radius sphere. This divergence was found to be $\sim 0.6 \cdot 10^{-3}$ rad, which was in fairly good agreement with the value deduced from measurements carried out in the far field of the laser radiation.

We shall now consider the question of the temporal coherence of laser radiation. This coherence depends not only on the construction of the laser but also on the conditions under which it is operated. A particularly complex situation is encountered in Q-switched lasers. The short duration of the laser pulses, high density of the radiation flux, and changes in the amplifying properties of the active medium during the deexcitation time result in considerable variation of the properties of the radiation during a pulse.

The main method used in the determination of the temporal coherence is the measurement of the contrast of a pattern obtained with an interferometer when the delay time between two interfering beams is varied. We shall consider the interference between two identical light pulses of duration T, one of which is delayed relative to the other by a time interval τ. The field of such a pulse is described by the expression

$$\left.\begin{aligned} E(t) &= AM(t)\exp j\,[\omega_0 t + \Theta(t)], \\ M(t) &= \begin{cases} 0 & \text{for } t \leqslant 0,\ t \geqslant T, \\ 1 & \text{for } 0 < t < T, \end{cases} \end{aligned}\right\} \tag{9}$$

where A is the field amplitude; ω is the frequency; $\Theta(t)$ is the phase shift during one pulse.

The total energy in the interaction between such pulses is

$$W(\tau) = \int_{-\infty}^{+\infty} |E(t) + E(t-\tau)|^2\,dt. \tag{10}$$

If the pulse radiation is completely monochromatic, integration of Eq. (10) gives

$$W(T) = 2A^2T + 2A^2(T-\tau)\cos\omega_0 T. \tag{11}$$

The interference pattern contrast is defined by the expression

$$\gamma = \frac{W_{max} - W_{min}}{W_{max} + W_{min}}. \tag{12}$$

Using Eq. (11), we find that

$$\gamma = (T-\tau)/T. \tag{13}$$

We shall now assume that the coherence of a laser pulse decreases rapidly from some finite value to zero. We shall denote the total duration of the coherent part of the pulse by T_c ($T_c \leq T$). We shall introduce the concept of the coherence coefficient of a pulse, defined as the ratio of the duration of the coherent part to the total pulse duration: $K = T_c/T$. The expression for the interference pattern contrast then becomes

$$\gamma = \frac{T-\tau}{T}K. \tag{14}$$

The maximum contrast for zero delay ($\tau = 0$) is equal to the coherence coefficient K.

We measured the temporal coherence using a Fabry–Perot interferometer formed by mirrors whose reflection coefficients were $R_1 = 10\%$ and $R_2 = 11\%$, respectively. The delay

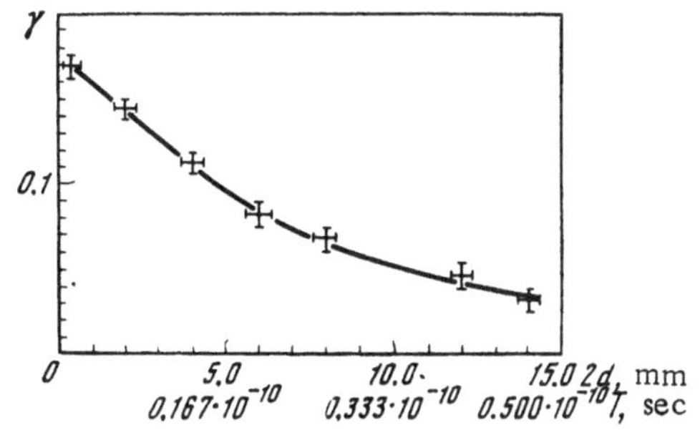

Fig. 14. Dependence of the interference fringe contrast on the delay time between interfering beams and on the path difference governed by the Fabry–Perot etalon base.

time τ was varied by altering the interferometer base d. We determined the dependence of the contrast on the delay time τ. This dependence on the delay time τ or the interferometer base d was of the type shown in Fig. 14. The maximum measured value of the contrast was 0.17 for d = 200 μ, i.e., the coherence coefficient of a pulse was 0.17. When the base was increased from 0.2 to 7 mm, the contrast decreased smoothly from 0.17 to 0.03 (Fig. 14). When the dependence in Fig. 14 was approximated by a linear function, measurements of the relative contrast yielded the coherence time τ_c:

$$\tau_c = \frac{\tau_1\gamma_1 - \tau_2\gamma_2}{\gamma_1 - \gamma_2}, \tag{15}$$

where γ_1 and γ_2 are the values of the contrast for the delay times τ_1 and τ_2, respectively. In our case the coherence time was $6 \cdot 10^{-11}$ sec, which corresponded to a coherence length $L_c \sim 1.8$ cm.

Thus, our investigation showed that the coherence of the radiation emitted by a Q-switched laser had properties quite different from those thermal sources and of cw lasers. The stimulated emission from such lasers was a combination of radiation of two types, one of which had a relatively high degree of coherence ($L_c \sim 1.8$ cm) and the other was practically noncoherent ($L_c \leq 10^{-2}$ cm), and the proportion of the coherent radiation could vary from one laser to another. The factors responsible for this nature of the laser radiation were not clear and further experiments would be required to determine such factors.

§ 4. Width of Emission Line

The coherence of laser radiation is linked directly to its monochromaticity. In plasma heating the monochromaticity of laser radiation is not essential but in complex laser systems it may be extremely useful if it is necessary to ensure reliable protection of the laser elements from the action of the radiation reflected by the plasma. The parameters of reflected light pulses (their duration and directionality) are close to the parameters of the incident laser pulses. The spectral characteristics are affected most strongly because of the Doppler shift of the emission line. This shift is due to the high velocities of an expanding plasma which may reach $\sim(3\text{-}5) \cdot 10^7$ cm/sec at temperatures of ~ 1 keV [26]. The Doppler shift at these velocities is approximately 5 Å. Thus, it should be possible to protect the laser from the reflected light pulses by the use of a narrow-band filter located at the exit from a laser emitting a narrow line. It follows that studies of the spectral composition of the laser radiation are of considerable interest.

We measured the width of the laser emission line using a Fabry–Perot interferometer. The reflection coefficients of the interferometer mirrors were 95 and 92%. The interferometer base was 1 mm and the dispersion range was ~ 5 Å. The interferometer resolution was deduced by recording the emission spectrum of an auxiliary gas laser emitting at $\lambda = 1.15\ \mu$; this resolution was at least 1/50 of the etalon order.

Fig. 15. Laser emission spectrum obtained with a Fabry–Perot etalon.

A typical interferogram is shown in Fig. 15. In addition to the main line, the radiation included also a weaker line which formed the second system of rings separated by $\Delta\lambda \sim 2$ Å from the main one. The intensity of the second line varied greatly from one flash to another but it remained considerably weaker than the intensity of the main line; in some cases the weaker line could not be recorded at all. The width of the main emission line, deduced from interferograms, was ~0.6 Å. The degree of monochromaticity $\Delta\lambda/\lambda$ was $5 \cdot 10^{-5}$. It should be noted that conventional lasers with the Q factor switched by a Kerr modulator subjected to a voltage applied to its plates were known to emit lines of width of at least ~10 Å. It was not clear why this considerable broadening occurred when voltage pulses were applied to the Kerr cell plates. The broadening could not be explained simply by the appearance of optical inhomogeneities due to the edge effects in the Kerr cell plates. It was clearly due to strong nonlinear effects which appeared in nitrobenzene in relatively strong electric fields.

§ 5. Amplifying Properties of Preliminary Stages and Parameters of Radiation Leaving These Stages

The amplification of a laser pulse in the preliminary amplifying stages (PAS) occurred at radiation flux densities across the beam close to the glass damage threshold. This gave rise to some special features in the amplifying properties of the system.

Figure 16 shows an oscillogram of a light pulse obtained at the exit from the modulator used in the pulse duration control system and of a pulse leaving the PAS. The shape of the pulses was recorded using a single coaxial photocell whose cathode received components of the laser radiation picked up from different points in the system separated by delay ensuring that the signals were resolved when scanned. It is clear from the oscillogram in Fig. 16 that the duration of the leading edge of a pulse decreased considerably as a result of amplification. This was evidence of nonlinear amplification in the PAS system.

Quantitative measurements of the gain of the exit stage in the PAS system, carried out for different values of the energy of the input signal, confirmed the nonlinearity of the amplification at high energy flux densities. Figure 17 shows the experimental dependences of the gain k on the energy of light pulses E_{in} at the amplifier input. The duration of the light pulses was different for curves 1 and 2 in Fig. 17. It is clear from these figures that the gain in this stage decreased rapidly when the input signal energy was increased from 3 to 12 J; this fall of the

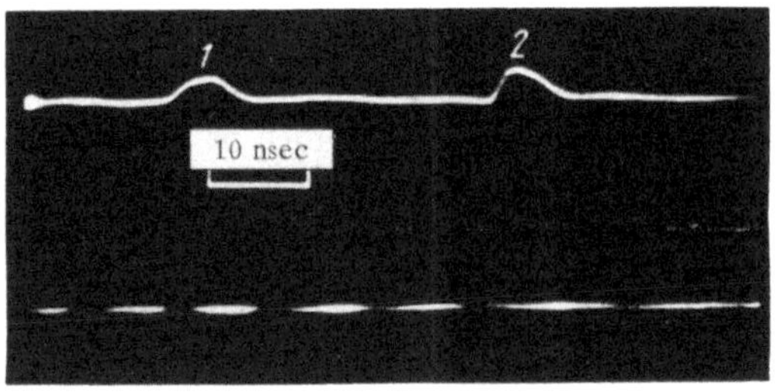

Fig. 16. Oscillogram of light pulses at the entry (first signal) and exit from the PAS system, illustrating the deformation of the leading edge by the nonlinear amplification. The lower trace represents timing marks of 10 nsec duration.

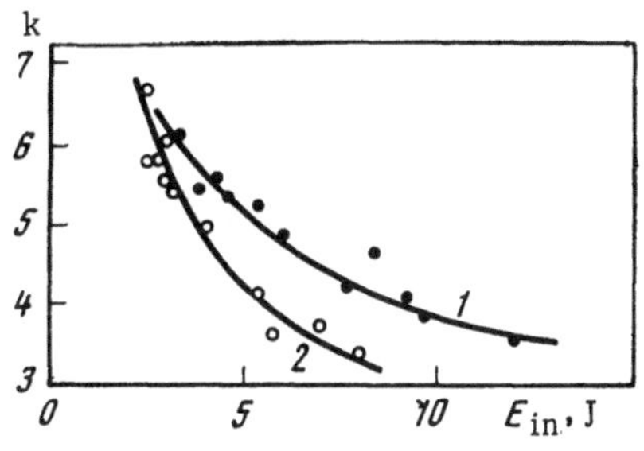

Fig. 17. Dependences of the gain k of the exit stage of the PAS system on the energy of laser pulses reaching this stage. Curve 1 corresponds to pulse durations $\tau_p \sim 4$ nsec, whereas curve 2 corresponds to $\tau_p \sim 2$ nsec. The pump energy is constant and amounts to 30 kJ.

gain was considerably faster for the shorter light pulses. Further increase of E_{in} in the range of input energies of ~20-50 J indicated that the gain varied slowly reaching ~3 for $E_{in} \approx 30$ J.

Table 1 lists the principal energy parameters of the radiation emerging from the PAS system pumped with diverging beams in the form of pulses of different durations. Here, E_1 and E_2 give the values of the energy at the exit of the three-pass amplifier and of the whole PAS system; k_1 and k_2 are the gains in the last three stages and in the whole system; q is the flux density; B is the brightness of the input radiation. It is clear from Table 1 that when the pulse duration was reduced from 16 to 1 nsec, the output energy fell by a factor of more than 3. This reduction in the gain of the system with decreasing pulse duration was due to the saturation of the lower nonradiative level of Nd, whose relaxation time was approximately 8 nsec for GLS-1 glasses [27].

Table 2 lists the radiation parameters determined at different points in the PAS system and representing the operation of the system as a whole. The values given were determined for diverging beams. Here, E_l and E_{pd} are the values of the energy at the exit from the master laser and from the pulse duration control system. Table 2 includes also the values of the en-

TABLE 1

τ, nsec	E_1, J	E_2, J	$k_1 \cdot 10^{-2}$	$k_2 \cdot 10^{-3}$	$B \cdot 10^{-17}$, $W/cm^2 \cdot sr$	$q \cdot 10^{-9}$, W/cm^2
1	—	35	—	7	—	—
2	0.9	60	0.7	5	1.2	4.0
4	0.9	90	1.0	3.7	0.95	3.0
8	1.0	110	1.1	2.4	0.58	1.8
16	1.1	140	1.3	1.5	0.37	1.2

TABLE 2

τ, nsec	E_l, J	E_{pd}, J	Stage No.	E, J	$\alpha \cdot 10^{-3}$, rad	S, cm^2	$B \cdot 10^{-16}$, $W/cm^2 \cdot sr$	$q \cdot 10^{-9}$, W/cm^2	k
2	0.8	0.025	1	0.9	0,53	0.2	0.25	2:25	36
			2	—	3.75	0,88	—	—	—
			3	21	3,75	2,5	—	4.2	—
			4	60	0,1	7.5	12	4.0	2,9
4	0,8	0,050	1	0,9	0.53	0.2	0,128	1,1	18
			2	5,8	3,75	0,88	—	1.65	6.5
			3	29	3,75	2,5	—	3.0	4.3
			4	90	0.1	7,5	9,5	3,0	3.2
8	0,8	0,075	1	1	0,53	0.2	0.071	0.62	13
			2	6.9	3.75	0.88	—	—	—
			3	34	3.75	2.5	—	1,7	—
			4	110	0.1	7,5	5.8	1,8	3,2
16	0,8	0.1	1	1.1	0,53	0.2	0.039	0.345	11
			2	—	3,75	0,88	—	—	—
			3	41	3.75	2.5	—	1.0	—
			4	140	0.1	7.5	3,7	1.2	3.4

ergy E, divergence α, beam cross section S, brightness B, flux density q, and gain in different stages in the PAS system.

Measurements of the radiation parameters at different points in the PAS system indicated that the coherence properties of the beam and the spectral composition of the radiation were practically unaffected by the amplification process.

CHAPTER III

HIGH-POWER AMPLIFYING STAGE WITH SERIES-PARALLEL CONFIGURATION

§ 1. Amplifying Stage Configuration

The laser beam leaving the preliminary amplifying stages (PAS) reached a high-power amplifier (HPA) in which a light pulse acquired the major part of its energy in spite of the relatively low value of the gain (k ~ 10). The optical system of the HPA was of the type shown schematically in Fig. 18. It consisted of two stages arranged in series. In the first stage the beam was divided into three components of approximately equal intensity and each of these was amplified additionally by one amplifier. In the second HPA stage each beam was again divided into three and amplified once more. Thus, nine beams appeared at the HPA exit and the parameters of each of them were governed approximately by the parameters of the radiation entering the HPA.

The HPA was constructed from four identical systems, each of which was a set of three parallel amplifiers whose parameters were similar to those of the exit part of the PAS. At the entry to each system there was a splitter for dividing the beam into three parts. The diameters of the active elements in the HPA was 4.5 cm and the length of the pumped part was ~600 mm. The ends of the rods were cut at the Brewster angle. The pump energy in each enclosure was 30 kJ and the maximum combined pump energy was ~900 kJ.

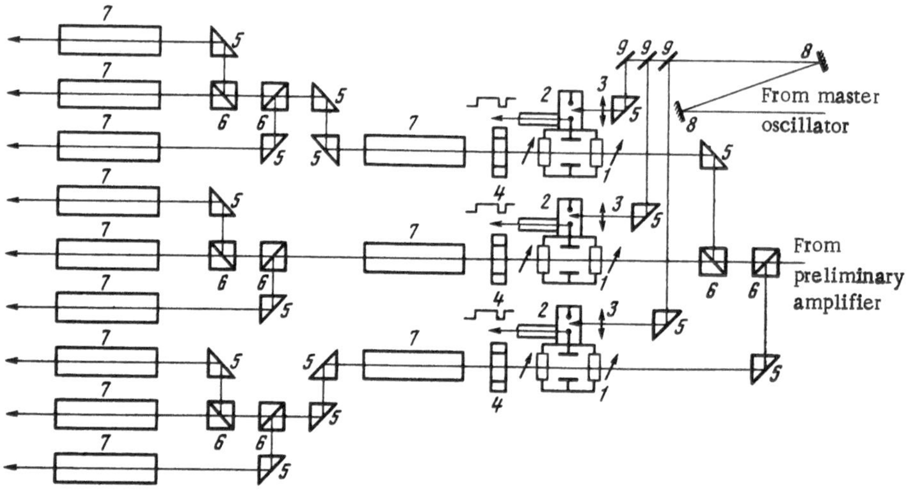

Fig. 18. Schematic diagram of the high-power amplifier: 1) Kerr switch; 2) laser-triggered discharger; 3) lens; 4) saturable filter; 5) total-internal-reflection prism; 6) beam splitter; 7) active elements of HPA; 8) optical delay mirrors; 9) plane-parallel plates.

§ 2. System for Dividing High-Power Beams

The division of a beam was performed in one plane throughout the HPA. The division factor was constant and equal to n = 3.

The division was performed by two different types of splitters. The operation of one of these splitters was based on the Fresnel reflection from plane-parallel glass plates with different refractive indices and use was made of the interference between the beams reflected from two parallel optical surfaces. Variation of the air gap between the two plane-parallel plates by amounts which were multiples of $\lambda/2$ made it possible to vary the reflection coefficient in the range $0-4R_0$, where R_0 was the reflection coefficient of one surface. This made it possible to achieve a relatively uniform division of the entry beam into three parts. However, elements of this kind had a number of important shortcomings. These included particularly the complexity of their use in the series division of linearly polarized laser beams in two orthogonal planes. Moreover, when these elements were used at the laser beam focus, the focal spot split into several smaller spots and the size of each of them was governed by the divergence of the arriving beam. We shall now consider this effect in some detail.

Let us assume that a laser beam of brightness B has a divergence angle α and that all the laser energy is concentrated within this angle. Subsequent propagation of the laser radiation can be considered by replacing the laser with an effective source point of light of equivalent brightness equal to the brightness of the original beam. The size of this source is given by

$$d=\frac{4}{\pi\alpha}\sqrt{\frac{E}{\tau B}}, \tag{16}$$

where E is the energy and τ is the duration of the light pulses. Reflection of a diverging beam from a plane-parallel plate of thickness l produces, in the image space, two sources whose axes are displaced relative to one another by a distance d given by

$$\delta=2l\frac{\sin\varphi\cos\varphi}{\sqrt{n^2-\sin^2\varphi}}. \tag{17}$$

If $\delta \geq d$, the effective brightness of the reflected beam decreases. We can ensure that this reduction does not exceed 10% of $1/n$ (n is the division factor) by using a configuration in which the gap thickness l does not exceed

$$l\leqslant 5\cdot 10^{-2}\frac{1}{\pi\alpha}\sqrt{\frac{E}{B\tau}}\frac{\sqrt{n^2-\sin^2\varphi}}{\sin\varphi\cos\varphi}. \tag{18}$$

If $B = 10^{17}$ W·cm^{-2}·sr^{-1}, $\alpha = 10^{-3}$ rad, E = 60 J, τ = 2 nsec, and $\varphi = 45°$, the permissible value of l is $\sim 2.4 \cdot 10^{-2}$ cm.

It would be difficult to construct an optical beam splitter with a gap of this thickness between plane-parallel plates, particularly in view of the fact that the diameters of the beams emitted by the currently available lasers can be as large as ~60 mm. Therefore, the effective brightness of the radiation at the exit of the HPA was increased using optical beam splitters based on the sudden change in the reflection coefficient at the boundary separating two media with different refractive indices; this change occurred when the angle of incidence was close to the total internal reflection angle [28]. The equivalent thickness of such a splitter can be reduced to ~50 μ, so that it could be used in beams of $\sim 10^{18}$ W·cm^{-2}·sr^{-1} brightness with a divergence angle $\sim 10^{-3}$ rad.

§ 3. Contrast of Radiation Emerging from High-Power Amplifier

The amplification in the HPA occurred at flux densities close to the limit set by the damage threshold of optical glass and, consequently, the gain in each stage did not exceed 3-4

[29]. Since under these circumstances the gain of weak signals could reach ~10, it was necessary to take special measures to maintain the required constrast. This was done by placing a special electrooptic Kerr switch of 4.5 cm aperture at the entry to the HPA. The switch was controlled as follows. A voltage pulse of ~10 μsec duration was applied simultaneously to the switch and to the triggering circuit of the pulse-discharge pump lamps used in the amplifying stages. Polarizers (glass piles) were parallel to one another so that the switch transmission was minimal for a given voltage. The voltage pulse applied to the switch was terminated ~5 nsec before the arrival of the light pulse. This was achieved using a laser-triggered discharger (LTD) which was controlled by a part of the pulse generated in the master oscillator. The required synchronization was achieved introducing an optical delay.

The parameters of the discharge circuit of the LTD were selected so that damped electrical oscillations were excited in the discharger. The oscillation period was ~20 nsec. The switch was then "open" for ~5 nsec, so that during the passage of the light pulse the switch transmission was maximal. This switch control scheme also provided protection of the master oscillator and of the LTD from the damaging influence of the radiation reflected from the target and amplified by the HPA system. Such protection was provided by matching the electrical oscillation period in the discharger to the time taken by the light pulse to travel twice from the switch to the target.

All these measures ensured that the contrast at the exit of the system was $\sim 10^7$. When fairly long light pulses (~8-16 μsec) were amplified in the HPA, the electrooptic switches were replaced by saturable filters with a low initial transmission.

The use of saturable filters in the exit stages of the HPA was undesirable in the case of amplification of short light pulses because the optical characteristics of these filters were not constant and because the contrast at the exit from the HPA decreased when the duration of the light pulses was reduced. A saturable filter with a reasonable initial transmission had practically no effect on the contrast of the pulse radiation relative to the background energy under Q-switching conditions.

§ 4. Parameters of Radiation at the Exit from High-Power Amplifier

The energy of the exit radiation, shape of the light pulses, and distribution of the intensity in the far zone (including the average value of the divergence angle) were monitored at different points in the HPA and recorded in each experiment.

The energy characteristics of the HPA obtained for different degrees of amplification and for light pulses of different durations τ are listed in Table 3. Here, E_1, E_2, and E_3 represent, respectively, the energy of a pulse entering the HPA and the total energy of three and none beams in the first and second stages of the HPA; B is the brightness of the radiation leaving the HPA; k is the total energy gain of two stages; q is the maximum density of the radi-

TABLE 3

τ, nsec	E_1, J	E_2, J	E_3, J	k	$B_1 \cdot 10^{-16}$, W/cm²·sr	$B_2 \cdot 10^{-16}$, W/cm²·sr	$q \cdot 10^{-16}$, W/cm²	$E_{el} \cdot 10^{-5}$, J	η, %
1	35	120	380	10.9	—	—	—	3.6	0.10
2	60	195	600	10.0	12	8.0	4	3.6	0.17
4	90	270	800	8.9	9.5	6.8	3.2	3.6	0.22
8	110	350	1000	9.1	5.8	5.0	2.5	3.6	0.28
16	140	400	1300	9.3	3.7	3.6	1.8	3.6	0.36

ation flux reaching the target; E_{el} is the electrical energy of a capacitor bank; $\eta = E_{ex}/E_{el}$ is the efficiency of the HPA. The results given in Table 3 indicate that the brightness of the radiation amplified by the HPA remained almost constant for pulses of 16 nsec duration and decreased somewhat (by a factor of about 2) for $\tau = 2$ nsec. The absolute brightness remained highest for the shortest duration and it was $\sim 4.3 \cdot 10^{16}$ W $\cdot$ cm^{-2} $\cdot$ sr^{-1}. The reduction in the brightness was due to a small increase in the divergence of the radiation caused by the nonlinear effects in the active medium and in nitrobenzene used in optical switches.

§ 5. System for Focusing Radiation on a Target

All nine beams emerging from the HPA were directed by a set of prisms to a focusing system (Fig 19). Each beam was focused by a separate objective each of which had two components. The long-focus entry components were mounted directly in the prism system. The exit components of the objectives were fixed to the walls of the vacuum chamber and suitably sealed. The geometric axes of the focusing objectives were oriented in such a way as to ensure a relatively uniform irradiation of a spherical target. The caustic diameter of each objective did not exceed 30 μ, which made it possible to irradiate targets of diameters ranging from 1 mm to 60 μ.

The prism collection system included also means for compensating for the differences between the optical paths in each beam. This compensation was achieved by an optimal distribution of the laser beams between the focusing channels and by using optical compensators each of which consisted of two right-angled total-internal-reflection prisms. A system of this kind ensured compensation to within 5.5 cm, which was of the order of the differences between the times of arrival of light pulses ($\sim 1.5 \cdot 10^{-10}$ sec) on the target.

The focusing system was aligned using combinations of gas lasers emitting at wavelengths of 0.63 and 1.15 μ. The correctness of the alignment was checked at the emission wavelength

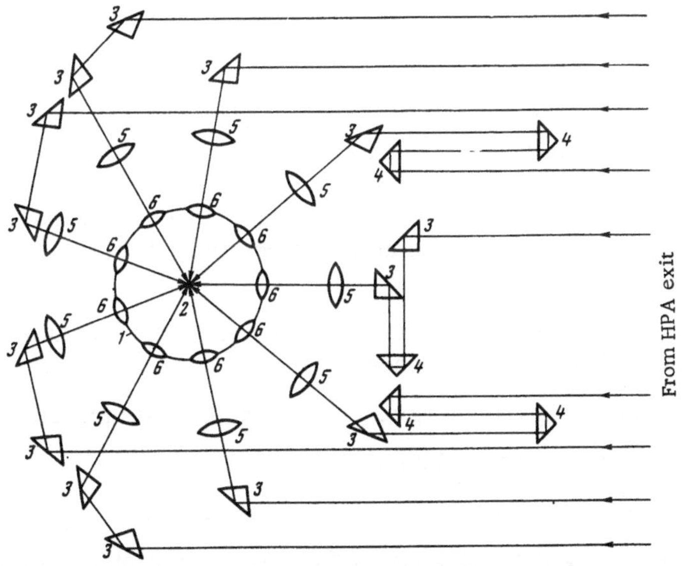

Fig. 19. Optical system for focusing HPA radiation on a target and compensating differences between optical paths (for the sake of simplicity the system is shown in one plane): 1) vacuum chamber; 2) target; 3) rotatable total-internal-reflection prism; 4) compensation prism; 5) first component of objective; 6) second component of objective.

of the neodymium laser. The construction of the focusing system and the control apparatus ensured that the focal plane coincided with the target to within $\pm 100\ \mu$ and that the optical axes of the objectives were positioned on the target to within $\pm 20\ \mu$.

The large number of the optical surfaces in the focusing system (~50) resulted in considerable losses of the radiation energy in the high-power laser because of the Fresnel reflection. The total transmission of the focusing system was 75%. When allowance was made for this transmission, it was found that in the case of pulses of 2 nsec duration the focusing system ensured a radiation flux density up to $1.5 \cdot 10^{16}$ W/cm^2 on the surface of a plane target and $2 \cdot 10^{15}$ W/cm^2 on the surface of a spherical target. In the spherical case the rate of evolution of the energy in the heated plasma reached 10^{18} W/cm^3 if the radiation was absorbed completely in the target. For a target of 60 μ diameter the experimentally achieved rate of evolution of energy was $\sim 5 \cdot 10^{17}$ W/cm^3.

CHAPTER IV

INVESTIGATION OF THE PARAMETERS OF A PLASMA FORMED AS A RESULT OF SPHERICAL IRRADIATION OF AN ISOLATED SOLID TARGET

§ 1. Vacuum Chamber. Diagnostic Apparatus

The laser radiation emerging from the exit of the high-power amplifier (HPA) system was focused on the surface of a solid target located at the center of a vacuum chamber. In the series of experiments under discussion we used spherical targets made of ordinary $(CH_2)_n$ and deuterated $(CD_2)_n$ polyethylene. The target diameter was selected in accordance with the experimental conditions and it could be varied within the range 55-850 μ. The maximum deviation of the target shape from an ideal sphere did not exceed 20%.

The target was suspended at the center of the vacuum chamber by a spherical positioning system in which the target was attached to a thin (5 μ diameter) filament made of $(CD_2)_n$; this filament was ~2 mm long and it was supported by a steel needle which was part of the positioning system. The laser focus was made to coincide with the center of the target by moving the target along three coordinates and this was checked with several microscopes. The vacuum chamber could be filled with various gases at pressures from 1 atm to 10^{-1} mm Hg. The lowest pressure in the chamber was 10^{-6} mm Hg.

Figure 20 shows schematically the vacuum chamber and the main diagnostic apparatus used in each experiment. The main components were:

apparatus for investigating the energy supply to the heated target;

unit for x-ray diagnostics of the plasma;

apparatus for investigating the neutron yield of the plasma;

apparatus for measuring the reflection of the laser radiation from the plasma;

interferometric unit for investigating the plasma density profile.

Moreover, Fig. 20 shows schematically two out of nine target irradiation channels (elements 3, 4, and 5). The interferometers are not included in Fig. 20.

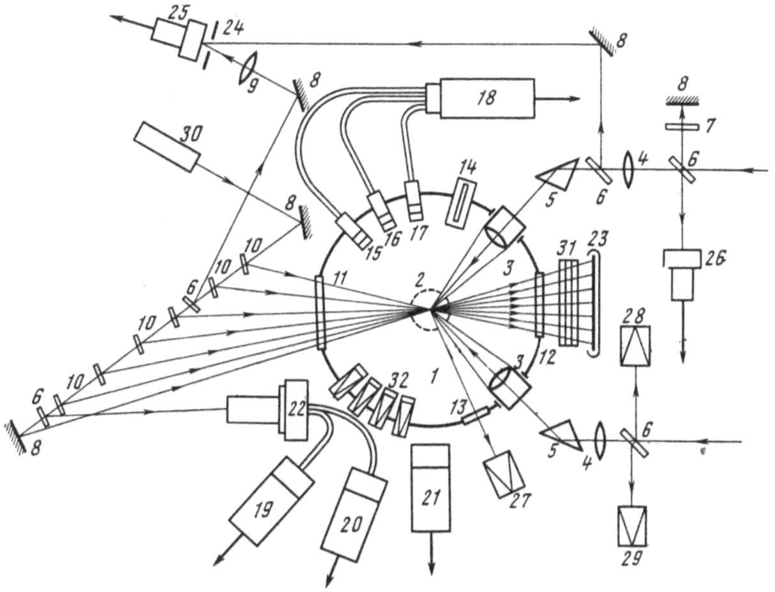

Fig. 20. Vacuum chamber and diagnostic apparatus: 1) vacuum chamber; 2) target; 3) short-focus component of focusing objective; 4) long-focus lens; 5) total-internal-reflection prism; 6) plane-parallel plate; 7) optical filter; 8) mirror; 9) lens; 10) two-component optical-delay mirrors; 11) diagnostic (entry) window of chamber; 12) exit window; 13) chamber window; 14) cassette with nuclear photographic emulsion; 15-17) x-ray "plasma thermometer" probes; 18) photomultiplier; 19-21) neutron scintillation counters; 22) cassette with filters and optical fiber; 23) cassette for shadow photography; 24) diaphragm; 25, 26) coaxial photocells; 27-29) calorimeters; 30) ruby laser; 31) cassette with filters; 32) four-channel x-ray calorimeter.

§ 2. Investigation of the Efficiency of Energy Supply to a Heated Target

When an isolated target is heated by the radiation of a multibeam laser, the energy balance can be represented in the form

$$E_l = E_{r1} + E_{r2} + E_{abs} + E_{tr}, \tag{19}$$

where E_l is the laser radiation energy; E_{r1} are the energy losses in the beam convergence system; E_{r2} is the energy reflected from the target; E_{abs} is the energy absorbed by the target; E_{tr} is the energy transmitted by the target without absorption.

A direct determination of the energy balance under spherical irradiation conditions is a difficult technical task. Special difficulties are presented by the measurements of E_{r2} and E_{tr}. In the spherical configuration when a target is irradiated with many beams, these measurements are practically impossible. Therefore, in the present investigation the energy absorbed in the target was determined by a method based on an investigation of the dynamics of motion of shock waves formed in the atmosphere of a residual gas as a result of expansion of a hot dense plasma. The propagation of shock waves was investigated using the method of high-speed multiframe shadow photography [30]. Typical shadowgraphs obtained under different experimental conditions are shown in Figs. 21 and 22.

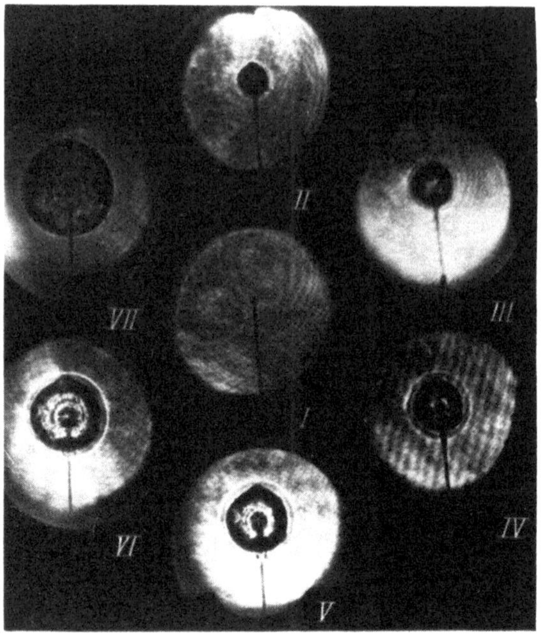

Fig. 21. Shadow photographs of shock waves obtained after the following delays relative to the beginning of the heating (nsec): I) 2; II) 20; III) 45; IV) 70; V) 140; VI) 210; VII) 460. The target was a $(CH_2)_n$ sphere of $r_0 = 250\,\mu$ radius. The energy of the laser radiation was 560 J. The duration of the laser pulses was 2 nsec. The pressure of the residual gas (air) was ~15 Torr. The diameter of each frame was 57 mm.

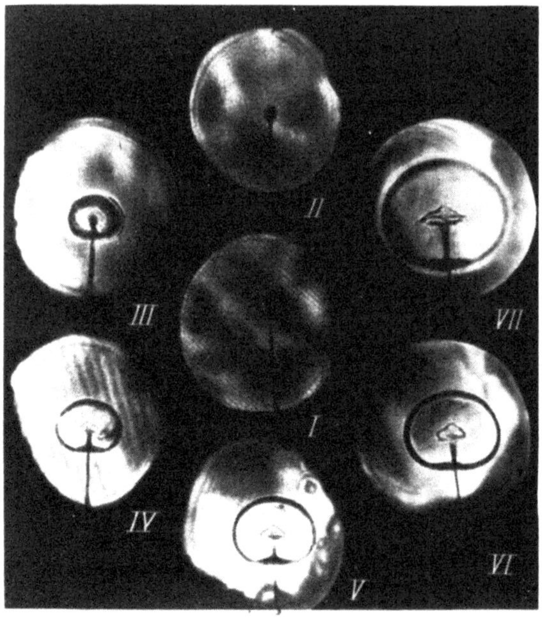

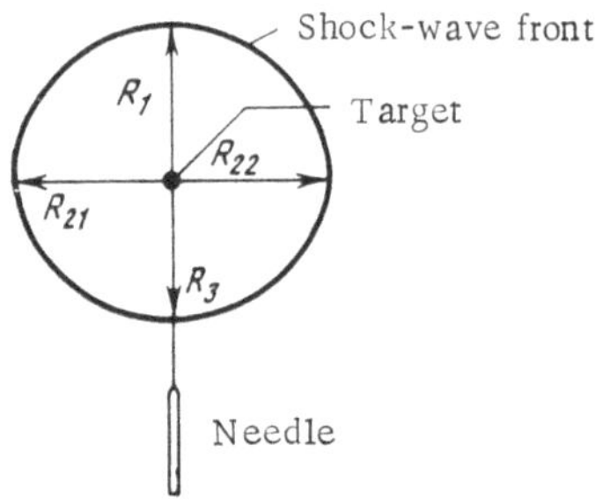

Fig. 22. Shadow photographs of shock waves in deuterium. The delay of the exposure of the first frame, relative to the beginning of the heating, was 22 nsec. The time intervals between the frames were the same as in the Fig. 21. The target was a $(CD_2)_n$ sphere of $r_0 = 125\,\mu$ radius. The laser energy was ~310 J and the duration of the laser pulses was 3 nsec. The residual gas pressure was ~12 Torr. The picture on the right shows schematically the arrangement used in the measurement of the radii of shock-wave fronts by averaging: $R_2 = (R_{21} + R_{22})/2$.

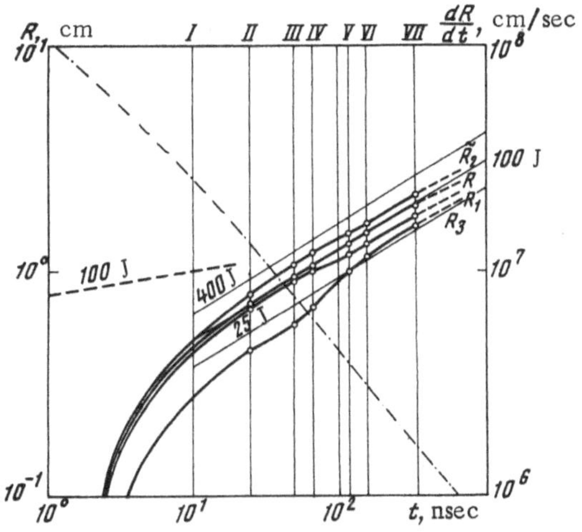

Fig. 23. Typical Rt diagrams of a shock wave in a deuterium atmosphere generated by laser radiation of 330 J energy in the form of pulses of 2 nsec duration. The target was a $(CD_2)_n$ sphere of 85 μ radius. The deuterium pressure was 10 Torr. The velocity of the shock-wave front is represented by the right-hand ordinate. The numbers above the curves represent the explosion energies in joules deduced from the point explosion model (on the right-hand side of the figure) and the thermal energy deduced from the instantaneous point heat evolution model (on the left). The chain curve is the velocity of the shock-wave front.

In an analysis of the shadow photographs the radius of a shock-wave front was taken to be an effective value R^* corresponding to experimentally determined parameters averaged over the volume of the gas affected by the shock wave. Figure 23 shows typical Rt diagrams of R_1, $\tilde{R}_2$, R_3, and R^* obtained by an analysis of a photograph shown in Fig. 22 (the notation is explained by a schematic diagram included in Fig. 22). The roman numbers on the abscissa in Fig. 23 denote the frame numbers. Continuous thin curves on the right-hand side of Fig. 23 represent the self-similar solutions of the problem of an instantaneous point explosion characterized by energies of 25, 100, and 400 J and they are used to determine the energy supply to the plasma.

It is clear from the curves in Fig. 23 that the early stages of the motion of a shock wave are far from self-similar. The front of a shock wave moves in accordance with a law close to the self-similar form only during the later stages, beginning from a moment when the mass of the gas affected by the shock wave becomes much greater than the mass of the evaporated target. This motion of a shock wave is due to the fact that in the case of shock waves formed by a laser jet during the earlier stages the energy is mainly dissipated by the gasdynamic motion of matter of the heated target and only after some time this energy is transferred to a shock wave. For example, our estimates indicated that at a moment t_i when the mass of the gas affected by the shock wave was 5 times as large as the target mass, the energy represented by the gasdynamic motion of the target material was $\sim 20\%$. The time t_i was governed by the experimental conditions and when the residual gas was deuterium (D_2), this time was given by

$$t_i \approx 1{,}4 \cdot 10^8 r_0^{5/2} p^{-1/2} E^{-1/2} \text{ nsec}, \qquad (20)$$

where r_0 is the initial target radius (cm), p is the gas pressure (Torr), and E is the explosion energy (J).

Another source of error in the measurement of the energy absorbed by a plasma is the intensive emission of x rays from a hot dense plasma in the course of its heating. According to the calculated and experimental results, presented in detail in the next section, the x-ray energy reach ~15-25% of the absorbed energy. X rays were absorbed weakly by the gas surrounding the target and did not influence significantly the process of formation of a shock wave.

Thus, the energy deduced from the experimentally obtained Rt diagrams represented the lower limit of the energy input. In the determination of the true value of the energy absorbed by a plasma one should make corrections allowing for the main loss mechanisms. In the present investigation the radiative losses were determined with x-ray calorimeters and the proportion of the energy represented by the gasdynamic motion of a plasma jet was estimated numerically. For example, in the case of the experiment represented by the Rt diagrams of Fig. 23 the shock-wave energy was ~100 J, the jet energy was ~15 J, and the x-ray radiation energy was ~65 J. Thus, in these experiments the total absorbed energy was ~180 J, which represented ~72% of the laser radiation energy reaching the target and amounting to ~250 J. The error in the measurement of the absorbed energy did not exceed 15%. The rest of the energy was probably carried away by the laser radiation reflected from the target or transmitted without absorption. Measurements indicated that in these experiments the reflection coefficient did not exceed 10% of the incident laser radiation energy.

In measurements of the energy input based on a study of a shock wave generated by an expanding plasma in a residual gas atmosphere it was important to select correctly the time interval during which the experimental Rt diagrams were compared with the law of motion of a shock wave produced by an instantaneous point explosion in a homogeneous atmosphere. The nature of motion of the shock wave during the initial stages was governed by the transfer of energy from the plasma to the surrounding gas and it could differ strongly from the self-similar law. A strong thermal wave could form initially as a result of the electronic thermal conductivity. The self-similar solution for spherically symmetric propagation of heat from an instantaneous point source was deduced in [31]

$$R_t \approx 0.59 \cdot E_t^{5/19} p^{-7/19} t^{2/19}, \tag{21}$$

where R_t is the radius of a thermal-wave front; E_t is the energy represented by heat evolution p is the gas pressure (Torr); t is the time (nsec). For $E_t = 10^2$ J, p = 10 Torr, and t = 1 nsec the radius of a thermal-wave front should be $R_t \approx 0.85$ cm. A thermal wave of this kind, traveling at a velocity $\dot{R}_t$ exceeding considerably the velocity of sound in a medium, could have an important influence on the motion of a shock wave overtaking the thermal wave during the later stages.

The law of propagation of a shock wave during later stages could also differ from that applicable to a point explosion because of the increasing losses due to the dissociation and ionization of the ambient gas molecules and also because of the emission of radiation from the gas affected by the shock wave. The total energy losses due to the ionization and dissociation at a moment t were

$$E_{id} = 1{,}4 \cdot 10^8 E^{3/5} p^{2/5} t^{6/5} \tag{22}$$

and for $E = 10^2$ J, p = 10 Torr, and $t = 3 \cdot 10^2$ nsec, the ionization and dissociation losses were $E_{id} \sim 25$ J.

The emission of radiation from a gas during the later stages could be estimated by assuming that all the mass of the gas was concentrated in the shock-wave front in a layer $\delta R = [\gamma - 1)/3(\gamma + 1)]/R$ thick. Applying a quasiclassical expression for the power of bremsstrahlung and recombination radiation emitted by a completely ionized deuterium plasma, we found that the time dependence of the energy of the continuous radiation could be represented in the form

$$E_{\text{cont}} = E^{ff}_{\text{cont}} + E^{fb}_{\text{cont}} = \frac{3 \cdot 10^7}{(\gamma - 1)^{1/2}} E^{1/5} p^{6/5} t^{8/5} + \frac{(\gamma + 1)^2}{(\gamma - 1)^{3/2}} 1.05 \cdot 10^{15} E^{-1/5} p^{8/5} t^{14/5}, \tag{23}$$

with the first term representing the bremsstrahlung and the second the recombination radiation. For the values of E and p given above and for $\gamma = 5/3$ the above expression gave $E_{\text{cont}} \approx 10$ J at t = 500 nsec; the bremsstrahlung represented ~1.9 J and the recombination radiation ~8.1 J.

Equation (23) is valid for a completely ionized plasma, which is true of deuterium at temperatures of several electron-volts. The plasma temperature in the radiating layer calculated using the point explosion approximation is given by

$$T = 4.92 \cdot 10^{-7} \frac{E^{2/5}}{p^{2/5} t^{6/5}} \frac{\gamma - 1}{(\gamma + 1)^2} \text{ eV}, \tag{24}$$

which shows that between 10^2 and 10^3 nsec the temperature should fall from 30 to 2 eV (at t = 500 nsec, T = 4.5 eV). At these temperatures and for a density governed by the maximum compression in the shock-wave front, the continuous radiation energy E_{cont} should represent ~50% of the line radiation E_{lin} [32]. Approximating, for pressures of ≤ 50 Torr and temperatures 3-30 eV, the ratio of the line radiation power to the total radiation power by a hyperbolic dependence, we obtain the following expression for the total radiation energy:

$$E_0 = E_{\text{lin}} + E_{\text{cont}} = \frac{3 \cdot 10^7}{\sqrt{\gamma - 1}} E^{4/5} p^{6/5} t^{8/5} \frac{[(\gamma - 1) E^{2/5} + 3.5 \cdot 10^7 (\gamma + 1)^2 p^{2/5} t^{6/5}]}{[(\gamma - 1) E^{2/5} - 6.1 \cdot 10^6 (\gamma + 1)^2 p^{2/5} t^{6/5}]}. \tag{25}$$

It is clear from Eq. (25) that the radiation losses should increase rapidly during the motion of a shock wave. This should alter the nature of a shock wave and result in increasing deviation of the motion from the self-similar law. Thus, Eqs. (22) and (25), together with Eq. (20), define the range of time during which the law of motion of a shock wave formed as a result of expansion of a laser jet into a residual gas atmosphere is in agreement with the point explosion theory.

For a fixed observation time the residual gas pressure is restricted by certain conditions and Eq. (20), on the one hand, and Eqs. (22) and (25), on the other, give, respectively, the upper and lower limits of the pressure. In the experiments described below the pressure at t_{max} = 334 nsec could vary between $p_{min} \approx$ 5 Torr and $p_{max} \sim$ 15 Torr for r_0 ~ 100 nsec and E_{abs} ~ 100 J.

We shall now consider the experimental results. The absorption of the radiation emitted by a dense plasma was investigated using a laser energy of ~300 J in the form of pulses of ~1.5-16 nsec duration. A target was made of deuterated polyethylene and its radius was within the range r_0 ~ 30-250 μ. We investigated experimentally the dependence of the efficiency of the transfer of the radiation energy to the laser plasma

$$\eta = \frac{E_{\text{abs}}}{E_{\text{inc}}} = \frac{E_{\text{abs}}}{E_l - E_{r1}} \tag{26}$$

(E_{inc} is the energy of the laser radiation incident on the target) on the parameters of the light pulses. Figure 24 shows the dependence of the efficiency of the energy input on the initial target radius. The duration of the laser radiation pulses τ_l is the parameter of the curves in this figure. It is clear from the curves that for targets with an initial radius $r_0 \geqslant 100$ μ an increase in the pulse duration reduced the energy transfer efficiency. For example, in the case of a target with r_0 = 120 μ irradiated with pulses of τ_l = 1.5 nsec duration the absorption efficiency was ~75% and when the duration was increased to 8 nsec, the efficiency fell to 40%.

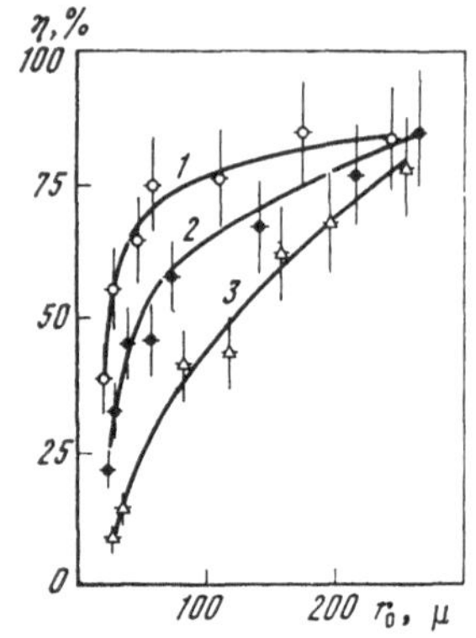

Fig. 24. Dependences of the efficiency of energy supply to a plasma on the initial target radius, plotted for different durations of the laser pulses τ_l (nsec): 1) 1.5; 2) 3; 3) 8. The pressure of the deuterium surrounding the target was ~8-12 torr. The energy of the laser radiation was ~290-350 J.

This dependence was most probably due to the target "bleaching" effect in which, beginning from a certain moment, the incident radiation passed through the plasma without significant absorption. This effect was due to a rapid reduction in the electron density in the plasma as a result of the gas dynamic expansion until the density fell below the critical value for the heating radiation frequency. The strong fall of the energy transfer observed for targets of $r_0 < 50\ \mu$ radius was clearly due to the poor focusing of the laser radiation on the target because in this case the target size became comparable with the size of the focal spot and the probability of some radiation bypassing the target was greater than for larger targets.

Figure 25 shows the dependences of the efficiency of absorption of the laser radiation on the duration of the laser pulses. The target size is the parameter of the curves. It was found that the absorption decreased more slowly for larger targets. This behavior was due to an increase in the characteristic gasdynamic time $t_{gd} = r_0/V_s$ (V_s is the velocity of sound in the plasma) when small targets were replaced with larger ones; this resulted in a weakening of the "bleaching" effect.

Information on the energy supplied to a plasma could be used to find the average energy per particle $\bar{\varepsilon}$ in a target of known composition and size. The results obtained in this way are plotted in Fig. 26 as the dependence of $\bar{\varepsilon}$ on the density of the laser radiation flux reaching the target. This flux density was varied by altering the target size and the sharpness of focusing of a beam ensuring a uniform irradiation of the target surface. The duration of light pulses in these measurements was fixed and it amounted to ~1.5 nsec. It is clear from the dependence in Fig. 26 that the average energy per particle increased approximately proportionally to the flux density $\bar{\varepsilon} \propto q$. The maximum value of this energy corresponding to $q = 10^{15}$ W/cm² was about 10 keV.

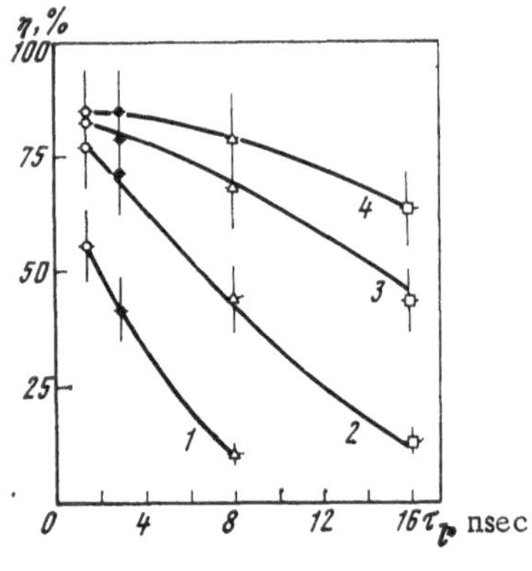

Fig. 25. Dependences of the efficiency of the energy input on the duration of the laser pulses plotted for targets of different radii r_0 (μ): 1) 3; 2) 100; 3) 200; 4) 250. The residual gas (deuterium) pressure was ~8-12 Torr. The laser radiation energy was ~300 J.

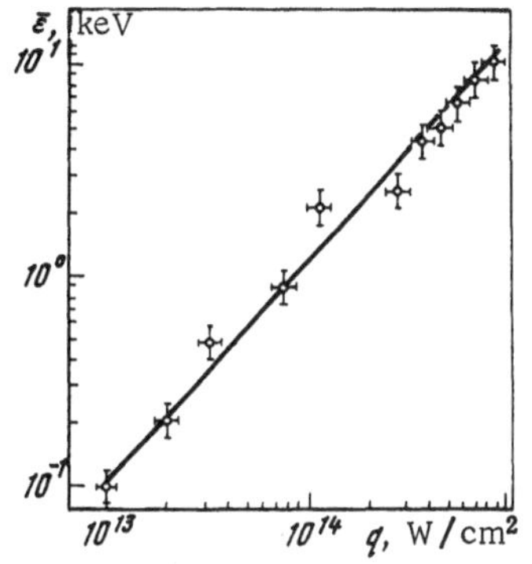

Fig. 26. Dependence of the average energy per particle in a plasma on the density of the laser radiation flux. The laser radiation energy was ~300 J. The deuterium pressure in the chamber was ~10 Torr.

Figure 27 shows the dependences of the efficiency of energy transfer on the laser radiation flux density, obtained using light pulses of different duration. An increase in the flux density resulted from a reduction in the target size and this led to a weaker energy transfer. It is clear from the dependence in this figure that although the absorption fell more strongly in the case of pulses of short duration, the heating of a plasma at high values of the flux density $\gtrsim 10^{14}$ W/cm² occurred most efficiently when a target was irradiated with pulses of minimum duration.

§ 3. Investigation of X-Ray Emission

The temperature of a plasma formed as a result of spherical irradiation of an isolated solid target made of $(CD_2)_n$ by laser radiation of $\sim 10^2$-10^3 J energy reached values of ~1-10 keV. At these temperatures the plasma was completely ionized and it emitted strongly in the x-ray part of the spectrum; in the case of an optically thin plasma such radiation originated from the whole volume and the intensity maximum was located at the wavelength $\lambda_{max} \propto 6.2T^{-1}$. The main emission mechanisms were the bremsstrahlung (*ff*) and recombination (*fb*) radiation, the latter involving electrons in the ground and first excited levels. The spectral density of the x-ray radiation power is described by the well-known expression

$$I_\lambda = 6 \cdot 10^3 \frac{\tilde{n}_e^2 (\tilde{z}^2 + 2)}{(\tilde{z} + 2)\lambda^2 T_e^{1/2}} \exp\left(-\frac{12.4}{\lambda T_e}\right)\left\{\bar{g}^{ff} + \frac{3.3 \cdot 10^{-3}\tilde{z}^2}{T_e}\left[\alpha \bar{g}_1^{fb} \exp\left(\frac{1.36 \cdot 10^{-2}\tilde{z}^2}{T_e}\right) + \beta \bar{g}_2^{fb}\left(\frac{3.4 \cdot 10^{-3}\tilde{z}^2}{T_e}\right)\right]\right\} \text{GW} \cdot \text{cm}^{-3} \cdot \text{Å}^{-1}, \quad (27)$$

where $\tilde{n}_e = 10^{-21}$ is the reduced electron density; $\tilde{z}$ is the average charge of ions in the plasma; λ is the emission wavelength (Å); T_e is the electron temperature (keV); $\bar{g}^{ff}$ is the averaged (over the Maxwellian distribution) Gaunt factor; $g_{1,2}^{fb}$ is the averaged (over the orbital quantum number l) Gaunt factor for n = 1 and 2; the coefficients α and β allow for the contribution of the

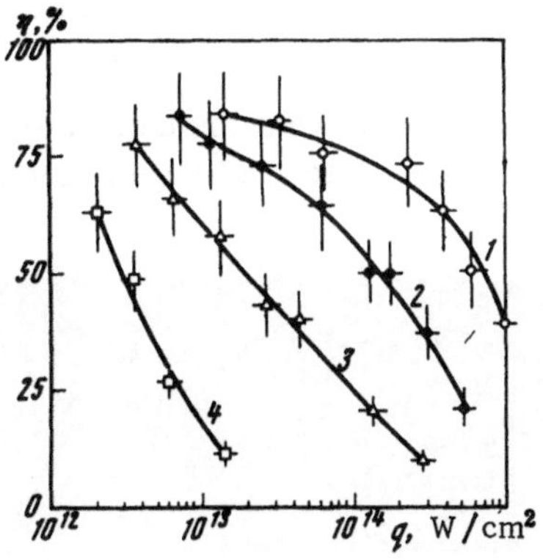

Fig. 27. Dependences of the efficiency of energy input on the density of the laser radiation flux plotted for different laser pulse durations τ_l (nsec): 1) 1.5; 2) 3; 3) 8); 4) 16. The deuterium pressure was 8-12 Torr. The laser radiation energy was ~300 J. The target radius was ~30-250 μ.

recombination radiation at wavelengths $\lambda \leq 912 n^2 \bar{z}^{-2}$ Å and they are, respectively,

$$\alpha = \begin{cases} 8 & \text{for } 0 < \lambda \leqslant 9.12 \cdot 10^2 \bar{z}^{-2}, \\ 0 & \text{for } \lambda > 9.12 \cdot 10^2 \bar{z}^{-2}, \end{cases} \qquad \beta = \begin{cases} 1 & \text{for } 0 < \lambda \leqslant 3.65 \cdot 10^{-3}, \\ 0 & \text{for } \lambda > 3.65 \cdot 10^3 \bar{z}^{-2}. \end{cases}$$

Thus, the electron temperature of a plasma can be determined by investigating the dependence of the intensity of the radiation it emits on the wavelength. The most widely used method in the x-ray wavelength range is the method of filters [33-36] which is based on a comparison of the intensities of the radiation transmitted by absorption filters with different transmission curves.

However, it should be pointed out that at high laser radiation flux densities $q \sim 10^{15}$ W/cm^2 the velocity distribution of electrons in a plasma formed by such radiation may differ from the Maxwellian distribution and this may give rise to considerable errors in the determination of the electron temperature T_e by this method.

If a target is sufficiently strongly compressed, which may be due to cumulation processes resulting from ablation, a plasma with a temperature of ~1-10 keV may be optically thick in relation to its own radiation, i.e.,

$$\tau = 1.1 \cdot 10^{-2} \frac{\bar{z}2}{\bar{z}+2} \lambda^2 r_0^6 r^{-5} T^{-3/2} \gg 1, \tag{28}$$

where τ is the optical thickness of the plasma for radiation of a given wavelength λ. The compression $\varkappa = \rho/\rho_0$ satisfying the condition (28) is given by the expression

$$\varkappa \gg 1.9 T^{2.1} r_0^{-0.6}. \tag{29}$$

In the case of a target of $2r_0 \sim 2 \cdot 10^{-2}$ cm diameter and a plasma of temperature T = 1 keV the compression at which the condition (29) is satisfied amounts to $\varkappa \gg 30$.

The spectral density of the power of the radiation emitted from the surface of a plasma is then given by the expression

$$S_\lambda = 3.74 \cdot 10^{11} \lambda^{-5} \left\{ \exp\left(\frac{12.4}{\lambda T_e}\right) - 1 \right\}^{-1} \text{GW} \cdot \text{cm}^{-2} \cdot \text{Å}^{-1}. \tag{30}$$

Thus, when the plasma temperature is determined from the x-ray radiation intensity, it is necessary to allow for the possibility of a change in the nature of the radiation itself.

The relative intensities of x-ray radiation passed by beryllium filters 100, 200, and 500 μ thick were determined experimentally. A single photomultiplier was used in the recording of light pulses generated in scintillators. The necessary synchronization of the light pulses was ensured by an optical delay line made of a fiber. A typical oscillogram of the pulses involved is shown in Fig. 28.

The absolute intensity of x-ray radiation at different wavelengths was measured at the same time. This was done using a four-channel x-ray calorimeter. Each calorimeter channel

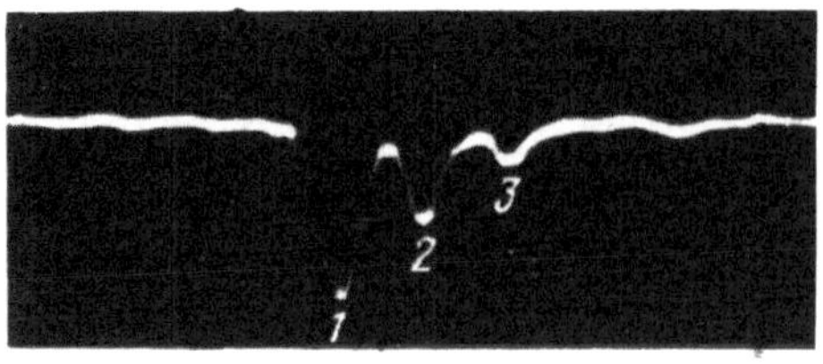

Fig. 28. Oscillograms of x-ray radiation pulses passed through Be filters of different thickness: 1) 100 μ; 2) 200 μ; 3) 500 μ. The lengths of the optical fibers used in these cases were: 1) 1 m; 2) 4 m; 3) 7 m.

recorded the energy of the radiation transmitted by beryllium filters 100, 200, 400, and 800 μ thick. These measurements enabled us to determine the total radiation energy E_{rad} emitted by the plasma

$$E_{rad} = \xi_c E_c, \tag{31}$$

where E_c is the energy measured in one of the channels; ξ_c is a factor allowing for the experimental geometry and the spectral composition of the radiation:

$$\xi_c = \frac{4\pi R_c^2 \int_0^\infty \int_0^t r^3(t') I_\lambda(t') d\lambda dt'}{S_c \int_0^\infty \exp[-\mu_c(\lambda)\delta_c] \int_0^t r^3(t') I_\lambda(t') d\lambda dt'}. \tag{32}$$

Here, R_c is the distance from the target to a radiation detector; S_c is the detector area; μ_c and δ_c are the mass absorption coefficient and thickness of a filter C.

The radiation power per unit volume of the plasma emitted in the continuous part of the spectrum with the spectral density given by Eq. (27) can be expressed in the form

$$P_{rad} = 4.86 \cdot 10^2 n_e^2 \frac{z^2+2}{z+2} T_e^{1/2} \left(1 + \frac{0.03z^2}{T_e}\right) \text{ GW} \cdot \text{cm}^{-3}. \tag{33}$$

The total radiation energy is proportional to the plasma compression: $E_{rad} \propto \varkappa$. Thus, measurements of the x-ray radiation energy can give some information on the degree of compresion of the heated plasma. With this in mind, calculations were made of the radiation emitted by a spherical plasma subject to the following assumptions: the plasma density and temperature are constant throughout its volume, $\rho(r)$ = const and $T(r)$ = const, whereas the gasdynamic expansion velocity varies linearly along the radius $(dr/dt)r = v(r/r_0)$. The system of equations describing the time dependences of the plasma parameters becomes

$$\left.\begin{aligned} &T = \alpha r \frac{d^2 r}{dt^2}, & &T(0, r) = 0, \quad \frac{dr}{dt}(0, r) = 0, \\ &\frac{dT}{dt} = -\frac{dr}{dt}\frac{d^2 r}{dt^2} + \beta(P_l - P_{rad}), & &r(0) = r_0, \end{aligned}\right\} \tag{34}$$

where T is the temperature in keV; P_l is the laser radiation power (GW); P_{rad} is the power of the radiation emitted by the plasma (GW); $\alpha = 3.02 \cdot 10^{-2}$; $\beta = 4.2 \cdot 10^{12} r_0^{-3}$. During the early stages of the expansion the plasma was assumed to be optically thick in relation to its own radiation and the power P_{rad} of the radiation emitted by the plasma was described by the expression

$$P_{rad} = 1.3 \cdot 10^9 r^2 T^4 \text{ GW}. \tag{35}$$

The plasma was assumed to absorb completely the heating radiation up to the moment when the range of the quanta in the plasma became comparable with the characteristic dimension of the plasma. In dealing with the later stages in the system (34) it was assumed that $P_l = 0$.

Figure 29 shows the time dependences of the power and energy of the radiation emitted by the plasma, which were obtained by numerical solution of the system (34). The radiation power was assumed to be constant for different durations of the light pulses, which were used as the parameters of the curves. The initial target radius was taken to be 10^{-2} cm. It is clear from the curves in Fig. 29 that the power of the radiation emitted by the plasma rose rapidly during the early stages and could reach 50% of the laser pulse power and the total energy losses due to the radiation represented ~25% of the energy supplied to the plasma.

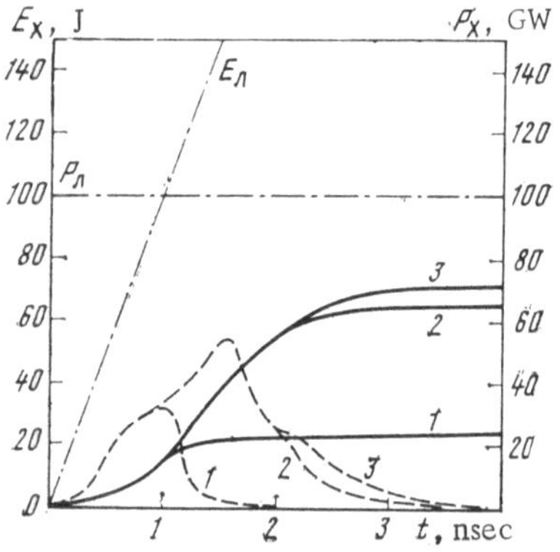

Fig. 29. Time dependences of the power (dashed curves) and energy (continuous curves) of x-ray radiation emitted by a plasma. The calculations were carried out for P_l = 100 GW, $r_0 = 10^{-2}$ cm, and different values of τ_l (nsec): 1) 1; 2) 2; 3) 3.

Figure 30 shows the dependence of the x-ray radiation energy emitted by the plasma on the target size for different durations of the laser pulses. For each fixed duration there was a definite value of the initial target radius for which the maximum amount of energy was emitted as radiation. The same figure shows the experimentally determined values of the x-ray radiation emitted from the plasma when excited with laser pulses of 1.5 nsec duration incident on a target of radius $r_0 \sim 130\ \mu$. This value (~30 J) was more than twice the value calculated for the same laser parameters and target size. For light pulses of real shape and with real temperature and density distributions in the plasma this discrepancy could be even greater.

The most probable cause of the high intensity of the x-ray radiation was probably the target material compression. The energy emitted as x-ray radiation from the plasma was proportional to the compression but one should remember that only a certain proportion of the initial target mass was concentrated in the hot core of the plasma where the density was high and the lifetime of such a core was considerably less than the total plasma lifetime. Bearing these points in mind, estimates were obtained of the compression coefficient and this was found to be $\varkappa \sim 30$ [40].

Measurements of the electron temperature of the plasma under these conditions were carried out by the absorption filter method and they gave ~1 keV, which was in a fair agreement with the calculated value. Similar temperature measurements were carried out also in all the other experiments. These measurements indicated that the electron temperature of the plasma fell rapidly with increasing size of the target and with decreasing laser radiation power. The maximum temperature was obtained by irradiating a target of $2r_0 \sim 60\ \mu$ diameter by laser pulses of $\tau_i \sim 1.5$ nsec duration and $E_l \sim 250$ J energy; this maximum temperature was $T_e \sim 6$ keV.

§ 4. Investigation of the Neutron Yield of Plasma

The neutron yield of a plasma was recorded using a system of three scintillation counters with photomultipliers placed at various distances from the target. Since the neutron pulses

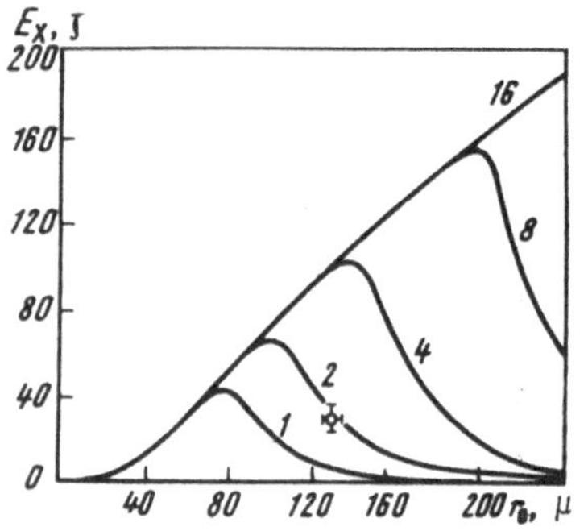

Fig. 30. Dependences of the energy radiated by a plasma on the initial target radius plotted for laser pulses of 1, 2, 4, 8, and 16 nsec duration (these durations are given alongside each curve). The symbol "+" denotes the experimentally determined energy radiated from a target of 130 μ initial radius heated by laser radiation of 250 J energy and 1.5 nsec duration.

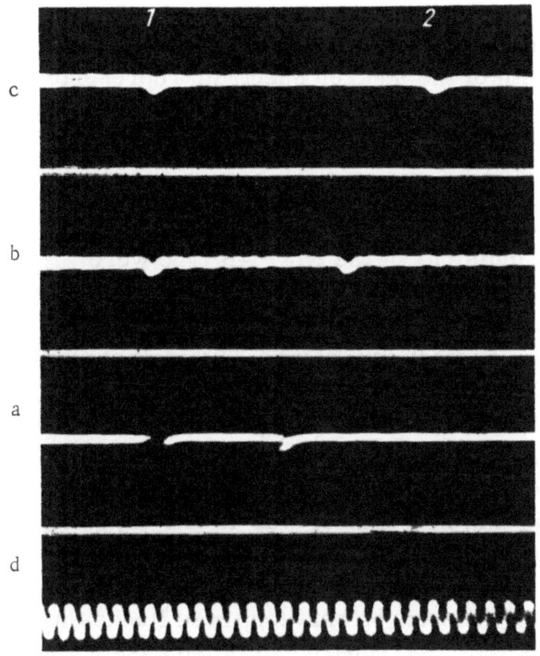

Fig. 31. Oscillograms of neutron pulses: 1) laser pulse; 2) neutron pulse; 2) pulse generated by a photomultiplier located at a distance of 1.15 m from the target; b) pulse generated by a photomultiplier located at a distance of 1.95 m; c) pulse generated by a photomultiplier located at a distance of 3 m; d) timing marks (10 nsec).

were short (the neutron pulse duration τ_n was considerably less than the plasma lifetime $\leqslant 1$ nsec) and the resolution of the scintillation system was relatively high (~5 nsec), it was possible to carry out time-of-flight measurements of the neutron energy. This was done by applying to each multiplier, by an optical fiber, a light pulse synchronized with the beginning of the target heating period. Typical oscillograms of the pulses obtained are shown in Fig. 31.

The neutron energy deduced by the time-of-flight method was ~2.45 MeV. This value agreed well with the energy of neutrons expected for a plasma with a temperature of the order of several kiloelectron-volts as a result of a (d, d) reaction. It should also be pointed out that the neutron radiation was isotropic. All this suggested that neutrons were generated by a thermal mechanism.

Measurements of the width of the neutron energy spectrum were carried out with a counter separated by $L = 3 \cdot 10^2$ cm from the target. The precision of the measurements of the spectral width was governed by the resolution time of the system and it amounted to

$$\Delta E_p = 2.74 \cdot \frac{E^{3/2}}{L} \Delta t_{tf} \text{ MeV}, \qquad (36)$$

where ΔE_p was the width of the spectrum; E was the neutron energy; Δt_{tf} was the resolution time (nsec) of the time-of-flight method. Under the experimental conditions the above quantity was ~170 keV. The experimentally determined width of the neutron spectrum made it possible to estimate the ion temperature of the plasma T_d which – for a Maxwellian distribution of the deuteron velocity – was related approximately to the width of the neutron spectrum by

$$\Delta E_p \approx 82.5 \sqrt{T_d} \text{ keV}. \qquad (37)$$

The measurements carried out indicated that the ion temperature of the plasma formed as a result of irradiation of a $(CD_2)_n$ target of ~110 μ diameter with a light pulse of ~250 J energy and ~1.5 nsec duration did not exceed 5 keV. This ion temperature demonstrated that neutrons were generated in the laser plasma by a thermal mechanism.

The ion temperature was not determined sufficiently accurately to draw any definite conclusions on the thermal equilibrium between electrons and ions in the plasma (the electron temperature measured under the same experimental conditions was ~1 keV), but estimates of the time needed to establish thermodynamic equilibrium suggested that the temperatures were equal. The ion temperature deduced from the width of the neutron spectrum represented only the hot dense core of the plasma with a very short lifetime $\tau \ll 1$ nsec, whereas the electron temperature deduced from the x-ray radiation measurements was averaged over space and time.

Quantitative measurements of the neutron yield were carried out using nuclear photographic emulsions and the proton-recoil method [14]. Emulsions of the NIKFI-R type were 300 μ thick and they were placed at a distance of 6 cm from the target. After three experiments in which scintillation detectors indicated the presence of thermonuclear neutrons, the irradiated and control emulsions were developed under identical conditions. The control emulsion belonging to the same batch was located in a different room during these experiments. The irradiated emulsion had 87 tracks per 1 cm^2 and these represented the protons which suffered recoil from the neutrons generated by the D(d, n) He reaction, whereas the control emulsion had 48 such tracks. The identity of the photographic properties of the emulsions was checked by determining the number of "stars" in 1 cm^2, and this number was 45 and 49, respectively. These quantities and the assumption that the neutron emission from the plasma was isotropic indicated that the total number of neutrons in three flashes was $\sim 10^7$. The minimal neutron yield was $\sim 3 \cdot 10^6$ neutrons/pulse, but in view of differences between these flashes in respect of the energy input, the yield should be taken as $\sim 10^7$ neutrons/pulse in the case when a target of 110 μ diameter was heated by a light pulse of ~3 nsec duration and ~215 J energy.

The generation of neutrons in the experiments described above was fairly irregular. For example, in some cases the absorbed energy was ~250 J, representing ~75% of the laser radiation energy reaching the target, but no neutrons were observed (within the limits of the experimental error) although the electron temperature of the plasma was found to be 1 keV. In other experiments we observed neutron pulses whose times of flight did not agree with the values expected for the thermonuclear neutrons. This irregularity of the neutron generation was due to the probabilistic nature of the effective compression of the plasma by high-power spherical irradiation resulting from deviations of the targets from sphericity and from inhomogeneity of the irradiation process.

§ 5. Reflection of Laser Radiation from Dense Plasma

An investigation of the proportion of the radiation reflected from a target during heating was carried out using two different channels of a multibeam laser. In one channel measurements were made of the amplitude-time characteristics of the reflected signals employing a coaxial photocell with a time resolution of ~0.5 nsec. This photocell received a small proportion (deflected by plane-parallel plates) of the incident and reflected radiation after a time delay which ensured that the signals were separated when displayed on the screen of an oscillograph (Fig. 20). A typical oscillogram of pulses of this kind is shown in Fig. 32. We can see that the pulse shape of the reflected signal differed somewhat from that of the incident signal.

a

b

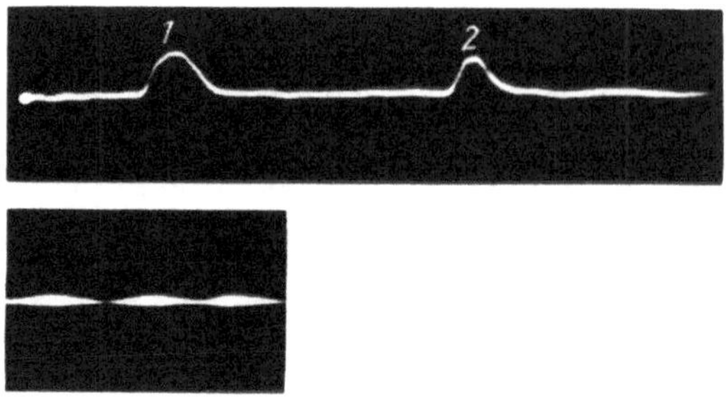

Fig. 32. Oscillograms of laser radiation incident on a target (1a) and reflected from it (2a); b) timing marks (10 nsec). The target was of $r_0 = 100\ \mu$ radius, the laser energy was 330 J, the pulse duration was 4 nsec, and the residual pressure of D_2 was 6 Torr.

The greatest change was observed in the duration of the leading edge of the incident signal. This was due to the fact that the reflection coefficient depended strongly on the electron density gradient in the plasma, whose value varied rapidly with time during the plasma heating process

Measurements of the absolute energy of the reflected radiation in a different channel of the laser system were carried out using a calorimeter and the results were used mainly in the determination of the efficiency of absorption of high-power lase radiation by the dense plasma. These measurements indicated that when the laser radiation flux density on the target surface was 10^{13}-10^{15} W/cm^2, the amount of radiation reflected into the solid angle supported by the objectives did not exceed 5-10%.

§ 6. Gasdynamic Pressure of "Corona" and Cumulative Compression of Plasma Core

We used high-speed interferometric methods to investigate the rapidly changing spatial distributions of matter in the outer region of the plasma. Two photographic methods were used: interferometric photography with an image-converter streak camera [37] and multiframe interferometric photography.

In the first case the illumination was provided by part of the radiation generated by the master laser, which was amplified and then transformed into the second harmonic of wavelength $\lambda = 0.53\ \mu$. The scan was synchronized with the investigated process by a laser-triggered discharger, which generated pulses that controlled the streak camera. The principal parameters of the system were as follows: the scan duration was ~30 nsec, the best time resolution was ~0.1 nsec, the spatial resolution was at least 200 lines/mm, and the maximum electron density, governed by the aperture of the optical system, was $\sim 10^{20}$ cm^{-3}.

In these experiments we used a Jamin interferometer, which allowed us to carry out measurements in the zeroth interference order. The interferograms were recorded for different orientations of the entry slit of the streak camera relative to the object. The symmetry of the outer region of the plasma ("corona") was checked by comparing the results obtained by analysis of interferograms for two opposite directions of the plasma expansion relative to the target center.

Fuller information on the plasma symmetry during the initial expansion stage was provided by multiframe interferometric photography. In these experiments measurements were also carried out in the zeroth interference order using the Jamin interferometer. The apparatus used enabled us to record seven consecutive frames. The exposure time of each frame was governed by the duration of a light pulse used to illuminate the plasma and it amounted to ~1.5 nsec. The interval between the frames was ~3 nsec and the synchronization was within 10^{-10} sec.

The results obtained in an analysis of the experimental data enabled us to determine the mass transfer in the gasdynamic plasma expansion process and to estimate the pressure pulse experienced by the surface as a result of ablation. When the laser radiation flux density was $\sim 10^{15}$ W/cm^2, the pressure was $\sim 2 \cdot 10^7$ bar. It should be noted that this was the minimum pressure exerted on the target surface because it was found ignoring the motion of the dense plasma layers.

These high pressures demonstrated that the core plasma was compressed to a density considerably higher than the density of a solid (~1 g/cm^2). This compression was most probably produced by a spherically converging shock wave formed by the strong pressure pulse.

A shock wave with a pressure amplitude in the front exceeding 10^7 bar was propagated through the condensed matter and the thermal energy resulting from the increase in the entropy of this matter was considerably greater than the binding energy of the solid, which was then converted into a strongly compressed plasma. This compression of matter gave rise to an electron density $n_e \gg (m_e e^2/\hbar^2)^3 Z^2$ (here n_e is the electron density, m_e is the electron mass, and Z is the atomic number) with a temperature $T \ll 40 Z^{4/3}$ eV, so that the electron component of the plasma formed a degenerate Fermi gas in which the pressure was [38]

$$p = \frac{(3\pi^2)^{2/3}}{5} \frac{\hbar^2}{m_e} (n_e)^{5/3} \approx 2.45 \cdot 10^{-33} n_e^{5/3} \text{ bar.} \quad (38)$$

This expression gave essentially the minimum pressure ensuring the required density in the plasma core. For example, a pressure of $\sim 10^9$ bar was needed to compress deuterated polyethylene by a factor of 40 ($n_e \sim 1.15 \cdot 10^{25}$ cm^{-3}).

The shock adiabat for an extremely strong shock wave in a solid was essentially identical with the shock adiabat of an ideal gas and the condensed matter was also characterized by a finite compression in the shock-wave front which was governed by the adiabatic exponent (for $\gamma = 5/3$, the compression was 4). A spherically converging shock wave moving toward the center was continuously accelerated and amplified. Close to the focusing moment the shock-wave motion reached a certain limiting regime for which a self-similar solution was found [39]. An analysis of self-similar equations of motion indicated that at the moment of focusing of a shock wave the density reached its maximum value

$$\rho_{max} = \varkappa_{max} \rho_0 \quad (\text{for } \gamma = 5/3 \; \varkappa_{max} \sim 15) \quad (39)$$

and it was constant in the radial direction. The plasma temperature in this region rose strongly, reaching values of the order of several kiloelectron-volts. Subsequent motion of the gas could be regarded as a strong explosion of a point charge except that the shock wave reflected from the center of symmetry traveled along a moving inhomogeneous gas. A numerical solution of this problem for $\gamma = 5/3$ predicted a further increase in the compression (to ~40) behind the front of the reflected shock wave.

In reality the pressure on the target surface varied continuously with time and the motion of matter toward the center could be regarded as a result of action of shock waves of increasing amplitude, each of which made its own contribution to the maximum compression. In this situation the optimal conditions for the strongest compression were provided by an isentropic process in which all the hydrodynamic perturbations reached the center of symmetry of the target at the same moment [12, 39]. It was suggested in [11] that this could be achieved by a special shaping of the laser pulses in time and the power of these pulses should be $\sim 10^{11}$-10^{15} W, which could not yet be achieved in practice.

It should also be pointed out that the maximum compression of a spherical target depended strongly on its initial symmetry and on the uniformity of radiation. If the deviation of the target shape from an ideal sphere was δ_0, then even an ideally uniform irradiation would result in a deviation $\delta \propto \delta_0 \varkappa^{1/3}$ of the dense core of the plasma from the spherical shape, which would give $\delta \sim 70\%$ for compression by a factor $\varkappa \sim 40$ and $\delta_0 \sim 20\%$. This asymmetry increased with the compression and it could reduce the maximum density of the plasma in such experiments. This was why the neutron yield of the plasma obtained as a result of spherical irradiation of a $(CD_2)_n$ target was irregular in our experiments.

The authors are deeply grateful to I. M. Buzhinskii, M. P. Vanyukov, and S. K. Mamonov for valuable advice, to B. I. Belov, V. M. Groznov, I. M. Divil'kovskii, D. V. Kovalevskii, and B. V. Kruglov for their help in assembling the laser system, and to V. A. Boiko, A. A. Erokhin, Yu. A. Mikhailov, S. N. Zakharov, N. N. Zorev, A. A. Rupasov, and A. S. Shikanov for their help and valuable advice in the course of our experiments.

Literature Cited

1. N. G. Basov and O. N. Krokhin, Zh. Eksp. Teor. Fiz., 46:171 (1964); Vestn. Akad. Nauk SSSR, No. 6, 55 (1970).
2. O. N. Krokhin, in: Physics of High Energy Density, Academic Press, New York (1971), p. 278.
3. H. Hora, in: Laser Interaction and Related Plasma Phenomena (Proc. Second Workshop, Rensselaer Polytechnic Institute, Hartford Graduate Center, Conn., 1971), Vol. 2, Plenum Press, New York (1972), p. 383.
4. P. P. Pashinin and A. N. Prokhorov, Zh. Eksp. Teor. Fiz., 60:1630 (1971).
5. F. Floux, J. F. Benard, D. Cognard, and A. Saleres, in: Laser Interaction and Related Plasma Phenomena (Proc. Second Workshop, Rensselaer Polytechnic Institute, Hartford Graduate Center, Conn., 1971), Vol. 2, Plenum Press, New York (1972), p. 409.
6. E. D. Jones, G. W. Gobeli, and J. N. Olsen, in: Laser Interaction and Related Plasma Phenomena (Proc. Second Worshop, Rensselaer Polytechnic Institute, Hartford Graduate Center, Conn., 1971), Vol. 2, Plenum Press, New York (1972), p. 469.
7. Yu. V. Afanas'ev, E. M. Belenov, O. N. Krokhin, and I. A. Poluéktov, ZhETF Pis'ma Red., 13:257 (1971).
8. S. Kaliski, Proc. Vibr. Probl., 12:231, 243 (1971).
9. N. G. Basov, O. N. Krokhin, and G. V. Sklizkov, in: Laser Interaction and Related Plasma Phenomena (Proc. Second Workshop, Rensselaer Polytechnic Institute, Hartford Graduate Center, Conn., 1971), Vol. 2, Plenum Press, New York (1972), p. 389.
10. J. W. Daiber, A. Hertzberg, and C. E. Wittliff, Phys. Fluids, 9:617 (1966).
11. J. Nuckolls, L. Wood, A. Thiessen, and G. Zimmerman, Abstracts of Papers presented at Seventh Intern. Conf. on Quantum Electronics, Montreal, 1972, in: Digest of Technical Papers, Institute of Electrical and Electronics Engineers, New York (1972), Post-Deadline Paper.
12. K. A. Brueckner, Preprint KMSF-NPS, KMS Fusion Inc. (1972).
13. N. G. Basov, O. N. Krokhin, G. V. Sklizkov, S. I. Fedotov, and A. S. Shikanov, Zh. Eksp. Teor. Fiz., 62:203 (1972); Preprint No. 123 [in Russian], Lebedev Physics Institute, Academy of Sciences of the USSR, Moscow (1971).
14. N. G. Basov, Yu. S. Ivanov, O. N. Krokhin, Yu. A. Mikhailov, G. V. Sklizkov, and S. I. Fedotov, ZhETF Pis'ma Red., 15:589 (1972).
15. N. G. Basov, V. A. Boiko, S. M. Zakharov, O. N. Krokhin, and G. V. Sklizkov, ZhETF Pis'ma Red., 13:691 (1971).
16. S. W. Mead, R. E. Kidder, and J. E. Swain, Preliminary Measurements of X-Ray and Neutron Emission from Laser Produced Plasmas (Preprint UCRL-73356), Lawrence Radiation Laboratory, University of California (1971).
17. C. Yamanaka, T. Yamanaka, T. Sasaki, et al., Plasma Generation and Heating to Thermonuclear Temperature by Laser (Preprint IPPJ-117), Institute of Plasma Physics, Nagoya University (1972).
18. F. Floux, D. Cognard, L. G. Denoeud, G. Pair, D. Parisot, J. L. Biboin, F. Delobeau, and C. Fauguignon, Phys. Rev. A, 1:821 (1970).
19. N. G. Basov, V. S. Zuev, P. G. Kryukov, V. S. Letokhov, Yu. V. Senatskii, and S. V. Chekalin, Zh. Eksp. Teor. Fiz., 54:767 (1968).
20. A. A. Mak, Yu. A. Anan'ev, and B. A. Ermakov, Usp. Fiz. Nauk, 92:373 (1967).
21. A. J. DeMaria, W. H. Glenn, J. Michael, et al., in: Laser Interaction and Related Plasma Phenomena (Proc. Second Workshop, Rensselaer Polytechnic Institute, Hartford Graduate Center, Conn., 1971), Vol. 2, Plenum Press, New York (1972), p. 11.
22. V. A. Gribkov, G. V. Sklizkov, S. I. Fedotov, and A. S. Shikanov, Prib. Tekh. Eksp., No. 4, 213 (1971).
23. G. V. Sklizkov and S. I. Fedotov, Prib. Tekh. Eksp., No. 2, 176 (1972).

24. N. N. Zorev, G. V. Sklizkov, S. I. Fedotov, and A. S. Shikanov, Preprint No. 56 [in Russian], Lebedev Physics Institute, Academy of Sciences of the USSR, Moscow (1971).
25. M. P. Vanyukov, V. A. Venchikov, V. I. Isaenko, V. A. Serebryakov, and A. D. Starikov, Opt. Spektrosk., 28:1008 (1970).
26. N. G. Basov, V. A. Gribkov, O. N. Krokhin, and G. V. Sklizkov, Zh. Eksp. Teor. Fiz., 54:1073 (1968).
27. A. A. Mak, D. S. Prilezhaev, V. A. Serebryakov, and A. D. Starikov, Opt. Spektrosk., 33:689 (1972).
28. N. N. Zorev and S. I. Fedotov, Preprint No. 123 [in Russian], Lebedev Physics Institute Academy of Sciences of the USSR, Moscow (1972).
29. N. G. Basov, O. N. Krokhin, A. A. Rupasov, et al., Preprint No. 47 [in Russian], Lebedev Physics Institute, Academy of Sciences of the USSR, Moscow (1973).
30. G. V. Sklizkov, in: Laser Handbook, Vol. 2, North-Holland, Amsterdam (1972), p. 1545.
31. Ya. B. Zel'dovich and Yu. P. Raizer, Physics of Shock Waves and High-Temperature Hydrodynamic Phenomena, 2 vols., Academic Press, New York (1966-7).
32. L. M. Biberman, V. S. Vorob'ev, and G. É. Norman, Opt. Spektrosk., 14:330 (1963).
33. F. G. Jahoda, E. M. Little, W. E. Quinn, G. A. Sawyer, and T. F. Stratton, Phys. Rev., 119:843 (1960).
34. K. Buchl, K. Eidmann, P. Mulser, H. Salzmann, R. Sigel, and S. Witkowski, Proc. Fourth Conf. on Plasma Physics and Controlled Thermonuclear Fusion, Madison, Wisconsin, 1971, publ. by International Atomic Energy Agency, Vienna (1971), p. 645.
35. N. G. Basov, V. A. Boiko, V. A. Gribkov, S. M. Zakharov, O. N. Krokhin, and G. V. Sklizkov, ZhETF Pis'ma Red., 9:520 (1969).
36. V. A. Boiko, O. N. Krokhin, and G. V. Sklizkov, present issue, p. 183.
37. N. G. Basov, V. A. Boiko, V. A. Gribkov, S. M. Zakharov, O. N. Krokhin, and G. V. Sklizkov, Zh. Eksp. Teor. Fiz., 61:154 (1971).
38. L. D. Landau and E. M. Lifshitz, Statistical Physics, 2nd ed., Pergamon Press, Oxford (1969).
39. K. P. Stanyukovich, Unsteady Motion of Continuous Media [in Russian], Nauka, Moscow (1971).
40. N. G. Basov, E. G. Gamalii et al., Preprint No. 15 [in Russian], Lebedev Physics Institute, Academy of Sciences of the USSR, Moscow (1974); E. G. Gamalii, ZhETF Pis'ma Red., 19:520 (1974).

INVESTIGATION OF THE PARAMETERS AND DYNAMICS OF A PLASMA OBTAINED BY SHARP FOCUSING OF LASER RADIATION ON SOLID TARGETS

V. A. Boiko, O. N. Krokhin, and G. V. Sklizkov

A review is given of the physical conditions in a plasma produced from solid targets in vacuum as a result of interaction with laser pulses of 10^{-11}-10^{-8} sec duration and q ~ 10^{10}-10^{15} W/cm^2 flux density. The results are given and an analysis is made of experimental determinations of the maximum electron temperature in a plasma, electron density in the plasma, lifetime of ions in the hot core of a laser plasma jet, gasdynamic velocities of plasma expansion, and maximum degree of ionization of the ions (T_e = 0.01-2.0 keV, $N_e \sim 10^{20}$ cm^{-3}, $\tau \sim 10^{-9}$ sec, v ~ 10^7 cm/sec, z = 20). These parameters are used in a discussion of the models of gasdynamic plasma motion and of ionization equilibrium in the hot core of a plasma jet.

§ 1. Introduction

The possibility of heating matter to thermonuclear temperature by high-power laser radiation [1] has stimulated numerous investigations of plasmas formed as a result of irradiation of condensed matter in vacuum (laser jet) or as a result of optical breakdown in gases (laser spark).

In addition to thermonuclear applications, there are also promising uses of the laser plasma as a source of multiply charged ions in spectroscopic investigations (which are of particular interest in astrophysics), in accelerator studies of the formation of new superheavy elements, and so on. The interest in the laser plasma is increasing with the temperatures achieved and these temperatures have become possible because of improvements in laser technology: in the last few years the output power of lasers has been increasing on the average by an order of magnitude per annum. A special feature of the laser plasma is its high rate of energy evolution, which is considerably greater than that which can be achieved using pulse electric discharges. The rate of energy evolution is now exceeding 10^{12} W with the energy density in a plasma of 10^7 J/cm^3 and the specific rate of energy evolution is ~10^{17} W/cm^3. The corresponding values for the most powerful "plasma focus" units are ~10^9 W, ~10^4 J/cm^3, and ~10^{11} W/cm^3; the specific energy density of explosives is also ~10^4 J/cm^3.

Much theoretical and experimental information has been published on the laser plasma. Experimental investigations of the physical conditions in a laser jet can be divided somewhat arbitrarily into two stages. The main result of the first stage (1964-1967) has been the estab-

TABLE 1

Year	Laboratory	Total number of neutrons formed	Laser parameters: energy, J	Laser parameters: duration, nsec	Target	Reference
1968	Lebedev Inst., USSR	~10	20	0.02	LiD	[17]
1969	Sandia Corp., USA	~10	50	0.01	LiD	[18]
1969	Limeil, France	10^2—10^3	30	7	D	[19]
1971	Lebedev Inst., USSR	~10^4	50	2	$(CD_2)_n$	[20]
1971	Livermore, USA	~10^4	20—70	2—5	$(CD_2)_n$	[25]
1971	Garching, W. Germany	~10^2	20	10	D	[24]
1972	Nagoya, Japan	~10^4	50	2	D	[23]
1972	Lebedev Inst., USSR	10^6—10^7	214	3	$(CD_2)_n$	[26]

lishment of the existence, near the surface of a target, of a dense and relatively hot plasma cloud capable of emitting ions of ~1 keV energy, generating strong shock waves in the ambient atmosphere, and so on [2-14]. The optical diagnostics methods used in these investigations (multiframe schlieren photography and interferometry with background pulse laser illumination), as well as probe and mass-spectrometric methods, have yielded information mainly on the consequence of the interaction of the heating radiation with a target because these methods have been used to study the structure of expanding plasma after the end of a laser pulse at fairly large distances from the target. A sufficiently detailed bibliography of the papers which appeared up to the end of 1968 can be found in [15, 16].

The next stage (beginning from 1967-1968) includes investigations of the parameters and dynamics of the hot phase of jets and of plasmas heated directly by the laser radiation near the target surface. The results obtained in investigations of this kind are the subject of the present review.

Systematic studies of the properties of high-temperature laser plasmas and improvements in heating and diagnostic methods have made it possible to use such plasmas as high-power sources of x rays and neutrons [17-26] (Table 1). These studies have made it possible to raise the neutron yield to 10^6-10^7 neutrons/pulse [26] and to generate x-ray radiation of energy up to 100 keV [20, 25]. We should mention also spectroscopic investigations of the structure of levels of multiply charged ions formed in laser plasmas and characterized by ionization potentials up to ~7 keV [27-31].

In the present review we shall try to give a general picture based on the published experimental investigations of the hot plasma in laser jets. We shall concentrate on the investigations of various spatial-temporal distributions of the principal laser plasma parameters (plasma temperature, density, linear dimensions, lifetime and effective charge of ions) and on the relationships between these distributions and the characteristics of the heating radiation. As a rule, the experiments which yielded these results were carried out using laser beams focused sharply on massive targets.

The results obtained so far do not provide a full picture of the phenomena involved. There are still many unexplained observations, particularly in the range of flux densities $q \geqslant 10^{14}$ W/cm^2 and pulse durations shorter than 10^{-9} sec. Nevertheless, the results reviewed in the present paper should help in the design of future experiments concerned with laser heating of plasmas.

§ 2. Characteristic Parameters of Laser Jets

2.1. Parameters of Laser Radiation. Focusing Conditions

Laser systems used for plasma heating consist of a master oscillator, which is a Q-switched or mode-locked laser, and a system of amplifiers. The use of special electrooptic devices in shaping the master oscillator pulses makes it possible to select separate spikes of $\Delta t_l \sim 10^{-11}$-10^{-12} sec duration in the case of mode-locked lasers [17, 18, 32], and to reduce the pulse duration to $\Delta t_l \sim 10^{-9}$ sec in the case of Q-switched lasers [19, 20]. Neodymium glass is used exclusively in high-power laser systems because large-volume active elements can be made from this material. Descriptions of the construction of such systems can be found, for example, in [33, 37].

Table 2 gives information on the parameters of laser radiation and ranges of variation of these parameters in the experiments discussed in our review, and also on the focusing of laser radiation on targets. An important parameter of a laser system is the ratio of the intensity of a laser pulse to the intensity of the preceding background (this is known as the contrast). The background radiation may give rise to a cloud of a relatively cold plasma in front of the target and this cloud may prevent the energy of the main pulse from reaching the target [11]. This can be avoided by ensuring the highest possible contrast. In modern laser systems it is possible to achieve a change in the intensity by a factor $>10^7$ in $\sim$1 nsec.

The light flux densities in the output stages of high-power laser systems have to be restricted because of the possibility of damage to neodymium glass or because of self-focusing which reduces the radiation brightness. One of the ways of increasing the energy supplied to a plasma is the use of multichannel laser systems. A nine-channel laser system with a variable pulse duration (2-16 nsec) and with an output energy up to 1000 J is described in [38].

Interaction of a laser beam with the surface of a condensed target in vacuum produces a plasma absorbing heating radiation only when the flux densities exceed a critical value q_0 [39, 40]. If $q < q_0$, the target is partly evaporated and the density jump separating the transparent vapor from the target surface is of the order of the atomic distances ($\sim 10^{-8}$ cm). If $q > q_0$, the energy supplied (per particle) exceeds the binding energy of atoms in solids and the ionization energy, so that a strongly absorbing dense plasma is formed. The gasdynamic motion of a plasma spreads the discontinuity at the target surface over a distance greater than the wavelength of the heating radiation ($\lambda \sim 10^{-4}$ cm). This has made it possible to determine experimentally the value of q_0 for many substances from the strong reduction of the intensity of the laser radiation reflected from a target when q is increased from 10^7 to 10^9-10^{10} W/cm^2 [41]. Thus, in the plasma heating experiments the range of flux densities of interest is $q > 10^{10}$ W/cm^2.

2.2. Heating Conditions and Formulas for Estimating Parameters of the Hot Core

Two situations are possible in laser heating of plasmas generated as a result of interaction of radiation with condensed matter [42]. In the first situation we have the inertial confinement regime when a laser radiation energy sufficient to heat a plasma is introduced in a time

TABLE 2

Parameter	Symbol	Range
Energy	Q	0,1—100 J
Divergence	ϑ	10^{-2}—10^{-4} rad
Duration at midamplitude	Δt_l	10^{-7}—10^{-12} sec
Duration of leading edge	Δt_{le}	10^{-8}—10^{-12} sec
Average power	$W = Q/\Delta t_l$	10^7—10^{12} W
Focal length of lens	f	1—10 cm
Diameter of focusing spot	$d = 2r_0 \approx \vartheta f$	50—500 μ
Area of focusing spot	$S \approx d^2$	10^{-3}—10^{-5} cm^2
Flux density on target surface	$q = \frac{W}{S}$	10^{10}—10^{16} W/cm^2

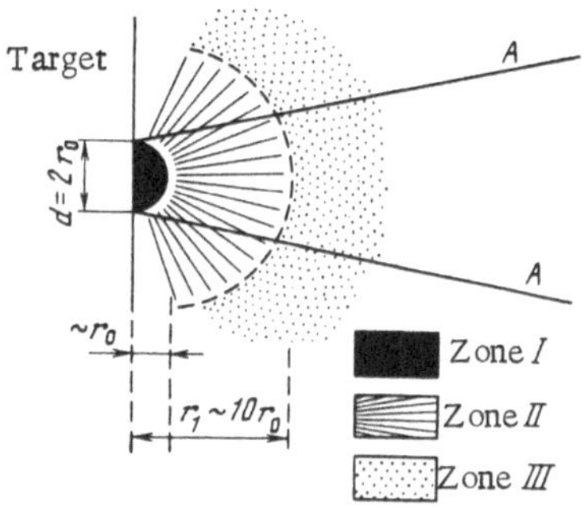

Fig. 1. Typical zones in a laser jet formed as a result of sharp focusing of radiation on the surface of a condensed target in vacuum: I) dense hot core; II) acceleration zone; III) zone of asymptotic motion with constant velocity; AA denotes the caustic of the heating radiation.

$\Delta t_h \sim 10^{-9}$ sec governed by the nature of gasdynamic expansion. This regime is obtained using mode-locked lasers emitting light pulses of duration $\Delta t_l \ll \Delta t_h$. The second situation is the gasdynamic motion regime which is obtained using Q-switched lasers emitting pulses of duration Δt_l exceeding the gasdynamic expansion time Δt_h. It should be pointed out that the major part of the information on laser plasmas has been obtained for the gasdynamic motion regime because the parameters of mode-locked lasers are unstable and difficult to control.

When laser radiation is focused sharply on a target, the principal parameter which governs the dimensions of the various jet zones with different types of gasdynamic motion is the radius of the focusing spot r_0 (Fig. 1). Zone I is the dense hot core of a jet where the laser radiation is absorbed because of the high density of the plasma. At distances from the target $r \ll r_0$ the gasdynamic motion of the plasma can be regarded as planar. At distances $r \gg r_0$ the lateral spreading reduces strongly the plasma density so that hardly any laser radiation is absorbed in this zone. In the gasdynamic motion case the dynamics of plasma expansion in a laser jet is described qualitatively but satisfactorily by the results of a theoretical paper [43], which deals with the case of a spherically symmetric steady-state irradiation of a target with laser light. According to the results of this paper, the gasdynamic plasma velocity u near the target surface is less than the velocity of sound c. At distances $r \sim r_0$ the velocity of the plasma flow is comparable with the velocity of sound but beyond this distance the flow becomes supersonic (this is similar to the flow of gas out of a Laval nozzle). The surface on which $u = c$ can be regarded arbitrarily as the boundary of zone I. The region of supersonic flow can be divided into two zones: II, which is the acceleration zone ($r_0 < r < r_1 \sim 10r_0$) and in which the thermal energy of an adiabatically expanding plasma is converted into the kinetic energy of directional gasdynamic motion; III, which is the zone of asymptotic motion ($r > r_1$) with a practically constant velocity in which the thermal energy of the plasma is less than the energy of directional motion.

In the gasdynamic regime the average values of the parameters of the hot core, which we shall assume to apply at $r \approx r_0$, can be conveniently estimated using the formulas given in [40] for a hydrogen target:

$$\varepsilon_H \approx 5.3 \cdot 10^{14} r_f W_f^{4/9} \text{ erg/g}, \tag{2.1}$$

$$T_H \approx 2.1 \cdot 10^6 r_f^{-2/3} W_f \ {}^\circ\text{K}, \tag{2.2}$$

$$\rho_H \approx 1.3 \cdot 10^{-3} r_f^{-1} W_f^{1/3} \text{ g/cm}^3, \tag{2.3}$$

$$N_H \approx 8 \cdot 10^{20} r_f^{-1} W_f^{1/3} \text{ cm}^{-3}, \tag{2.4}$$

$$M_H \approx 9 t r_f^{2/3} W_f^{5/9} \text{ g}. \tag{2.5}$$

Here, N_H is the concentration of hydrogen ions; T_H, ρ_H, and ε_H are, respectively, the temperature, density, and internal energy of the plasma; $r_f = r_0 \cdot 10^2$ cm; $W_f = W \cdot 10^{-9}$ W; M_H is the total amount of the target material heated in a time t.

A very important characteristic of the gasdynamic expansion is the lifetime of the plasma in the hot dense core Δt_h. The value of Δt_h can be estimated from

$$\Delta t_h \sim \frac{r_0}{c}, \tag{2.6}$$

where c is the velocity of sound in the plasma.

§ 3. Temperature Measurements

The dense core of a jet, where the plasma has the highest temperature, is of the greatest interest from the physical point of view and in the majority of practical applications. Thermonuclear reactions producing neutrons occur in the core of a jet when deuterium-bearing targets are irradiated; when heavy-element targets are used, multiply charged ions with z ~ 20-30 are formed in the same region. The range of temperatures which can be achieved in the core of a jet and, consequently, the nature of the phenomena which accompany the heating and expansion of a plasma depend strongly on the laser flux density reaching the surface of a target. Consequently, it is of interest to determine the dependence of the core temperature on the flux density q using the experimental data obtained for a wide range of such densities.

It should be stressed that we shall consider only measurements of the electron temperature T_e, because of the absence of reliable methods for the determination of the ion temperature T_i of the plasma in a laser jet. Nevertheless, in the majority of the experiments described below we may assume that $T_e = T_i$ in the core of a jet because the electron-ion relaxation time is shorter than the plasma lifetime.

3.1. First Measurements of T_e in Laser Jets [4, 5]

These measurements were carried out using relatively modest flux densities ($q \sim 10^{11}$ W/cm^2) and they established the existence of a high-temperature plasma. Absolute measurements of the intensity of the $\lambda = 2270.90$ Å line of the C V ions were used in [4] in estimating the lower limit of T_e, which was 10 eV. A comparison of the intensity of the $\lambda = 4500$ Å line of Li III ions with the intensity of the continuous spectrum, assumed to be close to the radiation of a black body, was carried out in [5] and it yielded $T_e \sim 20$ eV.

It should be stressed that measurements of the temperature of hot laser plasmas based on the radiation emitted in the visible and adjoining parts of the spectrum [44] are not reliable because the cold zones of a laser jet emit photons in numbers comparable with the hot core and the optical thickness of a jet in a continuous spectrum is, according to the estimates based on the formulas in [45], $\tau_\nu \geqslant 1$. This is illustrated clearly by a determination of the plasma temperature based on the absolute intensities of the continuous spectrum and reported in [44]. The temperature deduced from experimental monochromatic intensities and the Planck function was found to increase severalfold when the wavelength was increased from 4000 to 10,500 Å. This indicated that the spectral distribution of the radiation emitted by a laser jet did not correspond to the black-body situation. The maximum temperature of a beryllium jet, measured at $\lambda = 10{,}500$ Å for various values of q, is given later in this section in Fig. 4, which is a summary of the dependences T(q).

It is generally known that when the temperature is increased the maximum of the plasma bremsstrahlung radiation shifts toward shorter wavelengths [45]: $\lambda_{max} = 6200/T_e$ Å, where T_e is the electron temperature in eV; λ_{max} is the wavelength at which the spectral distribution of the bremsstrahlung radiation emitted by an optically thin plasma has its maximum. The first measurements of the temperature of laser jets based on the determination of the maximum of the bremsstrahlung radiation in the vacuum ultraviolet region were reported in [47, 48]. The maximum of the continuum was close to 200 Å for all the targets employed (Table 3) and this maximum corresponded to $T_e \sim 30$ eV.

TABLE 3

Designation of T_e in Fig. 4	Laser parameters		Target	Reference	Comments
	Q, J	Δt_l, nsec			
1	1		C	[4]	Lower limit of T_e is determined
2	3	40	LiH	[5]	
3		7.5	Be Al Pb	[44]	The value in Fig. 4 is given for Be
4	0.4	40	Be, Al, Ni Ta, W, Pt	[47, 48]	
5	5	17	C, $(CH_2)_n$, Ti Mn, Fe, Ni Cu, Zn	[27, 51, 52]	Use is made of spectral lines of Fe XVI and Ni XVIII ions
6	0.6—1	30	Al	[49]	
7	~1	30	Al	[60]	
8	20	15—20	Al Ca	[29, 30, 53]	Use is made of spectral lines of Ca XIII-XV ions
9	27	15	C	[55, 56]	
10	90	50	D	[58, 81, 82]	$\Delta t_{le} \sim 40$ nsec
11	7	7	C	[61]	
12	7	7	LiD	[61]	
13	20	10	D	[24]	$\Delta t_{le} \sim 2$ nsec; 10-10^2 neutrons observed
14	1—3	0.01—0.001	D	[24]	
15	~50	2	$(CD_2)_n$ D	[50]	Emission of ~10^4 neutrons and gamma rays to 100 keV energy
16	20—40	7—8		[19]	Emission of 10^3 neutrons; q in Fig. 4 estimated using $2r_0 \sim$ 150 μ [58]
17		40	CH_2	[64]	
18	0.1—0.3	0.01—0.001	LiD	[101]	

Similar measurements were also carried out at somewhat higher flux densities [27]. A strong continuum with a maximum at about 70 Å (T_e ~90 eV) was observed for all targets and it was recorded on photographic film. One should mention also an investigation [49] in which a careful analysis of the contribution of the various orders of the spectra and calibration of a spectrograph in the 50-2000 Å range yielded the position of the bremsstrahlung maximum at λ_{max} = 250 Å corresponding to a temperature T_e = 25 eV. The slope of the Lyman continuum could be used to determine the temperature of a plasma containing hydrogen and helium-like ions. This method was used in [50, 51], where the maximum temperature of a carbon jet T_e ~ 100 eV was the same as that obtained under identical conditions in [27]. Since the recombination radiation at these temperatures was considerably stronger than the bremsstrahlung radiation (because it was proportional to z^4 instead of z^2), it was possible to determine in [51] the distribution of T_e at distances up to 2 mm from the target surface (Fig. 2).

The determination of T_e from the line spectra of multiply charged ions is possible only if a specific model of the ionization equilibrium in the hot core of a laser jet is selected. This will be discussed later (§6) and here we shall simply mention [28, 29],

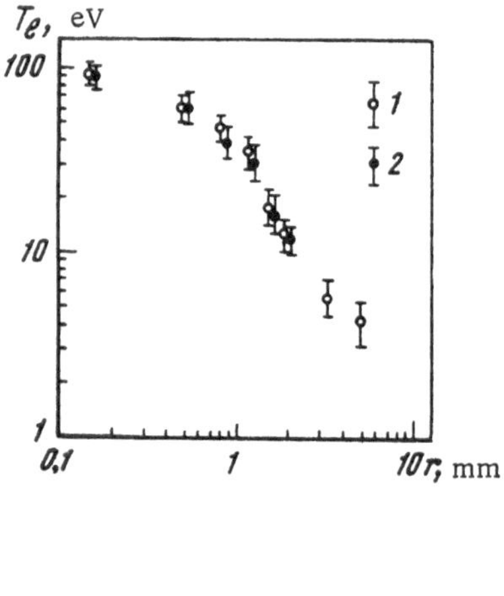

Fig. 2. Integrated (over the laser pulse duration) spatial distribution of the electron temperature in a carbon jet (Q ~ 5 J, Δt_l ~ 15 nsec) given in [51]. In the range r ≤ 2 mm the values of T_e were obtained from the slope of the Lyman continuum of C VI (1) and C V (2) ions. The values of T_e for r = 3.5 and 5 mm were obtained from the relative intensities of the λ = 3434 and 5290 Å lines of C VI ions.

where the coronal model and the presence of lines of Ca XIII-XV ions yielded a rough estimate $T_e \sim 100$-300 eV for different values of q, as well as [52], where a situation was similar and the relative intensities of the $2s^22p^n-2s2p^{n+1}$ transitions in Ca XIII-XV ions were used to find the temperature $T_e \sim 150$-200 eV.

The absorber method, first suggested for measurements of the electron temperature in theta pinches [53], is now seen to be most suitable for the determination of the time dependences of T_e for laser plasmas and of the correlation of these dependences with the shape of the heating pulses. This method is used widely in studies of laser jets [54-65] and laser sparks [66-68].

The essence of the absorber method (for details see [53, 69, 70]) is the measurement of the relative intensities of soft x-ray radiation emitted by a plasma and this is done using filters of different thickness. Large numbers of calculations of the transmission of thin films of beryllium, polyethylene, aluminum, nickel, gold, and some of their combinations have been made for the bremsstrahlung spectrum [53, 61, 69, 71-73]. The results of such calculations (for example those reported in [72, 73] in a convenient graphical form) can be used to determine, for a wide range of temperatures from 50 eV to 100 keV, the spectral distribution of the radiation transmitted by filters of different thickness and the proportion of this radiation relative to the bremsstrahlung integrated over the whole spectrum.

When calculations carried out for pure bremsstrahlung spectra are used in the diagnostics of laser plasmas containing multiply charged ions, which have high ionization and excitation potentials when z is fairly high, it is necessary to pay special attention to recombination radiation, line spectrum, and optical thickness of the plasma. For example, the applicability of the absorber method to carbon jets with $T_e \sim 200$ eV is discussed in [54]. Estimates obtained using the formulas in [74] show that the radiation passed through a beryllium filter 100 μ thick is dominated by the recombination component due to transitions to the ground state of C VI and, since jumps in the recombination radiation and line spectrum lie in the strong-absorption region of the filter, the recorded radiation has the same dependence on the wavelength as the bremsstrahlung spectrum.

The radiation transmitted by filters is usually recorded with scintillators, which transform x rays into visible radiation, and ordinary photomultipliers (see, for example, [75]). The deexcitation time of plastic scintillators based on polystyrene with admixtures of p-type terphenyl and POPOP is 2.0 ± 0.2 nsec [76]. The linearity of the conversion of x-ray energies into the radiation recorded by photomultipliers is a subject of special importance. At energies exceeding 20 keV, the energy yield of plastic scintillators is exactly proportional to the energy of hard quanta. In the 1 keV range there may be deviations from the linearity in the conversion process. However, according to the results reported in [77], in narrow intervals amounting to several kiloelectron-volts in the range $\lambda \leqslant 10$ Å the dependence of the light yield of a scintillator on the energy of the incident x-ray quanta can be regarded as linear.

Typical measurements of relatively low temperatures $T_e \sim 30$ eV of an aluminum plasma were reported in [60], where the radiation in the 20-100 Å range was recorded using an open-window photomultiplier (see, for example, [78]) with a tungsten photocathode. Painstaking calculations were carried out in [60] in order to allow for the recombination radiation of Al IV-IX ions and for the spectral dependence of the quantum efficiency of the photocathode.

3.2. Time Dependences of T_e in Laser Jets

The time dependences of the intensity of soft x-ray radiation and of temperature, carried out using the absorber method, yield information which is very important for the determination of the dynamics of laser plasmas. Measurements of this kind may yield the lifetime of the hot zone of a jet as a function of the parameters of the heating radiation and they can also give time dependences of the temperature which—together with the pressure and density—governs the transient state of a laser plasma.

The short duration of the heating radiation pulses and the brief lifetime of the plasma impose stringent requirements on the time resolution of the recording apparatus ($\sim 10^{-9}$-10^{-10} sec). Morevoer, it is necessary to record several processes simultaneously and to determine accurately the time relationships between them. A method described in [79] involves the use of six oscillographic recording channels for the determination of the time relations between various processes to within (2-3) $\cdot 10^{-10}$ sec. Fast-response high-current photomultipliers of the ÉLU type (linear current up to 2.5 A, gain 10^7-10^8, time resolution of edge 1.5 nsec, time resolution of width at midamplitude 2 nsec) ensure that the signals have sufficient amplitudes for the absorber method in the nanosecond range.

In the gasdynamic regime and for laser pulses of $\Delta t_l \sim 10^{-8}$ sec duration the shapes of the laser radiation and x-ray emission of the plasma are identical and the maxima coincide [55] (Fig. 3). The experimental results show that the x-ray radiation intensity "follows" the rate of supply of energy by the laser radiation. A similar time dependence is exhibited also by the visible radiation in the part of a jet where $0 < r < 0.1$ mm (see §4.1). This is manifested particularly clearly in the case of random fluctuations of the laser pulse shape (spreading of the leading edge, two humped pulses, etc.). Thus, the processes of evaporation, heating, and expansion of the target material occur quite rapidly and they can follow changes in the incident laser power W(t).

When the duration of laser pulses is reduced to $\Delta t_l \sim 2$ nsec ($q \sim 10^{14}$ W/cm^2) the x-ray radiation pulses also become shorter and comparable with the time resolution of photomultipliers (Fig. 3). In this case the electron temperature T_e of a plasma produced from a $(CD_2)_n$ targets is ~1 keV [56]. The absence of x-ray radiation after the end of a heating pulse shows that in the gasdynamic case the lifetime of the hot part of the plasma is $\Delta t_h \sim 10^{-9}$ sec.

A special feature of the experimental results reported in [55] is the constancy of T_e, to within ±15%, in a time interval of ~25 nsec, during which the power of the laser radiation W(t) varies by more than one order of magnitude. The apparent conflict between this observation and the theoretical dependence of the maximum temperature in the core of a laser jet on the incident flux density q under gasdynamic conditions, which is $T \propto q^{4/9}$ [see Eq. (2.2) and Fig. 4], is eliminated by the use of the results of measurements of the spatial-temporal struc-

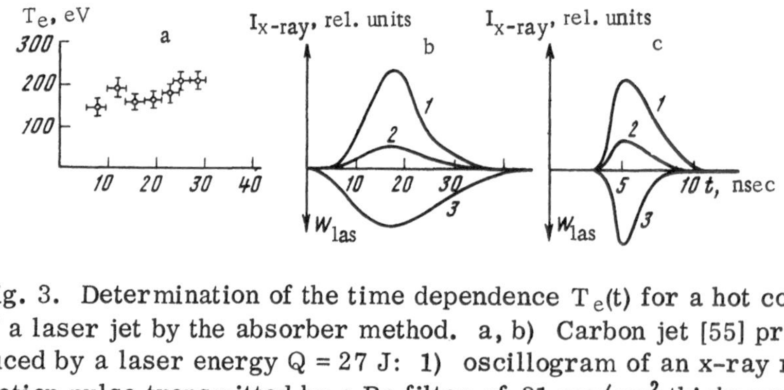

Fig. 3. Determination of the time dependence $T_e(t)$ for a hot core of a laser jet by the absorber method. a, b) Carbon jet [55] produced by a laser energy Q = 27 J: 1) oscillogram of an x-ray radiation pulse transmitted by a Be filter of 31 mg/cm^2 thickness; 2) same oscillogram, Be filter of 15.5 mg/cm^2 thickness; 3) time dependence of the laser power W(t). c) $(CD_2)_n$ target, $T_e \sim 1$ keV [56]: 1) oscillogram of an x-ray pulse transmitted by a Al filter of thickness 17.2 mg/cm^2; 2) same oscillogram, Al filter of thickness 34.4 mg/cm^2; 3) W(t), $W_{max} \approx 10$ GW.

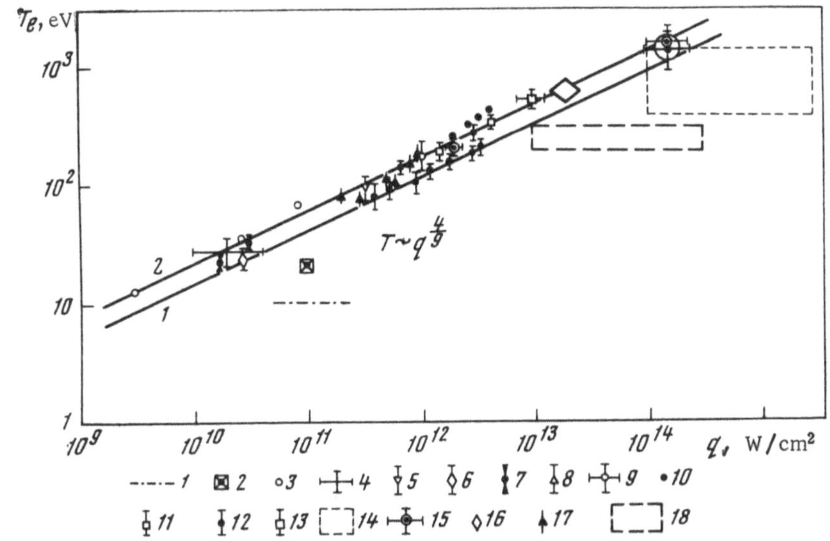

Fig. 4. Summary of experimental determinations of the electron temperature T_e of laser plasmas. The designations and references are given in Table 3. The theoretical dependences $T_e(q)$ are the continuous lines [see also Eq. (2.2) in the present paper)]: 1) hydrogen target [40]; 2) carbon target [64].

ture of the radiation emitted by the laser employed in [80]. The divergence of the laser radiation during a pulse is such that the flux density q remains constant over a considerable part of the focusing spot and this makes T_e constant. The bell shape of the x-ray pulses (and of the visible-radiation pulses including the continuous spectrum near the target and the line spectrum at greater distances) and the corresponding change in the area of the focusing spot show that the amount of the hot plasma in direct contact with the target surface varies with time.

Differences between the time dependences of the divergence of the heating radiation, which are related to the construction of the laser systems, may give rise to different time dependences of T_e. For example, it is reported in [65] that 20-nsec laser pulses give rise to bell-shaped time dependences of T_e with maxima coinciding with the laser radiation density maximum. An analysis of the experimental results yields $T_e \propto q^{4/9}$, which shows that the quasi-steady model of the gasdynamic motion of plasmas is valid.

When a plasma is heated by laser pulses of $\Delta t_l \sim 100$ nsec duration, the time dependence of T_e may be affected by the gasdynamic effects associated with the motion of the absorption zone of laser radiation into the target to a depth exceeding the diameter of the focusing spot (for details see § 4). For example, it is reported in [57, 58, 81, 82] (for heating pulses with a leading edge of $\Delta t_l \sim 50$ nsec duration) that the maximum values of the intensity of x-ray radiation emitted by a plasma formed from solid deuterium and the corresponding maximum temperatures occur 20-30 nsec ahead of the laser pulse maximum. Similar results were obtained in [83], where an interference method was used to determine the time dependence of the gasdynamic pressure exerted by a plasma on the target and the magnitude of this pressure was governed by the temperature of the hot core of the laser jet.

In interpreting such results we must bear in mind that the maximum temperature in a laser jet, measured in experiments just mentioned, may depend not only on the laser flux density q but also on the electron density gradient ∇N_e near the point of reflection of laser radiation where the electron density has the critical value N_{cr} (for neodymium laser radiation this value is $N_{cr} = 10^{21}$ cm^3). This is due to the fact that when the bremsstrahlung absorption (inverse bremsstahlung) coefficient is propotional to N_e^2, the rate at which energy is supplied to one particle governs the temperature rise and is proportional to N_e. A reduction in ∇N_e [a less steep profile $N_e(r)$] corresponds to an increase in the optical thickness of the plasma so that the radiation is absorbed at lower values of N_e and, therefore, it heats the plasma to a lower temperature [84]. This explanation is supported by the experimental results in [83], which show that ∇N_e decreases before reduction in the pressure of the plasma on the target. Clearly, the fall in ∇N_e before the laser pulse maximum is observed also under the conditions used in [57, 81, 82] because during the leading edge of a laser pulse the absorption zone of the heating radiation is displaced into the target to a depth of ~1 mm. This depth is considerably greater than the diameter of the focusing spot $2r_0 \sim 0.1$-0.2 mm, as a result of which the motion of the plasma in a narrow channel bounded by relatively cold dense walls in the target may become one-dimensional and the profile $N_e(r)$ may become less steep and this reduces T_e.

3.3. Deviations from Maxwellian Equilibrium

The absorber method is based on the concepts which are usual for a thermalized plasma and in which the plasma temperature is related to the Maxwellian distribution of electrons. In principle, when the heating radiation density is sufficiently high, we may encounter such transient phenomena as the deviation of the electron distribution function from the Maxwellian form and departure of the average thermal energy of electrons from that corresponding to the ionization state of the plasma [85-88].

The first experimental data on the occasional deviation of the electron distribution function from the Maxwellian form are reported for laser plasmas in [24, 62]; these data were obtained by the four-channel method of recording the x-ray spectrum in the range ~1-4 keV. Use was made of plastic scintillators and beryllium filters 25, 50, 100, and 200 μ thick. The time resolution of the system was ~10 nsec. The experimental points for a carbon jet fitted well straight lines corresponding to an equilibrium Maxwellian plasma with $T_e \sim 120$ eV and $T_e \sim 160$ eV. A selection of six pairs out of four recorded relative intensities yielded six values of T_e and in the case of carbon all these values differed by not more than 10% from the average.

The situation was found to be quite different for deuterium (and hydrogen): a straight line corresponding to a definite temperature could not be drawn through the experimental points and the different pairs of intensities yielded temperatures increasing from 200 to 700 eV with the filter thickness. The presence of hard (x-ray) quanta indicated that the distribution function of electrons had a considerable non-Maxwellian "tail." Thus, reliable diagnostics of deuterium laser plasmas requires the use of multichannel variants of the absorber method. Nevertheless, using two filters transmitting the radiation near the maximum of the continuous spectrum, we can determine the average energy of the majority of electrons, as was done in [24] (beryllium filters 25 and 75 μ thick, deuterium plasma, $q \sim 10^{13}$ W/cm^2, $T_e \sim 500$ eV, $\Delta t_l \sim 10$ nsec).

The spectral distribution of the radiation emitted by (CD_2) and Fe plasmas in the photon energy range E = 1-12 keV was analyzed in [56] for laser parameters $Q \sim 30$ J, $\Delta t_l \sim 2$ nsec, $q \sim 10^{14}$ W/cm^2 (the absorbers were stacks consisting of 1-8 layers of aluminum foil 16 μ thick). In the case of the Fe plasma the radiation of all energies corresponded to a Maxwellian electron distribution with $T_e \sim 0.8$ keV. In the case of the $(CD_2)_n$ plasma there was a deviation from the equilibrium case: when E was increased from 2 to 10 keV the spectral temperature $T_{sp}(E)$, deduced from the slope of the derivative of the spectral distribution of the plasma radiation (plotted using semilogarithmic coordinates), increased from 0.45 to 8.0 keV.

These results are still tentative and a quantitative interpretation must await additional experiments and theoretical studies. We shall now discuss the current state of the knowledge of this subject.

The distribution of electron energies F(ε, t) in a laser plasma at a moment t cannot be described by the usual formula for the "maxwellization time" of a free plasma due to electron–electron collisions [89]

$$t_{ee} = \frac{0.26T_e^{3/2}}{N_e \ln \Lambda}, \tag{3.1}$$

which gives $t_{ee} = 2.6 \cdot 10^{-13}$–$8 \cdot 10^{-12}$ sec for the case $N_e = 10^{20}$ cm^{-3} of interest to us; the electron temperature is $T_e = 10^6$–10^7 °K and the Coulomb logarithm is $\Lambda \sim 10$. Thus, in the presence of strong laser radiation in a plasma the transport equation for F(ε, t) includes not only the terms due to the electron–electron collisions but also the contribution of the laser radiation field and of inelastic collisions between electrons and ions. A characteristic parameter of this problem is the rate of loss of the electron energy due to ionization and the rate of acquisition of energy from the optical field [86]

$$\beta_0 = \frac{\chi(z)\,\nu_i(z)}{\varepsilon_0 \nu_{\text{eff}}(z)}, \tag{3.2}$$

where $\chi(z)$ is the ionization potential of the ions present in the plasma, $\varepsilon_0 = 4\pi e^2 q/m\omega^2 c$ is the energy of an electron oscillating in the field of a light wave of frequency ω; $\nu_i(z)$ and $\nu_{\text{eff}}(z)$ are the frequencies of inelastic and elastic collisions.

If $\beta_0 \gg 1$, i.e., in the range of low values of q, the electron distribution function terminates at energies ε close to $\chi(z)$ because the ionization processes are due to the "tail" of the distribution function. Then, the influence of the terms due to the laser radiation and inelastic collisions on the transport equation for F(ε, t) is compensated and, because of the electron–electron collisions, the function F(ε, t) is nearly Maxwellian. Clearly this is the situation observed experimentally in [62] for carbon because for $q = 10^{12}$ W/cm^2, $T_e = 150$ eV, $\chi(z) = 500$ eV, $z = 6$, $\varepsilon_0 = 0.2$ eV, and $N = 10^{20}$ cm^{-3} we find that Eq. (3.2) yields $\beta_0 \sim 1$.

If $\beta_0 \ll 1$, i.e., in the range of relatively high values of q, targets consisting of heavy elements whose atoms have a sufficient "reserve" of electrons exhibit nonequilibrium ionization for which the electron distribution function is nearly Maxwellian.

Nonequilibrium radiation emitted by a deuterium jet for $q \sim 10^{12}$ W/cm^2 [24] corresponds to a completely ionized plasma for which $\beta_0 = 0$. There is as yet no published theoretical analysis of this case and in this analysis the transport equation for F(ε, t) should include contributions due to electron–electron collisions and acquisition of energy from the optical field. The published work is confined to a solution of the transport equation for a growing electron avalanche during the initial stage of formation of a plasma when allowance for the acquisition of energy from the optical field is sufficient because of the initially low electron density [85]. In this case the distribution function has a considerable non-Maxwellian "tail."

Various mechanisms of the absorption of radiation in plasmas which are not associated with the inverse bremsstrahlung effect have been considered recently in the literature [90, 96]. These effects are particularly important at high flux densities q. In some cases allowance for these effects alters the nature of the transport-equation terms describing the acquisition of energy from the optical field and the distribution function may then deviate from the Maxwellian form [95, 96].

We shall conclude this subsection by pointing out that irradiation of a solid deuterium target with a single picosecond pulse of density $q \sim 10^{15}$–10^{16} W/cm^2 produces a plasma which emits x-ray radiation that corresponds (in contrast to the case when $q \sim 10^{12}$ W/cm^2) to a Maxwellian electron distribution function [24, 62].

We should note also that hard x-ray quanta of energies 100 keV or higher may be formed. These quanta were recorded in [20] using a scintillation probe covered by thick aluminum

(1.5 cm) and copper (0.5 cm) filters [$q \gg 10^{14}$ W/cm^2, carbon and (CD_2) targets]. In all probability these x-ray (strictly speaking γ-ray) quanta were knocked out from the chamber walls by fast electrons since there was no significant reduction in the radiation intensity when a thick lead filter (known to absorb γ-rays) was placed between the detector and the focusing spot. Similar results were also reported in [25].

3.4. Dependence $T_e(q)$

Figure 4 and Table 3 summarize the experimental determinations of the electron temperatures T_e of laser plasmas. The wide range of heating radiation densities (extending over almost five orders of magnitude) makes it possible to compare the experimental results with the main conclusions that follow from various theoretical models.

Since all the plasma heating experiments have been carried using Q-switched lasers, the information presented in Fig. 4 applies largely to the gasdynamic regime. This figure includes also the theoretical dependences T(q): curve 1 represents a hydrogen target illuminated with light focused into a spot of $r_0 = 10^{-2}$ cm radius [40] [see Eq. (2.2) in the present paper]; curve 2 represents a carbon target with $r_0 = 1.5 \cdot 10^{-2}$ cm [Eq. (1) in [61]]. The agreement between the absolute values of the theoretical and experimental temperatures can be regarded as good particularly in view of the following two points. First of all, the temperatures deduced from the gasdynamic laws of conservation under steady-state conditions represent averages over the volume of the hot core. This approach ignores the experimental time dependences of such quantities as the electron density profile $N_e(r)$, diameter of the focusing spot, etc. The second point is related to the failure of most of the methods employed to give any spatial resolution and even the temporal resolution may be poor. The use of the short-wavelength parts of the emission spectra in the measurements ensures that the experimental values of T_e apply to the dense hot part of a jet but again these values are subject to temporal and spatial averaging. In addition to these methodological errors, there is also some indeterminacy in the values of q which is due to the variation of q during a laser pulse and also due to the fact that the diameter of the focusing spot is not always given in the published literature. Therefore, some of the values of q given in Fig. 4 are estimates based on the usual values of the divergence of laser radiation.

The experimental points for the gasdynamic regime fit well the dependence $T \propto q^{4/9}$ throughout the range of variation of q, which shows that Eqs. (2.1)-(2.5) give reasonable values of the parameters of the hot core.

In some investigations [58, 64, 82] the experimental values of the electron temperature (these values are included in Fig. 4) are interpreted using a one-dimensional model of gasdynamic motion of the plasma in which the zone of absorption of laser radiation moves into the target following a shock wave. This model predicts the dependence $T \propto q^{2/3}$ and it clearly applies to the case of long laser pulses (~50-100 nsec), when the absorption zone moves into the target to a depth several times greater than the diameter of the focusing spot, which ensures that the gasdynamic motion is one-dimensional. However, in the case of shorter pulses (this applies to the majority of the results given in Fig. 4 and Table 3), the dependence is $T \propto q^{4/9}$.

When q is increased in the gasdynamic case, the dependence $T \propto q^{4/9}$ remains valid as long as the electron density at the boundary of the hot core of a jet, given by Eq. (2.4), is below the density which is critical for given heating radiation [97, 98]. The value of q corresponding to this case is ~10^{14} W/cm^2 for $r_0 = 10^{-2}$ cm. As reported in [97, 98], when $q > q_{cr}$, the dependence should be $T \propto q^{2/3}$. This result is obtained by analyzing the balance of the energy flow across the boundary of the hot core employing the relationship $q \propto N_{cr} T^{3/2}$, which applies when the absorption coefficient of the plasma depends weakly on temperature but this is not generally true in the cases considered (see, for example, [99]). When $q > q_{cr}$, we have to make a more careful theoretical analysis allowing for the reflection of the heating radiation from the surface where the density is critical. Preliminary experimental results obtained in [98] by methods which give indirect estimates of the temperature demonstrate that the dependence T(q) changes when q is increased but no further information is yet available.

We shall now make some comments about the results plotted in Fig. 4. The temperatures obtained in [58, 81, 82] for flux densities $q \sim 10^{12}$ W/cm^2 are approximately 1.52-2 times higher than the average values for the same range of q. This may be due to the nonequilibrium nature of the distribution function of electrons in a deuterium jet because the beryllium foils of density 10.2 and 19.75 mg/cm^2 used in these investigations [81] remove a part of the spectrum quite far from the maximum.

Another interesting point is the fact that the temperatures are basically the same for targets ranging from hydrogen to tungsten. At first sight this seems to be strange because an increasing proportion of the laser energy should be used in the ionization of ions with higher values of the charge z: for example, in the case of carbon at $T_e \sim 100$ eV these losses are ~1 keV per one completely bare nucleus. However, one should note that these losses are compensated by an increase in the energy of laser radiation per ion because of the proportionality of the absorption coefficient of the plasma to z^2. Estimates obtained for the gasdynamic regime in [100] on the basis of the energy balance in the hot core of a jet demonstrate that T_e is not very sensitive to the charge and mass of ions formed in the core.

The experimental data (see [24, 101]) on the inertial confinement regime ($\Delta t_l \ll \Delta t_h$) are clearly insufficient for plotting any dependences or for making comparisons with the theory. Instability of the operation of mode-locked lasers and the absence of reliable information on the time profiles of picosecond pulses make the values of the flux density uncertain (see Table 3 and Fig. 4). Moreover, the temporal resolution of the modern methods of measuring T_e is insufficient for investigating the processes of transfer of the electron energy from that relatively small portion of the plasma which is heated directly by laser radiation in $\Delta t_l \sim 10^{-12}$-10^{-11} sec to the surrounding target material by electronic heat conduction or the processes of energy transfer from electrons to ions in a time $\Delta t_h \sim 10^{-9}$ sec.

§ 4. Gasdynamic Motion of Laser Plasmas

Since the gasdynamic motion is the main cause of cooling of laser plasmas, investigations of this motion are of great importance in the determination of the nature of the phenomenon of cooling as a whole.

The results of experimental investigations of the kinetics of formation and gasdynamic motion of the hot core of a laser jet given in the present section can be used to determine the spatial-temporal distribution of the principal gasdynamic parameters (temperature T, velocity u, density ρ) and the degree of ionization of expanding plasma. It should be stressed that information of this kind can be obtained only using methods with a sufficient spatial-temporal resolution for the observation of the different phases of motion and acceleration of plasma (including the initial expansion stage) and this resolution is required first during the action of the heating radiation, and second at distances close to the target comparable with the diameter of the focusing spot. The ultrafast spectroscopic methods are of special importance because they make it possible to follow the "fate" of the various parts of a jet with different temperatures, which cannot be done using schlieren or interference methods.

4.1. Spatial-Temporal Structure

The spatial and temperature structure of an expanding plasma in a laser jet formed as a result of gasdynamic motion was investigated in greatest detail in the flux density range $q \sim 10^{11}$-10^{12} W/cm^2 in [51, 79, 102, 103] using the radiation emitted by the plasma in the form of spectral lines of carbon ions with different degrees of ionization. The target materials were solid carbon and polyethylene CH_2 because at temperatures $T \sim 100$ eV, which were achieved using lasers with an output power of 0.5-1 GW, the carbon atoms lost all their electrons becoming hydrogen-like ions C VI. The considerable information available on the spectra of carbon ions of lower degrees of ionization made it possible to investigate also the colder parts of a jet because of the large difference between the ionization potentials χ_i (for example, in the case of C VI the potential was $\chi_i \sim 490$ eV, whereas for C III it was $\chi_i \sim 48$ eV) ions with difference degrees of ionization formed at very different temperatures.

The characteristic geometric dimension—the radius of the focusing spot—was $r_0 \sim 0.1$ mm for the experiments discussed in the present section. At distances $r > 2$ mm the time of appearance (relative to the moment of arrival of a heating radiation pulse on the target) and the shape of a pulse representing emission of spectral lines of C I-VI ions in the visible part of the spectrum were recorded using photomultipliers (see [51, 79]). The use of an image amplifier [102, 103] ensured a spatial resolution of ~0.05 mm in the plane of the jet and a temporal resolution of 0.5 nsec; the use of this amplifier made it possible to observe the various phases of the motion and acceleration of the plasma at distances from the target comparable with r_0 (Fig. 5). The results of an analysis of oscillographic and image-amplifier measurements are plotted in Fig. 6 in the form of space-time (rt) diagrams showing the expansion of ions of different degrees of ionization. Figure 6 shows not only the time dependence of the intensity of the lines and the continuous spectrum but also the corresponding dependences of the power of the laser radiation absorbed by the plasma.

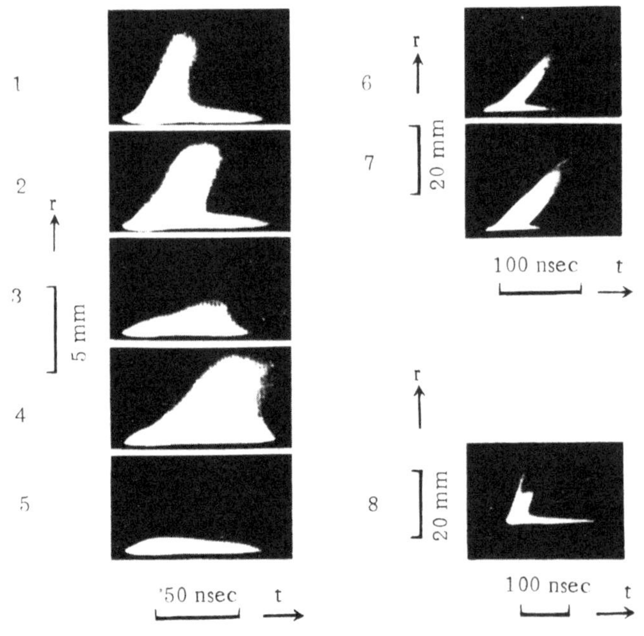

Fig. 5. Typical time scans of the radiation emitted by a carbon jet obtained in the light of the carbon lines [102, 103]: 1) C VI, λ = 5292 Å; 2) C V, λ = 4952 Å; 3) C IV, λ = 5801.51 Å; 4) C III, λ = 4663 Å; 5) continuous spectrum, λ = 5000 Å; 6) C VI, λ = 5292 Å; 7) C V, λ = 4952 Å; 8) C VI, λ = 5292 Å. Frame 6, corresponding to the leading edge of a laser pulse, shows a shock wave in a carbon vapor atmosphere recorded in the light of the C VI ions.

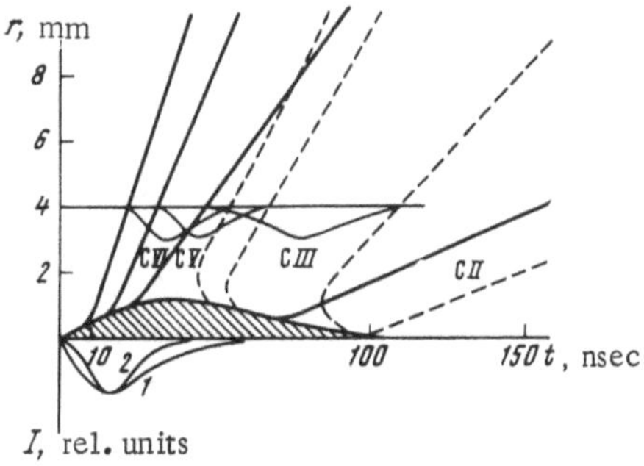

Fig. 6. Expansion of a laser jet illustrated by the rt diagrams obtained in the light of the C II-VI lines (compare with Fig. 5) [102, 103]. The continuous curves correspond to the beginning of the emission whereas the dashed curves correspond to the end of the emission. The figure includes also the time dependences of the intensities of the lines at a distance r = 4 mm and of the continuum in the region $0 < r < 0.1$ mm (1), and of the laser radiation power (2). The intensity of the continuum in the shaded region exceeds the line intensities.

It should be noted that the intensity of the continuous spectrum emitted by the hot core of a jet ($0 < r < r_0 \sim 0.1$ mm) in the visible region was fully correlated with the time dependence of the laser pulse power W(t) (in exactly the same way as the x-ray emission—see § 3 and Fig. 3). This result [79, 102, 103] was in conflict with the results in [44, 51], which indicated that the maximum of the continuous visible radiation occurred at the end of a laser pulse. This was due to the fact that the plasma radiation was recorded in [44, 51] with a poor spatial resolution, i.e., the radiation examined was generated in a large part of the jet. A special experiment [59] confirmed this explanation. We have discussed this point in order to stress the presence of a strongly emitting core of $r \sim 0.1$ mm size, whose parameters are governed in practice by the instantaneous value of the laser pulse power.

The luminous front of the hot plasma near the target was found to travel at a velocity $\sim (0.5\text{-}0.6) \cdot 10^7$ cm/sec. During the next ~10 nsec, over distances up to $r \sim 1$ mm from the target, the front accelerated to an asymptotic velocity of $\sim 3 \cdot 10^7$ cm/sec. The high optical thickness of the jet in the continuous visible part of the spectrum at distances of $0 < r < 0.75\text{-}1$ mm suppressed the line spectrum. However, at greater distances ($r > 1$ mm) the rt diagrams of the flight of different ions (for example, C V and C VI) could be followed right up to $r \sim 20$ mm. The velocities of propagation of the beginning, maximum, and end of emission of the C V and C VI lines were practically identical over all distances $r > 1$ mm, and the shapes of the pulses representing the C V and C VI lines were identical (like the radiation emitted by the jet core) with the shape of the laser pulses when accidental fluctuations occurred in the latter. The rt diagram passing through the points of the maximum intensity of the continuum near the target surface and of the maximum of the radiation represented by the C VI lines for all values of r repeated exactly the rt diagram of the front, confirming that the hot phase of the jet traveled in a quasisteady manner.

The asymptotic velocity of the radiation front of the C III ions was considerably less than that of C VI and it amounted to $\sim 1.3 \cdot 10^7$ cm/sec; the C III ions escaped the region with $r < 1$

mm after the end of the laser pulse. The C II ions appeared ~100 nsec after the end of the laser pulse and they represented the relief of the load on the target after the expansion of the hot phase of the jet.

An analysis of the ionization equilibrium indicated that the rt diagram of the radiation emitted by ions represented the mass motion of the plasma layers consisting mainly of the ions in question. Time scans of the C VI and C V lines, shown in frames 1 and 2 in Fig. 5, represented the flight of the bare C^{6+} nuclei and of the C VI ions, respectively. This was due to the fact that the upper levels of the transitions of the corresponding C VI and C V lines were located very close to the ionization limit so that they were in equilibrium with the continuum [104] and their populations were governed by the densities of ions of the next higher degree of ionization (ionic charge). However, the rt diagrams of the radiation emitted by ions with a lower charge (Fig. 6) represented the flight of the ions themselves. A notable point was the fast fall of the intensity of the λ = 5801.51 Å line of C IV with increasing r (frame 3 in Fig. 5). This occurred because the upper level of the $3s^2S-3p^2p^0$ transition corresponding to this line had a very short lifetime (~0.2 nsec) due to the high probability of the decay to the ground state $2s^2S$ [105]. However, the population of the level $3p^2p^0$ from higher levels and from the continuum (i.e., from the ground state of C V) was a much slower process because the energy gap between this level and higher levels was considerably greater than in the case of C VI and C V.

4.2. Quasisteady Nature of Gasdynamic Expansion of Plasmas

If the laser pulse duration exceeds considerably Δt_h (which is true if $\Delta t_l \sim 10^{-8}$ sec, $q \sim 10^{12}$ W/cm^2, $r_0 \sim 0.1$ mm), a nearly steady motion is observed in the gasdynamic regime during certain stages of heating.

The quasisteady motion is understood in the sense that a plasma leaves the hot zone absorbing laser radiation in a time Δt_h much shorter than the laser pulse duration Δt_l. A hot plasma begins its motion at the surface of a target with a velocity of $u \sim 0.6 \cdot 10^7$ cm/sec and the lifetime of a plasma in the hot core of a jet can be found from Eq. (2.6): $\Delta t_h \sim 1.5$-3 nsec $\ll$ $\Delta t_l \sim 40$ nsec. It follows from the experimental results that when a plasma leaves the region $r < r_0$, it is accelerated and its motion becomes supersonic. Therefore, its subsequent "fate" does not affect in any way the region of subsonic flow (the low density of zones II and III, due to the lateral expansion of the plasma, makes the absorption coefficient extremely small so that laser radiation reaches the hot core of a jet without any significant absorption on the way). The rt diagrams representing the flight of "elementary volumes" of the plasma are similar.

The quasisteady nature of the plasma flow can also be described differently because there is only a slight change in the mass during the time that gasdynamic conditions exist in a given "elementary volume" (i.e., the thermal energy of the plasma is comparable with the energy of the directional motion and the transformation of the energy into the kinetic energy can change the latter significantly).

We shall now establish the relationship between the experimental observations discussed in the preceding subsection and the complex spatial-temporal structure of laser radiation in the far zone, which governs the variation of the error of the focusing spot $S \approx d^2$ on the target surface. As pointed out in § 3, this variation may ensure a practically constant laser radiation flux density in spite of the bell-shaped time dependence of the laser power W(t), and this is responsible for the constancy of the maximum temperature ($T_e \sim 100$-120 eV for $q \sim 10^{11}$ W/cm^2). When the temperature of the hot core varies weakly throughout a laser pulse, ions with a charge corresponding to this temperature are formed in the core. At $T_e \sim 100$ eV these ions are C^{6+} and C VI. Then, the time dependence of the rate of formation of the C^{6+} and C VI ions and of the flux of these ions from the hot core should be analogous to W(t), i.e., should have the same bell shape as the laser pulse. Since these ions do not recombine in the region where the plasma is accelerated, the bell-shaped radiation of C V and C VI is observed experimentally at distances r > 2 mm. The constancy of the shape of the radiation pulse emitted by these ions, observed with the aid of photomultipliers up to distances of r ~ 10 mm, is evidence that the time dependence of the mass remains constant at these distances, i.e., the spherical flow of mass through surfaces bound by a single solid angle remains constant.

Experimental confirmation of the adiabatic nature of the expansion in zones II and III is provided by measurements of the distributions of the density N_e and temperature T_e of electrons in the parts of a jet containing C^{6+} and C VI ions [51]. According to the results of these measurements the expansion of a plasma from distances r ~ 0.5 mm to r ~ 5 mm is characterized by $T_e \cdot N_e^{-2/3}$ = const, which corresponds to adiabatic expansion of the electron gas.

4.3. Anisotropy of Flight of Ions Out of a Laser Jet

Figure 7 gives information on the angular distribution of ions with high charges escaping from a carbon jet [59]. In the angular range $\varphi = 0$–$60°$ (measured from the normal to the target) the leading edge of the radiation emitted by the C VI ions coincided exactly with the edge for $\varphi = 0$, which demonstrated that the flow of the C^{6+} ions was symmetric during the first half of the laser pulse (at least in a sector of ±60°). However, the duration of the radiation of the C VI ions observed at distances r > 1 mm from the target decreased smoothly between $\varphi = 0$ and 60°. The trailing edge of the radiation was similar to the leading edge. The radiation of the C III ions, which appeared after the end of the laser pulse, was the same in the range φ = 0–90° as in frame 4 of Fig. 5, which was evidence of the spherical symmetry of the flight of the C III ions.

Similar results were reported in [106-108] where an investigation was made of the asymptotic structure of a jet after the end of a laser pulse. According to spectroscopic investigations [106, 107], the radiation emitted by the C VI and C V ions was localized in a cone whose vertex was located at the focusing point and whose axis was the laser beam. Mass-spectrometric investigations [108] also demonstrated that the higher the ionic charge, the smaller was the solid angle within which the flow of the ions took place.

We shall now discuss the experimental results of investigations of the expansion of a jet along directions oriented at various angles relative to the target normal. Zones with temperatures considerably lower than the maximum value appear during a laser pulse at the periphery of the focusing spot in the plane of the target. Formation of fairly large numbers of ions with low charges (for example, C III ions with ionization potentials an order of magnitude lower than the ionization potential of the bare carbon nuclei C^{6+}) in these zones prevents the escape of ions with higher charges from the hot central region at angles $\varphi > 60°$ over distances $r \gg r_0$

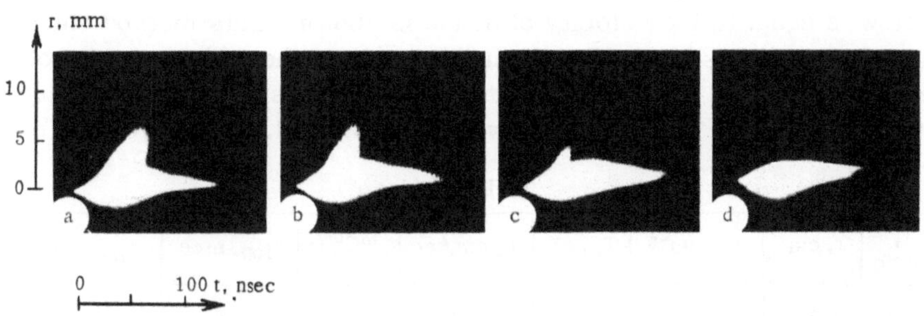

Fig. 7. Typical time scans of the expansion of a carbon jet observed in the light of the C VI line (λ = 5292 Å) for different angles φ between the monochromator slit and the normal to the target: a) $\varphi = 0°$; b) $\varphi = 40°$; c) $\varphi = 60°$; d) $\varphi = 80°$.

for which the whole focusing spot can be regarded as a point source. Then, the spatial distribution of, for example, the C^{6+} and C III ions on the target surface transforms into a distribution between the flight sectors. A displacement of the hot zone into the target during a laser pulse may play a definite role. After the end of a laser pulse the temperature of the hot core falls strongly and the C III ions escape in a spherically symmetrical manner. By this time the C^{6+} and C VI ions have already departed from the target surface and have become localized in a conical cluster moving at a constant velocity of $\sim 3 \cdot 10^7$ cm/sec.

The appearance of a low-temperature "halo" at the periphery of the focusing spot may be due to the fall of the flux density of the heating radiation caused by aberrations of the focusing lens. It follows from estimates of the diffusion of ions and electronic thermal conductivity [89] that in this case ($q \sim 10^{11}$ W/cm^2, $T_e \sim 100$ eV) the characteristic escape time of ions from the hot zone is too short for a significant equalization of the temperature and mixing of ions with different charges in different parts of the focusing spot.

We shall conclude this discussion by pointing out that investigations of the flight of highly charged ions provide information which is very important in the applications of laser jets as sources of multiply charged ions in accelerator technology.

4.4. Kinetics of the Relaxation of the Ionization State of Expanding Plasmas

According to [109, 110], the ionization state of a plasma expanding in a vacuum is "frozen" as a result of time dependences of the gasdynamic parameters of an "elementary volume" of the plasma which ensure that the recombination time of ions is considerably greater than the characteristic times of the gasdynamic expansion process.

The kinetics of the relaxation of the ionization state can be followed in our case by considering an "elementary volume" of a plasma and obtaining the time dependences of the distance from the target $r(t)$, velocity of expansion $u(t)$, electron temperature $T_e(t)$, and electron density $N_e(t)$ [59]. Table 4 gives the parameters of such an elementary volume as a function of time from the moment of evaporation and heating of the target surface. The dependences $r(t)$ and $u(t)$ are based on the experimental data plotted in Figs. 5 and 6. The time dependence of the electron density $N_e(t)$ is obtained from the equation of conservation of mass in steady-state spherical flow of an ideal gas; carbon plasma is assumed to be fully ionized with an ionic charge $z = 6$. The profile $N_e(r)$ can be related to the experimental results by substituting $N_e = 4 \cdot 10^{17}$ cm^{-3} at $r = 0.096$ cm, which corresponds to a hemispherical escape of mass at a rate of $\mu = 1.8$ g/sec. In principle, the time dependence of the temperature $T_e(t)$ can be obtained from the supersonic branch of the solution for steady-state expanding adiabatic flow (see, for example, [111]) quite simply if the coordinate of the surface on which the velocity of flow is equal to the velocity of sound is known. This method cannot be used in our case because it is not possible to determine accurately where the velocity of flow becomes equal to the velocity of sound. The experimental data simply show that at distances $r \sim 0.1$-0.2 mm the velocity of the plasma is close to the velocity of sound. This information is clearly

TABLE 4

t, nsec	r, cm	N_e, cm^{-3}	T_e, eV	u, cm/sec	$M_C u^2/2$, eV	τ_{ion}, nsec	τ_{rec}, nsec
2.4	0.013	$8 \cdot 10^{19}$	125	$0.55 \cdot 10^7$	—	0.8	6
4	0.021	$3.3 \cdot 10^{19}$	110	$0.55 \cdot 10^7$	—	4	20
6	0.033	$1.3 \cdot 10^{19}$	90	$0.55 \cdot 10^7$	187	40	40
7.5	0.046	$3.5 \cdot 10^{18}$	60	$1,1 \cdot 10^7$	775	500	110
9	0.066	$1.1 \cdot 10^{18}$	45	$1.66 \cdot 10^7$	1720	—	250
10.5	0.096	$4 \cdot 10^{17}$	35	$2.2 \cdot 10^7$	3000	—	350
12	0,133	$1.6 \cdot 10^{17}$	15	$2.8 \cdot 10^7$	4850	—	400

insufficient because the error in the determination of the velocity of sound by, for example, 20%, would result in a 40% error in the determination of temperature and since the ionization and recombination coefficients depend strongly on the temperature T_e, this approach may result in a basically incorrect description of the relaxation of the ionization state of a plasma. This difficulty can be bypassed using the experimental dependence of T_e on the distance r [32] (see Fig. 2). The values quoted in Table 4 for periods beginning from t = 6 nsec are based on the data plotted in Figs. 5 and 6 and on the results reported in [52]. It is assumed in this table that at t = 2.4 and 4 nsec the velocity is constant and equal to $u = 0.55 \cdot 10^6$ cm/sec. At a distance r ~ 0.01 cm the value of T is based on the absorber method measurements reported in [103]. The acceleration process is represented in Fig. 4 by the kinetic energy of the directional gasdynamic motion per one ion, regarded as a function of time and distance. The ionization time τ_{ion} and the recombination time τ_{rec} for the C VI $\rightleftarrows$ C^{6+} process are taken from [104]. It is clear from Table 4 that during the motion along the trajectory of an elementary volume the recombination time of the C^{6+} ions becomes much greater than the characteristic gasdynamic expansion time.

An experimental confirmation of the absence of the recombination of the absence of the recombination of the C^{6+} ions in this case is provided by the constancy of the ratio of the densities of the C^{6+} and C VI ions in the range r = 1-5 mm; this ratio is determined from the relative intensities of the C VI and C V spectral lines (see § 4.1) [51].

It should be stressed that in spite of the purely illustrative nature of the data in Table 4, the electron density profile given in this table describes correctly the dependence $N_e(r)$ for r → 0.1 mm; this profile is obtained by assuming that expansion is a steady-state process and using the experimental value at r ~ 1 mm. This follows from a comparison of the results in Table 4 with those given in § 5 for the $N_e(r)$ distribution measured near the target surface. Thus, the approach described above fits the real situation encountered in the expansion of a plasma jet.

4.5. Total Number of "Hot" Ions, Mass Loss, Plasma Pressure on Target, Energy Balance

We shall use the ideas put forward above in estimating several parameters of a laser jet and comparing some of them with the results of independent measurements. This not only gives information on the characteristics of a laser jet but may also be used as a check of the model of gasdynamic plasma expansion.

We shall carry out estimates for typical conditions in a carbon jet formed by illumination with laser radiation of $q \sim 5 \cdot 10^{11}$ W/cm² density and Q ~ 5-10 J energy per pulse. The total number of "hot" ions (i.e., of those ions which compose the directly heated core of a jet: in our case, C^{6+} and C VI) can be estimated from the flux of these ions crossing the boundary of the hot core (r ~ 0.1 mm) and the total duration of the laser pulses $\Delta t_l = 4 \cdot 10^{-8}$ sec:

$$n_{tot} = N_i u S \Delta t_l \sim 4 \cdot 10^{15}, \tag{4.1}$$

where $S = 10^{-3}$ cm² is the area of the focusing spot; N_i (r = 0.1 cm) = $N_e/6 = 1.7 \cdot 10^{19}$ cm^{-3} is the density of the C^{6+} ions ($N_e = 10^{20}$ cm^{-3} is taken from Fig. 14); $u = 6 \cdot 10^6$ cm/sec is the gasdynamic velocity at the boundary of the hot core.

Since the results plotted in Fig. 14 are deduced from the Stark broadening of the C VI lines which – together with the C^{6+} ions – form the hot core, we can definitely state that the value $n_{tot} = 4 \cdot 10^{15}$ applies to the hot phase of the jet. Similarly for r = 1.5 mm ($N_e = 1.5 \cdot 10^{17}$ cm^{-3}, $N_i = 2.5 \cdot 10^{16}$ cm^{-3}, $u = 3 \cdot 10^7$ cm/sec, $S = \pi r^2$), we again obtain $n_{tot} \sim 4 \cdot 10^{15}$. We must bear in mind that because of the anisotropy of the expansion as well as because of the temporal and spatial averaging, etc., the absolute values given in the present section are only illustrative and are subject to an error of ± 100%. Nevertheless, the agreement between the total number of the "hot" ions at r = 0.1 mm and r = 1.5 mm justifies the use of the steady-state equations in an approximate description of the gasdynamic plasma motion [43]. Independent measurements of the number of the C^{6+} and C VI ions based on the absolute intensities of the spectral lines of the C VI and C V ions in the range r > 2 mm, reported in [51] for Q ~ 5 J, also give $n_{tot} \sim 2 \cdot 10^{15}$. The mass of the "hot" ions is

$$M = n_{tot} M_C \sim 8 \cdot 10^{-8}\ g, \tag{4.2}$$

where $M_C = 2 \cdot 10^{-23}$ g is the mass of a carbon ion.

The rate of mass transfer across the focal spot in the hot zone of a jet, averaged during a laser pulse, is

$$\mu = \frac{dM}{dt} \sim \frac{n_{tot}}{\Delta t_l} \sim 2\ g/sec. \tag{4.3}$$

It is interesting to compare this value with the results in [83], where a high-speed interferometric method was used to determine the rate of mass transfer which was 4 g/sec at the laser pulse maximum. Bearing in mind that these are estimates of the absolute values, we can regard the agreement as quite satisfactory.

The pressure exerted by the hot core of a jet on the target is estimated to be

$$p = \frac{M u_\infty}{\Delta t_l S} \sim 0.75 \cdot 10^{12}\ dyn/cm^2 \sim 10^6\ atm \tag{4.4}$$

($u_\infty = 3 \cdot 10^7$ cm/sec is the asymptotic velocity of ions in zone III), which is again close to the independent results given in [83].

The pressure of a hot plasma in a laser jet [6] produces a shock wave in condensed matter and this wave travels into the target. The shock-wave velocity D can be estimated from the formula (see Chap. XI in [110])

$$p = \frac{D^2}{V_0}\left(1 - \frac{V}{V_0}\right), \tag{4.5}$$

where $V_0 = 1/\rho_0$ is the volume of a unit mass behind the shock-wave front.

If we assume that a carbon target is compressed so that $V_0/V \sim 1.5$, we find that when $p \sim 0.75 \cdot 10^{12}$ dyn/cm^2 the shock-wave velocity is $D \sim 10^6$ cm/sec, which is close to the rate of burning of a polyethylene foil $(CH_2)_n$ determined in [112] for different laser flux densities (according to the results reported in that paper, the rate of burning is $v_b = 7.5 \cdot 10^5$ cm/sec for $q \sim 5 \cdot 10^{11}$ W/cm^2). Using this value of v_b, we can compare the number of the "hot ions" with the total number of atoms in a target participating in the gasdynamic flow. During a laser pulse the shock wave perturbs a cylindrical region (in the polyethylene target) whose volume is

$$\Delta V = S v_b \Delta t_l \sim 3 \cdot 10^{-5}\ cm^3 \tag{4.6}$$

and which contains $1.3 \cdot 10^{18}$ carbon atoms, a number considerably greater than the number of the "hot" highly charged carbon ions ($n_{tot} \sim (2\text{-}4) \cdot 10^{15}$ or the total number of carbon ions of any kind, found experimentally in [12] and amounting to $\sim 1.3 \cdot 10^{16}$. This large difference can be explained by the model [113] according to which the hot zone of a jet moves into the target at a velocity v_b pushing aside the target material away from the laser beam. The hot phase flows outside through the cylindrical channel formed in this way and the mass of this phase is considerably less than the mass of the condensed matter perturbed by the shock wave.

In the adiabatic expansion of a plasma in the acceleration zone (zone II) a considerable proportion of the absorbed laser pulse energy is converted into the kinetic energy E_{kin} of the directional gasdynamic motion of highly charged ions. Our estimates for "hot" ions give

$$E_{\text{kin}} = n_{\text{tot}} \frac{M_{\text{C}} u_{\infty}^2}{2} \sim 3.2 \text{ J}. \quad (4.7)$$

A calculation of the kinetic energy of carbon ions in all charge states, carried out on the basis of the experimental data reported in [52] for laser pulses of $Q \sim 5$ J energy, gives a similar value of ~ 3.5 J which is subject to an error of $\pm 70\%$. Collisions and cumulation of laser plasma jets [114] may transform the energy of gasdynamic motion directly into the energy of the thermal motion of ions. A streak pattern shown in Fig. 8 and obtained using a method described in [103] (see § 4.1) illustrates the possibility of an increase in the laser plasma lifetime as a result of such processes.

4.6. Time Dependence of the Pressure p(t)

Ultrahigh-speed interferometric photography can be used to determine the time dependence of the plasma pressure p(t) in the hot zone of a jet during a laser pulse [83].

The experiments described in [83] were carried out using flux densities $q \sim 10^{11}$ W/cm^2 and a direct determination was made of the spatial-temporal distribution of the plasma density $\rho(r, t)$ from interferometric data and of the velocity profile u(r, t) from the spectroscopic data (Table 4). Since plasma was accelerated in the direct vicinity of the target, the value of the momentum of the plasma governed uniquely the pressure exerted by the plasma on the target. It was assumed that the plasma was spherically symmetric and that the heating process was gasdynamic. The projection of the plasma momentum on the normal to the surface of the target was assumed to be

$$F(t) = \int\limits_V \rho(r, t) u_n(r, t)\, dV, \quad (4.8)$$

where $\rho(r, t) = N_e M_i z^{-1}$; M_i is the mass of an ion; z is the effective charge; $u_n(r, t)$ is the projection of the velocity on the normal; r is the distance from the center of the focusing spot. The integration was carried out over the whole volume occupied by the plasma.

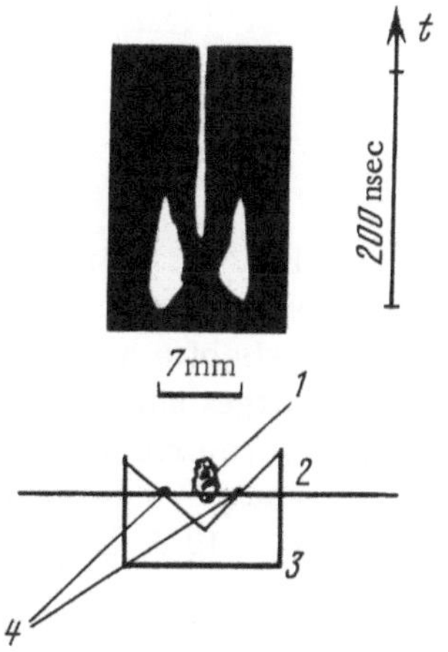

Fig. 8. Time scans showing a collision between two laser jets observed in the light of the C VI line at $\lambda = 5292$ Å; 1) collision region; 2) monochromator slit; 3) carbon target; 4) focusing spots of Q = 2 J laser radiation.

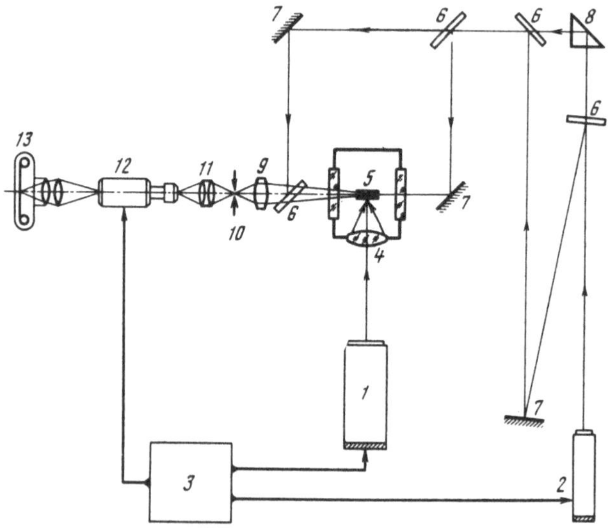

Fig. 9. Schematic diagram of the apparatus used for time scanning of interferograms of a jet with an image converter: 1) neodymium laser; 2) ruby laser; 3) control unit; 4) focusing objective; 5) target; 6) beam splitter; 7) mirrors; 8) rotatable prism; 9) objective for projecting interference image on a slit; 10) slit; 11) objective for projecting image of a slit on the photocathode of an image converter; 12) scanning image converter; 13) photographic camera.

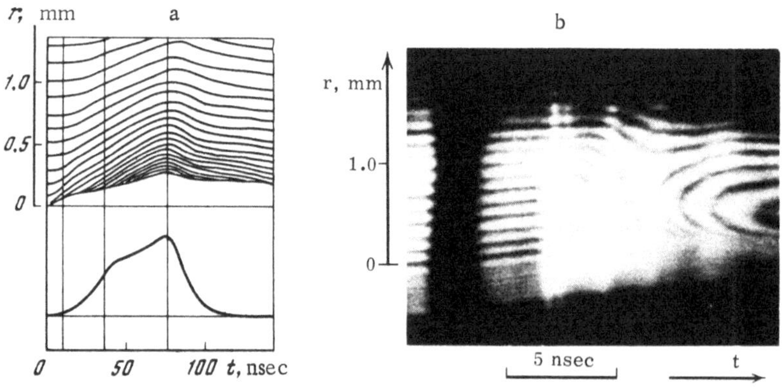

Fig. 10. Time scans of an interferogram of a laser jet. The slit of an image-converter scanning camera coincided with the axis of a neodymium laser beam and was perpendicular to the target surface. Interference fringes in the zeroth position were oriented parallel to the target; r is the distance from the target and r = 0 corresponds to the target surface. a) Carbon target, $Q \sim 10$ J, $\Delta t_l \sim 70$ nsec [83]; b) $(CD_2)_n$ target, $Q = 20$ J, Δt_l = 2 nsec [56].

Hence, the pressure in the heated zone was found from

$$p(\zeta) = (\pi r_0 \tau)^{-1} \frac{dF(\zeta)}{d\zeta}, \tag{4.9}$$

where r_0 is the radius of the focusing spot; τ is the heating time; $\zeta = t\tau^{-1}$ is the dimensionless time.

The apparatus used in these measurements is shown schematically in Fig. 9. A target was heated by a neodymium laser and the plasma was probed with a ruby laser emitting elongated pulses. A slit image of a Mach-Zehnder interferogram was scanned with an FÉR-1 image-converter streak camera. A typical time scan of an interferogram is shown in Fig. 10 and the profiles of the electron density $N_e(r)$ obtained as a result of an analysis of the interferogram are plotted in Fig. 14.

Figure 11 gives the time dependences of the momentum, force, and pressure exerted by the plasma on the target. We can see that the pressure maximum occurred at the beginning of a laser pulse because at low intensities the divergence was weak and the diameter of the focusing spot was governed by the caustic of the focusing lens. At the end of the laser pulse the electron density profile became flatter and the plasma zone where the absorption took place moved away from the target and became larger. The mass of the gas heated directly by laser radiation also increased. The fall of the pressure indicated that a characteristic screening of the laser radiation took place. The temperature in the hot region fell and an increasing proportion of the radiation energy was converted directly into the kinetic energy of the expanding matter. The occurrence of a pressure peak (and, consequently, of a temperature peak) at the beginning of a laser pulse was explained by the results obtained in [57, 58, 81, 82], where a strong x-ray radiation peak emitted from a jet was observed at the very beginning of a laser pulse when the laser radiation power was well below the maximum value (see also § 3.2).

4.7. Absorption Lines in Laser Plasmas

Figure 12 shows a spectrogram obtained in [115] using a normal-incidence vacuum ultraviolet spectrograph. This spectrogram illustrates clearly the features of the gasdynamic motion of a laser jet plasma discussed above. The absorption lines in the spectrogram were due to resonance transitions in the C IV ions and were analogous to the Fraunhofer lines in the solar spectrum. A continuous spectrum was produced by the dense hot core of the jet which was of ~0.1 mm size and which emitted radiation across a cloud of a low-density cold plasma characterized by a high kinetic energy of the directional motion of ions. Since the ions traveled away from the emitting core at a velocity u, the radiation reaching these ions was shifted (because of the Doppler effect) in the direction of longer wavelengths by an amount $\Delta\lambda = u\lambda/c$ and that the emission spectrum of the core did not include photons which, having been shifted toward longer wavelengths by $\Delta\lambda$, corresponded to the resonance wavelength λ_0. Consequently, the absorption lines in the spectrogram were shifted in the direction of shorter wavelengths

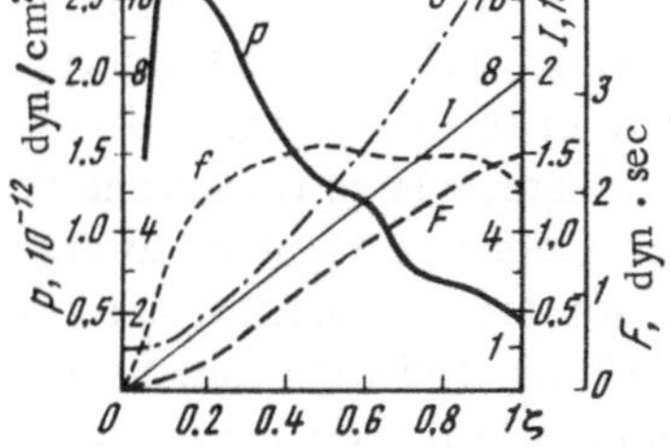

Fig. 11. Time dependences of the plasma pressure p in the hot zone, of the force f acting on the target, and of the momentum f carried away by the plasma. The figure includes also the time dependences of the focusing spot area S and of the radiation intensity T approximated by a straight line at the front of the pulse.

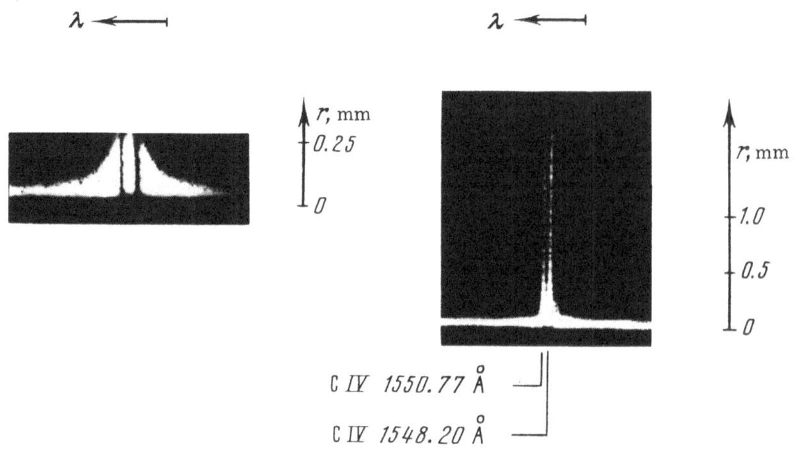

Fig. 12. Spectrogram of a carbon jet recorded using a DFS-29 vacuum spectrograph (the spectra are shown on different scales). Here the coordinate r is the distance from the target. It is clear from the spectrogram that the absorption lines of the cold outer region are shifted by $\Delta\lambda \sim 1.2$ Å relative to the emission lines. The rate of expansion corresponding to this shift is $2.3 \cdot 10^7$ cm/sec [115].

compared with the undisplaced emission lines of the cold part of the plasma ($r > 0.2$ mm), which were due to the C IV ions traveling at right-angles to the line joining the jet core to the spectrograph slit (for details of the experimental geometry the reader is referred to [116]).

At $\lambda_0 = 1549$ Å, when the Doppler shift was $\Delta\lambda \approx 1.2$ Å, the velocity of the directional motion of the C IV ions was $u = 2.3 \cdot 10^7$ cm/sec. This value was in good agreement with the results of measurements of the rate of expansion obtained using photomultipliers [51] (see § 4.1).

§ 5. Distribution of Electron Density in Laser Plasmas

In contrast to the many independent temperature measurements carried out over a wide range of the laser radiation flux densities q (see § 3), there is as yet little information on the distribution of the electron density in laser plasmas. Frame-by-frame photography of interferograms involving an exposure of ~2 nsec [12] made it possible to determine the distribution $N_e(r)$ during the later stages of plasma expansion (approximately 100 nsec after the end of a laser pulse) and at large distances r from the target (~10 mm). The dynamics of changes in the $N_e(r)$ profiles in the hot parts of a jet in the immediate vicinity of the target was investigated in [83] using an image converter with a time resolution of ~0.5 nsec.

Among spectral investigations of laser jets one should mention [51, 116, 117] (one should also cite measurements of the time-average electron density $N_e \sim 2 \cdot 10^{18}$ cm^{-3} in laser sparks based on the Stark broadening of spectral lines in the visible range [118]). The electron densities reported in [117] were obtained by measuring the profile of the C VI line at $\lambda = 3434$ Å (6-7 transition). Since this line was in the quartz ultraviolet range, where the intensity of the continuous radiation of a jet exceeded the intensity of the line spectrum right up to distances of 0.9 mm from the target surface, the internal part of the laser jet was outside the investigated range. The values of N_e for $r > 0.9$ mm reported in [117] agreed with those given in [79] and with the results of high-speed interferometric measurements [83] (see Fig. 14).

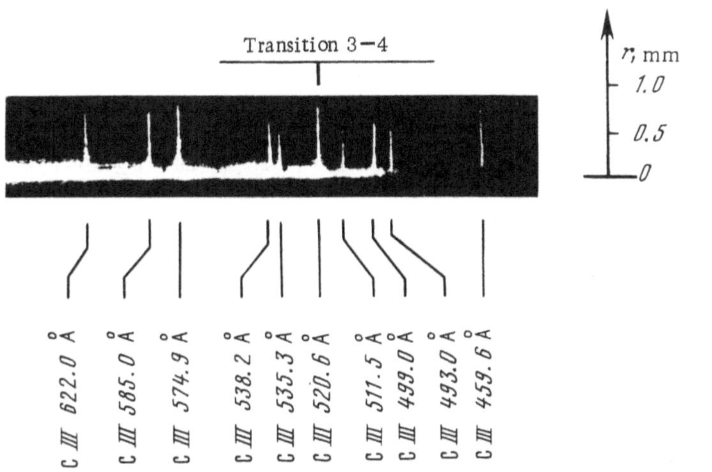

Fig. 13. Spectrogram of the radiation emitted by a carbon jet containing the C IV λ = 520.6 Å ion used in measurements of the plasma density [116]. The coordinate r is the distance from the target.

An estimate of the electron density in a laser jet near the target surface was obtained in [51] from the broadening of the spectral lines of the O VI and K IX ions lying in the range of ~150 Å. The value obtained was $N_e \sim 10^{21}$ cm^{-3}; the spatial resolution of these measurements was ~0.5 mm. The results reviewed in § 4 suggested that these measurements could not represent the hot core of the jet. In fact, since the ionization potentials of the O VI and K IX ions were low (only 138 and 176 eV, respectively), these ions should exist at temperatures much lower than the maximum value, which in these experiments was ~100 eV. Therefore, the value of N_e given in [51] was applicable either to the peripheral parts of the focusing spot or to the plasma near the target surface sometime after the end of a laser pulse.

In view of the characteristic spatial-temporal structure and the distribution of ions with different charges (see § 4), the selection of the C VI ion with a sufficiently high ionization potential (~500 eV) made it possible to use the time-integrated spectrograms in estimating the distribution $N_e(r)$ in the hot core of a laser jet during the action of a heating pulse [116]. Figure 13 shows a spectrogram of the radiation emitted by a laser jet in the wavelength range 500–700 Å obtained in [117] for a heating radiation flux density $q \sim 3 \cdot 10^{11}$ W/cm^2. The astigmatism of the image in the optical system employed made it possible to determine the half-width of the λ = 520.6 Å line of the hydrogen-like ion C VI (3-4 transition) at different distances from the target surface: the spectrogram showed clearly that the line could be observed right down to the shortest distances from the target where it appeared against the continuous radiation background.

The broadening of the spectral lines in high-temperature high-density plasmas may be due to the Doppler and Stark mechanisms. The relationship between these mechanisms in the case of hydrogen-like ions and n → n' transitions can be estimated from the formula [117]

$$\frac{\Delta\omega_S}{\Delta\omega_D} = 1.7 \cdot 10^{-5} \frac{n^2 n'^2}{z^2 \langle v_i \rangle} N_i^{2/3}, \tag{5.1}$$

where $\Delta\omega_S$ and $\Delta\omega_D$ are the Stark and Doppler half-widths, respectively; z is the ionic charge; N_i is the density of ions; $\langle v_i \rangle$ is the average thermal velocity. In the case of the C VI λ = 520.6 Å line (n = 4, n' = 3) and for $N_i \sim 10^{19}$ cm^{-3}, $T_i \sim 100$ eV, we have $\Delta\omega_S/\Delta\omega_D \sim 80$, i.e.,

TABLE 5

r, mm	Interferometric results [83]		$\Delta\omega_i$, sec^{-1} (Holtsmark broadening)	γ_{el}, sec^{-1}	δ	$\Delta\lambda_{theo}$, Å	$\Delta\lambda_{exp}$, Å	N_e, cm^{-3}, deduced from $\Delta\lambda_{exp}$
	N_e, cm^{-3}	N_i, cm^{-3}						
0.1	$6\cdot10^{19}$	10^{19}	$4.1\cdot10^{14}$	$3.8\cdot10^{18}$	2	4.6	4.9	$6.7\cdot10^{19}$
0.15	$3.5\cdot10^{18}$	$5.8\cdot10^{18}$	$2.6\cdot10^{14}$	$2.6\cdot10^{13}$	3.2	3.14	4.3	$5.5\cdot10^{19}$
0.2	$2.5\cdot10^{19}$	$4.2\cdot10^{18}$	$2.3\cdot10^{14}$	$1.7\cdot10^{13}$	3.2	2.5	3.3	$3.6\cdot10^{19}$
0.25	$1.9\cdot10^{19}$	$3.2\cdot10^{18}$	$1.9\cdot10^{14}$	$1.3\cdot10^{13}$	3.6	2.06	2.5	$2.5\cdot10^{19}$
0.8	—	—	—		30	—	19	$3.4\cdot10^{17}$
1.1	—	—	—		30	—	13	$2.0\cdot10^{17}$
1.4	—	—	—		30	—	10.4	$1.3\cdot10^{17}$

the broadening is due to the Stark effect. Table 5 gives not only the interferometric and spectrometric experimental data but also the estimates (see, for example, [74]) of the statistical ionic (Holtsmark) broadening $\Delta\omega_i$, contribution of the impact electronic broadening γ_{el} [119] and Ecker parameter δ (the C VI λ = 520.6 Å line was used in the range r = 0.1-0.25 mm and the C VI λ = 3434 Å line in the range r = 0.8-1.4 mm). The values of $\Delta\lambda_{theor}$, obtained from interferometric results making simultaneous allowance for the ionic and electronic broadening, are in satisfactory agreement with the experimental values of the half-width $\Delta\lambda_{exp}$ of the C VI λ = 520.6 Å line at different distances from the target. The profiles obtained from the spectroscopic [116] and interferometric [83] measurements are shown graphically in Fig. 14.

We shall conclude by pointing out that the selection of spectral lines of ions with sufficiently high ionization potentials ensures that the results give the density of the hot core of a jet, whereas in the case of interferometric measurements the temperature of the investigated plasma is not clear unless additional information is available. On the other hand, an undoubted advantage of interferometric techniques is the time resolution down to the nanosecond or subnanosecond range [12, 83], which makes it possible to investigate the dynamics of changes in the electron density profiles. The upper limit of the laser plasma densities which can be studied interferometrically is $\sim10^{20}$ cm^{-3}. This restriction is due to the refraction of the probing radiation in regions of a jet with high electron-density gradients. There is no such upper limit in the spectroscopic measurements and, in principle, densities up to 10^{23}-10^{24} cm^{-3} can be measured in the soft x-ray part of the spectrum.

§ 6. Ionization State of Multiply Charged Laser Plasmas

The choice of the model of ionization equilibrium is basic in the use of the results of experimental investigations of the spectra of multiply charged ions in plasma diagnostics.

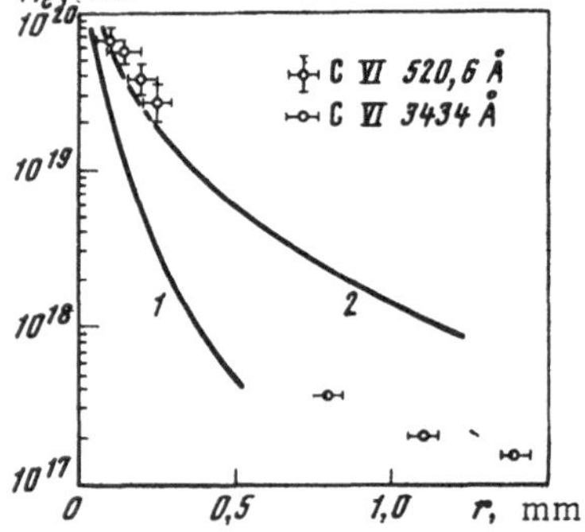

Fig. 14. Comparison of the results of measurements of the averaged (over 30-nsec laser pulse durations) profiles of the electron density $N_e(r)$, based on the Stark broadening of the C VI lines [116], with the results of interferometric studies [83]. The maximum possible errors are given for the experimental points. Curve 1 corresponds to 10 nsec after the beginning of a laser pulse and curve 2 to 30 nsec.

Consequently, theoretical and experimental investigations of possible ionization state models are of considerable interest. The development of the models of the transient state of plasmas in the presence of a strong laser radiation field, gasdynamic motion, and heat conduction [87, 88, 120] is necessary for estimating the possibility of using laser plasmas as sources of multiply charged ions in spectroscopy and accelerator technology, and as high-power pulse sources of neutrons [120].

6.1. Possible Models of the Ionization State of Laser Plasmas for Flux Densities $q < 10^{13}$ W/cm^2

The models used to describe the ionization equilibrium and the distribution of ions between the levels depend on the relationship between the rates of radiative and collisional processes, which are governed by the temperature T_e and density N_e of electrons and by the characteristic linear size of a plasma L. The following models are usual (see, for example, [45, 70, 121]): 1) the model of local thermodynamic equilibrium (LTE); 2) the coronal model (CM) (the formulas for the cross sections of the elementary processes are given in [122-126] and the numerical calculations of the ionization equilibrium curves are in [127-129]); 3) the collisional-radiative model (CRM) (for numerical calculations see [104]). A very important parameter in the description of any specific situation by these models is the ratio of the time needed to establish a steady-state ionization state to the plasma lifetime. At low plasma densities the coronal model is appropriate, whereas at high densities the LTE model is more suitable. The CRM approach is intermediate, as illustrated in Fig. 15 by the $C^{6+} \rightleftarrows$ C VI ionization equilibrium curve plotted as a function of N_e for T_e = 100 eV using numerical values of the collisional-radiative recombination and ionization coefficients, whose tables are given in [104] for a wide range of T_e and N_e. The experimental data obtained so far on the plasma parameters during heating radiation pulses make it possible to consider possible models of the ionization equilibrium in the hot core of a jet. Since the choice of a model for fixed values of the electron temperature T_e and density N_e and of the characteristic linear size L depends on the structure of the energy levels of the ions, we shall consider specific examples of the processes which occur in the laser jet for $q \sim 10^{12}$ W/cm^2.

Hydrogen-like Ions. We shall consider the C VI $\rightleftarrows C^{6+}$ process in a plasma with the following parameters: $N_e = 10^{20}$ cm^{-3}, T_e = 100 eV, L = 10^{-2} cm.

Under these conditions a plasma has a large optical thickness τ in respect of the Lyman lines of the C VI ion, which is a very important point in judging the model [59, 130]. If we estimate the optical thickness τ_{mk} and the probability of spontaneous emission A_{mk} due to the transition k → m using the formulas in [74, 131] and the oscillator strengths given in [104], and if we assume that the line profile is of the Doppler type and corresponds to the ion temperature T_i = 100 eV, we readily find the following values: $\tau_{1,2} = 94$, $\tau_{1,3} = 15$, $\tau_{1,4} = 5.5$, etc.

The large optical thickness for the resonance transitions in the C VI ion, which results in the "trapping" of the line radiation in the volume occupied by the plasma, reduces the rate

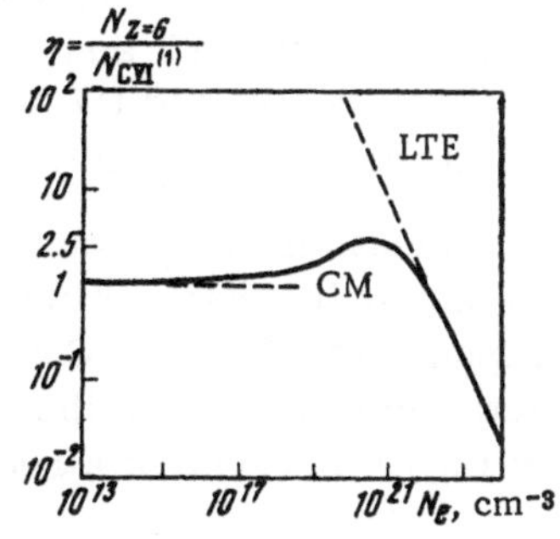

Fig. 15. Curve representing the $C^{6+} \rightleftharpoons$ C VI ionization equilibrium as a function of N_e at T_e = 100 eV, plotted using the tables given in [104]. The dashed curves represent the limiting case of the coronal model (CM) for $N_e \to 0$ and the local thermodynamic equilibrium (LTE) model for $N_e \to \infty$.

of radiative decay of the levels in accordance with the formula [121]

$$A^{*}_{mk} = A_{mk} G(\tau), \tag{6.1}$$

where A^{*}_{mk} is the "effective probability" of the radiative decay of a level k; $G(\tau)$ is the Holstein factor, known as the radiation escape coefficient.

Using the numerical values of $G(\tau)$ from [121] and the formulas for the rates of collisions in hydrogen-like ions [105], we can estimate the contribution of the radiative and collisional processes controlling the populations of the C VI ionic levels important in the ionization equilibrium. A comparison of such results [59] shows that:

1) the populations of the C VI levels are controlled only by the collisional processes because even in the case of a resonance transition the rate K(2, 1) of the 2 → 1 collisional transition is over an order of magnitude higher than the effective probability of the spontaneous decay $A_{1,2}$;

2) the ionization and recombination involving the ground state of the C VI ion are controlled by cascade collisional transitions via bound states because the rate of the 1 → 2 excitation K(1, 2) exceeds the rate of ionization from the ground state;

3) the characteristic times of the ionization and recombination processes are governed by the excitation and deexcitation of the level with n = 2;

4) the equilibrium of the levels with n > 2 with the continuum is established in a time much shorter than the ionization time; the populations of these levels are governed by the instantaneous value of the density of electrons and bare nuclei, and can be calculated using the Saha formula (compare with the results discussed in § 4.1).

The time δt needed to establish a steady ionization state is [121]

$$\delta t = \{N_e [S(T_e, \text{C VI}) + \alpha(T_e, C^{6+})]\}^{-1}, \tag{6.2}$$

where $S(T_e, \text{C VI})$ and $\alpha(T_e, C^{6+})$ are the ionization and recombination coefficients, respectively.

The number of electrons leaving the ground state to the continuous spectrum in 1 sec can be expressed in the form $K(1, 2)N_{\text{C VI}}(1)$, whereas the number of electrons returning to the ground state is $K(2, 1)N_{\text{C VI}}(2) = K(2, 1) \cdot 5.3 \cdot 10^{-3} N_{C^{6+}}$ [here, $N_i(m)$ is the population of the level m]. Assuming that $N_e S = K(1, 2)$ and $N_e \alpha = 5.3 \cdot 10^{-3} K(3, 1)$, we find that $\delta t \sim 0.5$ nsec, which is less than the lifetime of ions in a region of size $L = 10^{-2}$ cm, which is being considered here.

Thus, under the above conditions the ionization equilibrium of the hot core of a carbon jet can be described by the steady-state LTE model.

Ions with $2s^2 2n^n$ Ground-State Electronic Configuration. In the case of multiply charged ions with a more complex energy level structure than that found in the hydrogen atom the model can be selected in a different way. We can demonstrate it by considering the example of the Ca XV ⇌ Ca XVI ionization equilibrium in a calcium plasma with $N_e = 10^{20}$ cm^{-3} and $T_e = 200$ eV (the estimates which will be given below apply also to the Ca XIII ⇌ Ca XVI and Ca XIV ⇌ Ca XV processes because the ions in question have similar structures of the terms).

In contrast to the hydrogen-like ions, when the resonance transition corresponds to one line, the radiative transition between the $2s^2 2p^2$ and $2s^2 2p3s(d)$ electronic configurations of Ca XV, which is important in the ionization equilibrium, corresponds to 17 lines. Consequently, for the same order of the overall transition probabilities and total number of ions the radiation absorption cross section is considerably smaller in the case of Ca XV because of a reduction in the populations of the lower levels belonging to the $2s^22p^2$ configuration, i.e., because the optical thickness is less. For the $2s^22p^2-2s^22p3s(d)$ transition lines in Ca XV the optical thickness is ~ 1 and the corresponding Holstein factor is $G(\tau) \sim 0.5$, which is in contrast to, for example, the 1-2 transition in C VI.

A comparison of the effective probabilities of the radiative decay of the $2s2p^3$ and $2s^22p3s$ groups of levels with the rates of the collisional (excitation, ionization) processes and photorecombination, estimated using the formulas given in [125, 126], leads to the following conclusions:

1) the populations of the levels of the $2s^22p^2$ and $2s2p^3$ configurations are controlled by the collisional processes so that a Boltzmann distribution between these levels is established in $\sim 10^{-11}$ sec;

2) the radiative decay of the $2s^22p3s(d)$ group of levels is much faster than the collisiona deexcitation to the lower levels and the excitation to the higher levels;

3) the ionization of the Ca XV ions occurs mainly from the $2s^22p^2$ and $2s2p^3$ levels;

4) the recombination of the Ca XVI ions occurs because of the photorecombination terminating at the $2s^22p^2$ levels.

Estimates of the relaxation time of the ionization state obtained using a formula similar to Eq. (6.2) allowing for the ionization from the $2s^22p^2$ and $2s2p^3$ levels give values $\delta t \sim 2$-3 nse which is comparable with the lifetime of calcium ions in the core of a jet.

It follows from the reported results that the ionization equilibrium in a calcium plasma is far from the LTE model and from the classical coronal case, and it should be described by the collisional-radiative model (CRM).

6.2. Experimental Check of the Selected Ionization Equilibrium Model for $q \sim 10^{12}$ W/cm^2

In this subsection we shall use the experimental data in judging the validity of the above selection of the ionization equilibrium model for a calcium jet and we shall do this by considering the measured electron temperature T_e.

Estimates in the preceding subsection show that when the heating flux density is $q \sim 10^{12}$ W/cm^2, a calcium jet with $T_e \sim 200$ eV and $N_e \sim 10^{20}$ cm^{-3} is formed and the levels of the Ca XIII-XV ions belonging to the $2s^22p^n$ and $2s2p^{n+1}$ electronic configurations ($n = 4, 3, 2$ for Ca XIII, Ca XIV, and Ca XV, respectively) are populated in accordance with the Boltzmann distribution. The time needed to establish an equilibrium distribution between these levels ($\sim 10^{-11}$ sec) is much shorter than the lifetime of ions in the hot core of the jet ($\sim 10^{-9}$ sec). Consequently, the intensity of one line due to the $m \rightarrow k$ ($LSJ \rightarrow L'SJ'$) transition in an ion of charge i, which is one of a large number of lines (for example, one of the 18 lines in the case of Ca XV) resulting from the $2s^22p^n-2s2p^{n+1}$ transitions, can be related to the total number of ions i which are located effectively at the $2s^22p^n$ and $2s2p^{n+1}$ levels. Then, the relative intensities of two lines of ions i and i' can be used to find the relative concentrations of these ions. This procedure is applied in [52] to the Ca XIII-XV lines lying in the range 130-170 Å. If we consider the $m \rightarrow k$ and $m' \rightarrow k'$ lines of ions i and i' and if we use the experimentally determined intensities I_{km}, wavelengths λ_{km}, statistical weights g_m, and oscillator strengths f_{km} of these lines, we can find the relative concentrations of the ions η from

$$\eta(i, i') \equiv \frac{N_i}{N_{i'}} = \frac{I_{km}}{I_{k'm'}} \frac{B_{i'}}{B_i} \frac{\lambda^3_{km}}{\lambda^3_{k'm'}} \frac{g_{m'}}{g_m} \frac{f(m'k')}{f(mk)}, \tag{6.3}$$

where B_i are the partition functions of all the states of the corresponding ions which consists solely of the contributions of the $2s^2 2p^n$ and $2s2p^{n+1}$ levels.

For each of the m → k lines of the $2s^2 2p^n \to 2s2p^{n+1}$ transitions the oscillator strength f (mk) can be calculated using the formula

$$f(mk) = \frac{g(2s^2 2p^n)}{2J'+1} \frac{\tilde{S}(mk)}{\Sigma \tilde{S}(mk)} f_{abs}(2s^2 2p^n,\ 2s2p^{n+1}), \tag{6.4}$$

where $g(2s^2 2p^n)$ is the statistical weight of the $2s^2 2p^n$ configuration. The oscillator strength $f_{abs}(2s^2 2p^n, 2s2p^{n+1})$ of all the terms of these configurations was calculated on a computer using the wave functions from [132] and a program described in [133]. The relative oscillator strengths of the S lines in the LS-coupling case were taken from [131, 134].

In selecting the ionization state model we have to compare the values of T_e deduced from the experimentally obtained relative concentrations using the coronal and LTE models with the results of measurements of T_e by an independent method based on the x-ray recombination radiation and the absorber method (see § 3), which in this case give $T_e \sim 140$ eV [59].

We shall now give the ranges of the relative concentrations of calcium ions deduced from the experimentally obtained relative intensities of four Ca XV lines, five Ca XIV lines, and eight Ca XIII lines using Eq. (6.4), as well as the corresponding values of T_e obtained from the ionization equilibrium curves for the coronal model [129]. In spite of the considerable scatter of the ratio of the ion concentrations η, due to the large errors in the determination of the intensities in the far vacuum ultraviolet range and due to inaccuracies in the oscillator strength because of the deviations from the LS-coupling case (the interaction between the configurations has to be allowed for in the case of the Ca XIV and Ca XV ions [135]), the exponential dependence gives quite accurate values of T_e:

$\eta(i, i')$	$T_e \cdot 10^{-6}$, °K
η(Ca XIV, Ca XIII)=0.2–0.8	(1.3–1.6)
η(Ca XV, Ca XIII)=0.1–0.7	(1.6–1.8)
η(Ca XV, Ca XIV)=0.35–0.8	(1.8–2.1)

Using the Saha formula in the LTE model we find from the above relative concentrations of the Ca XIII, Ca XIV, and Ca XV ions and from $N_e = 10^{20}$ cm^{-3} that the electron temperatures are $T_e < 70$ eV.

Thus, the agreement between the electron temperature determined from the coronal model with the value of T_e deduced independently from the continuous x-ray radiation demonstrates the validity of the coronal model in the case of an ionization equilibrium in the dense core of a laser jet formed from elements with the nuclear charge $z \sim 20$-30 using laser radiation flux densities of $q \sim 10^{12}$ W/cm^2.

6.3. Nonequilibrium Ionization in Laser Plasmas for $q > 10^{14}$ W/cm^2

The above ionization equilibrium models apply to laser plasmas subjected to relatively low densities q of the heating radiation when the condition $\beta_0 \geq 1$ is satisfied [see Eq. (3.25)], i.e., they apply when the rate of electron energy losses due to ionization is less than the rate of acquisition of energy by electrons from the laser radiation. In this case the average energy of electrons is considerably less than the ionization potential of the ions present in the plasma: in the LTE model the difference is a factor of 8-10, whereas in the coronal model it is a factor of 4-5.

At high laser radiation densities q, such that $\beta_0 \ll 1$, we may expect a considerable deviation from the ionization equilibrium because of the rapid diffusion of electrons (in the energy space) to the region where the energy exceeds considerably the ionization potential χ_i of the ions present in the plasma. Then, in contrast to the equilibrium conditions, the electron temperature is $\gg \chi_i$. The flux density q_1 corresponding to $\beta_0 \sim 1$ can be estimated using the formula [88]

$$q_1 = 2.8 \cdot 10^{12} \frac{A^z}{\ln \Lambda} \left(\frac{\nu}{10^{14}}\right)^2 \left(\frac{I_0}{I_H}\right)^2 \ \mathrm{W/cm^2}, \tag{6.5}$$

where the numerical values of the parameter $A_z = A[Q/(2l_0 + 1)]$ are given in [125]; $\nu = 3 \cdot 10^{14}$ sec is the emission frequency of a neodymium laser; $\chi_z = I_0 z^2_{eff}$ is the ionization potential of ions whose effective charge is z_{eff}; $\ln\Lambda$ is the Coulomb logarithm; I_H is the ionization potential of the hydrogen atom.

If we use Eq. (5) and substitute the parameters from the preceding subsection (calcium plasma, $z \sim 15$, $\chi_z \sim 900$ eV, $I_0 \sim 4$ eV, $A_z \sim 10$, $\ln\Lambda = 6$), we find that $q_1 \sim 5 \cdot 10^{13}$ W/cm².

Thus, in discussing the physical conditions in a laser plasma containing ions of sufficiently heavy elements we find that if $q > 10^{14}$ W/cm², we have to allow for the nonequilibrium ionization. The gasdynamics of laser plasmas under nonequilibrium ionization conditions is discussed in detail in [88].

In investigations of nonequilibrium ionization processes in laser plasmas it would be interesting to determine the intensities of the lines emitted by multiply charged ions and at the same time to record the continuous x-ray emission of the plasma (in order to find the effective temperature of electrons) using laser radiation densities exceeding $q = 10^{14}$ W/cm². Experiments of this kind are reported in [31]: a laser with an output power $W \sim 10$ GW was used and a study was made of the spectral lines ($\lambda \sim 10$-20 Å) of the Fe XXIV ions with $\chi_i \sim 2$ keV. Measurements of the electron temperature based on the continuous spectrum and the absorber method, carried out using similar laser pulses in [56], gave $T_e \sim 0.8$ keV.

§ 7. Laser Plasma as a Source of Multiply Charged Ions

Laboratory sources of multiply charged ions are needed in various physical investigations. These include the classical study of the optical spectra of multiply charged ions (see, for exaple, the reviews [136, 137]), which have become very topical in the diagnostics of hot laboratory plasmas and in extraterrestrial spectroscopic investigations of the solar corona carried out using space satellites and rockets [138, 139]. Many of the solar lines have not yet been identified because information is still lacking on the energy levels and the corresponding transitions lying in the vacuum ultraviolet and soft x-ray parts of the spectrum for ions whose ionization potentials exceed 200 eV.

Efficient sources of multiply charged ions are needed also in nuclear-physics studies of the formation of new superheavy elements [140]. Sources of multiply charged ions with a sufficiently high ratio of the charge to the atomic weight $\varepsilon = z/A$ would make it much easier and economical to accelerate ions to the required energies of ~10 MeV/nucleon.

7.1. Spectroscopic Investigations

Until quite recently the traditional sources of the atomic spectra have been sparks in air with an electron temperature of the plasma amounting to $T_e \sim 3$-4 eV [141] and sparks in vacuum with $T_e \sim 20$-30 eV [142]. Sparks in air can be used to excite the spectra of ions with one or two ionic charges, whereas vacuum sparks can be used to produce ions which have lost 10 or 12 electrons. However, in astrophysical studies one needs laboratory-obtained information on the spectra of ions with all degrees of ionizations formed from elements with a nuclear charge z ~ 20-30 and with an ionization potential of several kiloelectron-volts. The highest ionization potential, $\chi_i \sim 9.2$ keV, among the ions found in the solar corona is exhibited by Fe XXVI [143].

Multiply charged ions with an ionization potential χ_i can form in an equilibrium plasma if at least the following conditions are satisfied: 1) the electron temperature of the plasma is high enough so that $T_e \sim (0.1\text{-}0.2)\chi_i$; 2) the plasma lifetime τ is comparable with the ionization time δt. An analysis of the rates of ionization of multiply charged ions yields $\tau > 10^{12}/N_e$ as the condition for attaining equilibrium ionization [121]. This criterion is 100 times weaker than the Lawson thermonuclear reaction criterion ($n\tau > 10^{14}$) and the former has been satisfied in many plasma experiments. Recent thermonuclear research has made available new sources for atomic spectroscopy with temperatures of ~100-1000 eV: these sources are magnetic-confinement systems, plasma focus, and laser plasma. Moreover, a reduction of the inductance of a line supplying a vacuum spark has made it possible to generate helium-like Cu XXVIII ions with $\chi_i \sim 11$ keV [144, 145]. A comparative analysis of such sources and their bibliography can be found in [146].

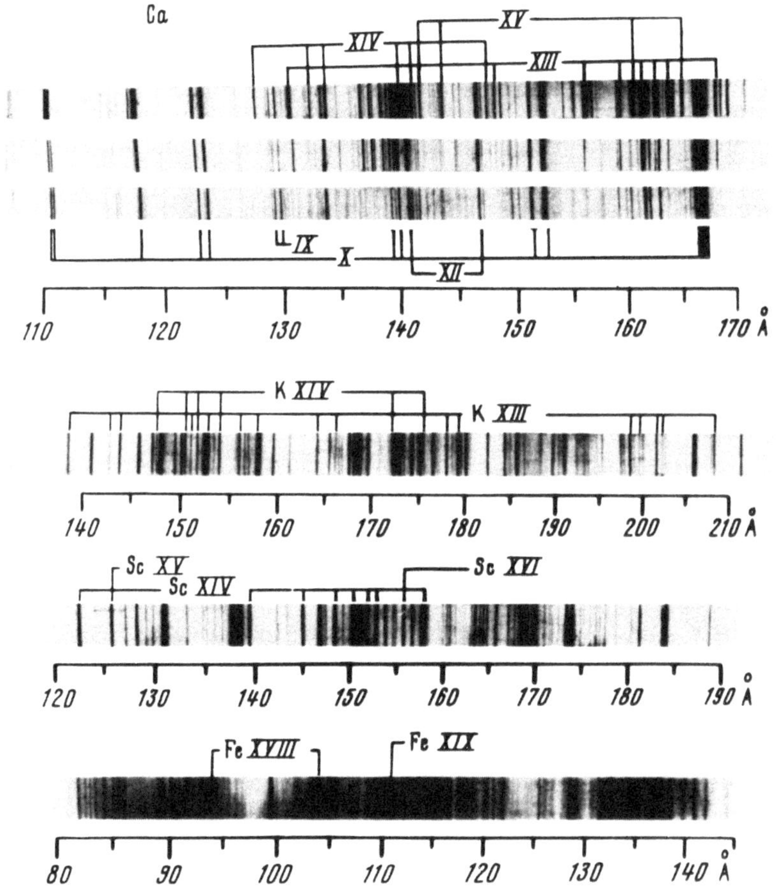

Fig. 16. Spectrograms of a laser plasma obtained using a DFS-26 spectrograph [29, 130, 155] and a laser emitting pulses of Q ~ 20 J energy and $\Delta t_l \sim 15$ nsec duration. The spectrograms of calcium were obtained using different laser radiation flux densities ($5 \cdot 10^{12}$, $2 \cdot 10^{11}$, and $2 \cdot 10^{10}$ W/cm^2 going in the downward direction [29]).

Laser plasma occupies a special place among the sources of multiply charged ions. This is due to the fact that the concentration of energy necessary for plasma heating is achieved using a simple lens in the direct vicinity of a spectral slit. Since solid targets can be used, impurity-free spectra can be excited in a laser plasma for practically any element in the periodic system and this can be done in the vacuum ultraviolet and soft x-ray range. The high energy concentration in a laser plasma makes it possible to heat very small amounts of matter to high temperatures and this gives rise to steep gradients so that the spectra are primarily due to highly ionized ions. This unique feature of the laser plasma facilitates greatly the identification of new spectral lines. The high brightness of the laser plasma, which can be regarded as a point radiation source, ensures a satisfactory density on photographic films after just a few laser flashes, whereas in electric-discharge devices the required number of flashes is usually several hundreds.

Spectroscopic determinations of the wavelengths of new lines of multiply charged ions emitted from laser plasmas have been carried out so far at relatively low flux densities $q \sim 10^{12}$ W/cm^2 ($T_e \sim$ 100-200 eV) [27-30, 147-155]. These investigations have been carried out using diffraction-grating spectrographs in the wavelength range $\geqslant$ 100 Å, and they have been mainly concerned with transitions of the $2s^2 2p^n - 2s2p^{n+1}$ type in ions with ionization potentials up to 1 keV (Fig. 16; see also tables in [156]).

Shorter wavelengths, 1-20 Å, are of particular interest because transitions involving changes in the principal quantum number in multiply charged ions with ionization potentials exceeding 1 keV lie in this range. The use of an x-ray spectrograph with a convex mica crystal [31] and a laser system ensuring that the heating radiation flux density was $\sim 10^{14}$ W/cm^2 yielded spectra of the hydrogen-like ions Mg XII and Al XIII and of the helium-like ions Mg XI and Al XII [157, 158] (Fig. 17). A characteristic feature of these spectra was the presence of satellites of the $1s^2 nl - 1s2pnl$ resonance lines (see, for example, [159]) of the Mg XI and Al XII ions and also of satellites of the $1snl - 2pnl$ lines of the Mg XII and Al XIII ions. The identification and measurements of the wavelengths of the resonance-line satellites of ions with $z > 10$ are of considerable interest because they can give information on the physical conditions in

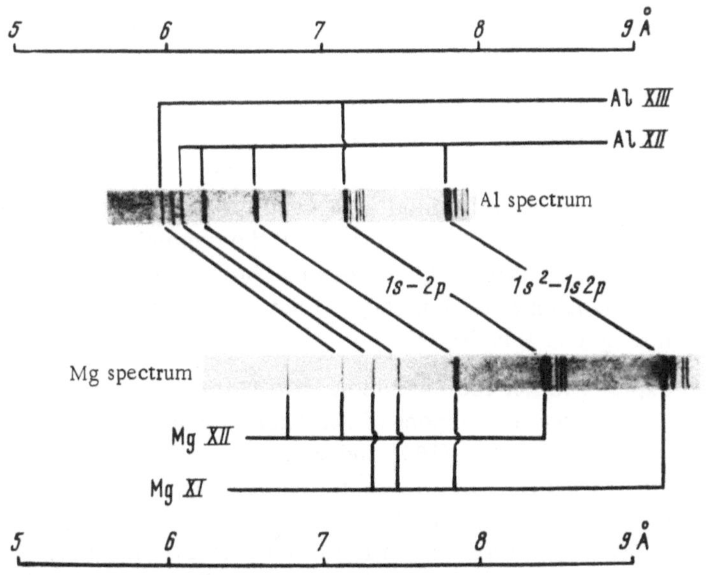

Fig. 17. Spectrograms of aluminum and magnesium laser plasmas obtained using an x-ray crystal spectrograph of [157] and a laser producing pulses of Q = 20 J energy and $\Delta t_l \sim$ 20 nsec duration.

high-temperature laboratory plasmas and in the solar corona so that independent measurements can be carried out of the electron temperature and density and studies can be made of the mechanisms of recombination of multiply charged ions.

7.2. Possible Uses of Laser Plasmas in Accelerator Technology

In spectroscopic investigations the laser plasma is one of many other sources of multiply charged ions, whereas in the accelerator technology the choice of sources is much narrower. This is primarily due to the basic restrictions such as the availability of systems for magnetic confinement of the plasma, low degree of ionization in the final stages of a discharge, etc.

Gas-discharge sources are used in heavy-ion accelerators (see, for example, [160]) and these produce ions of a very limited number of elements with fairly low values of the ratio of the charge to the atomic weight $\varepsilon = z/A \sim 0.1$. In nuclear physics applications an ideal source of ions should produce elements with $\varepsilon \sim 0.5$ and a flux of 10^6-10^{13} particles/sec [140].

The laser plasma is the most promising source in which these conditions can be achieved. The use of lasers with an output in the gigawatt range makes it possible to generate plasmas with electron temperatures of several hundreds of electron-volts containing $\sim 10^{15}$ ions with ε close to 0.5 (see §§ 4 and 6 in the present paper). It follows from the experimental results given in § 4 that ions with higher degrees of ionization have higher velocities so that, in principle, they can be separated spatially from ions with lower degrees of ionization. The absence of any cumbersome devices in the immediate vicinity of an expanding laser jet plasma should make it possible to solve readily the problem of "injection" of ions into an accelerator (see, for example, [161]). An important aspect is the efficiency of the recombination of ions in an expanding plasma. In a carbon jet the ions with the highest degree of ionization, C^{6+} and C VI, are almost completely "frozen." Experimental information on targets made of heavy elements is not yet available. Mass-spectrometric investigations of the flight of the ions at large distances from a target (of the order of tens of centimeters), where the conditions are no longer of the gasdynamic type, should be helpful [162-167]. The maximum degree of ionization of the ions recorded with the aid of a time-of-flight mass spectrometer for an expanding cobalt jet was 25 and the total number of such ions was 10^5 [165].

A theoretical analysis of the kinetics of recombination in an expanding laser plasma is very difficult to carry out because of the complexity of the selection of a suitable gasdynamic model. For example, the time dependences of the electron temperature $T_e(t)$ and density $N_e(t)$ are used in [161] and these dependences are obtained from approximate equations for the gasdynamic expansion of a plasma formed as a result of irradiation on small particles (see [168]). Since in this model we can determine only the dependences $T_e(t)$ and $N_e(t)$ averaged over the plasma volume, it follows that the results obtained in [161] give a qualitative idea of the "freezing" of the ionization state but they do not give a correct picture of the real situation obtained when a bulk target is irradiated.

A qualitative explanation of the characteristic features of the energy spectra of ions (i.e., of the distribution of ions in respect of the energy considered as a function of z) at large distances from the target is given in [169] and this explanation deals with the results obtained by mass spectrometry [166, 167]. In this explanation allowance is made for the recombination of ions in "elementary volumes" of a plasma with the coordinates r(t), for which the time dependences $T_e(t, r)$ and $N_e(t, r)$ are taken from the self-similar solutions gasdynamic equations for the spherical expansion of a plasma cloud of a fixed mass. This model again does not describe the expansion of the plasma in a laser jet whose mass increases with time because of the evaporation of the target material in accordance with Eq. (2.5). Nevertheless, the approach used in [169] is constructive and if the gasdynamic expansion model is refined, this model may explain the features of the "freezing" of ions in a laser jet.

Conclusions

The major part of the available experimental data on laser plasmas has been obtained by sharp focusing of heating radiation under gasdynamic conditions when the density of a hot plasma does not exceed 10^{20}-10^{21} cm^{-3}.

In thermonuclear applications the case of spherical radiation of a condensed target is more promising [26, 38, 170]. The reactive pressure of an expanding plasma should make it possible to generate plasmas with temperatures of several kiloelectron-volts and densities 10^3 higher than in solids, i.e., $\sim 10^{26}$ [171-173], which should ensure that the Lawson criterion ($N\tau > 10^{14}$) of thermonuclear fusion is satisfied with a margin to spare. Compression of the target material by a factor of ~30 has already been achieved experimentally by spherical irradiation [174].

In this connection we may mention two possible directions of experimental investigations of processes occurring in laser plasmas when the density of the heating radiation is 10^{14}-10^{16} W/cm^2 and the pulse duration is 10^{-10}-10^{-9} sec. One type of investigation should be concerned with the spatial region occupied by a plasma with an electron density $N_e < N_{cr}$ ($N_{cr} = 10^{21}$ cm^{-3} for a neodymium laser), where the heating radiation is absorbed by free electrons. Comprehensive measurements should be made of the electron energy distribution functions, electron density profiles, degree of ionization of the plasma, "effective" electron temperature, gasdynamic parameters of the jet, energy and spectral composition of the reflected and scattered laser radiation and of its harmonics (see, for example, the preliminary results given in [175-180]), and so on. This should make it possible to determine the mechanism of absorption of strong electromagnetic waves in dense plasmas and the characteristics of laser plasma heating at high values of the heating flux density. There is as yet no agreed theoretical view on these phenomena (see [90-96]).

The other type of investigation should be concerned with the properties of matter heated to densities above the critical value. Studies of the parameters of the hot cores obtained as a result of plasma cumulation in spherical irradiation of targets would require methods for measuring ion temperatures T_i and densities of 10^{23}-10^{25} cm^{-3}. The most promising techniques are those involving x-ray spectrometry of multiply charged ions. By way of example we can quote the method for the determination of the electron density from the intercombination ($2^3P \rightarrow 1^1S$) and resonance ($2^1P \rightarrow 1^1S$) lines of helium-like ions. This method has been used to measure the electron density $N_e \sim 10^{16}$ cm^{-3} in the θ pinch from the intensities of the lines of ions with $z \sim 6$ [181], in laser plasmas with $N_e \sim 10^{19}$-10^{21} cm^{-3} and $z \sim 10$-14 [182]; in principle, it should be possible to measure $N_e \sim 10^{23}$ cm^{-3} using the lines of helium-like ions with $z \sim 18$. Another possible way of obtaining the same result is the technique of measuring the profiles of spectral lines of multiply charged ions in the soft x-ray region (see, for example, [183]). In this case the ion temperature T_i can be determined using lines whose broadening is due to the Doppler effect and the electron density can be found employing lines due to transitions between levels with high values of the Stark constant.

The authors regard it as their present duty to thank N. G. Basov for his interest, and E. V. Aglitskii, A. V. Vinogradov, S. M. Zakharov, and E. A. Yukov for valuable advice.

Literature Cited

1. N. G. Basov and O. N. Krokhin, Zh. Eksp. Teor. Fiz., 46:171 (1964).
2. W. I. Linlor, Appl. Phys. Lett., 3:210 (1963).
3. N. R. Isenor, Can. J. Phys., 42:1413 (1964).
4. E. Archbold and T. P. Hughes, Nature (Lond.), 204:670 (1964).
5. R. V. Ambartsumyan, N. G. Basov, V. A. Boiko, V. S. Zuev, O. N. Krokhin, P. G. Kryukov, Yu. V. Senatskii, and Yu. Yu. Stoilov, Zh. Eksp. Teor. Fiz., 48:1583 (1965).

6. N. G. Basov, V. A. Boiko, V. A. Dement'ev, O. N. Krokhin, and G. V. Sklizkov, Zh. Eksp. Teor. Fiz., 51:989 (1966).
7. A. F. Haught and D. H. Polk, Phys. Fluids, 9:2047 (1966).
8. Yu. V. Afanasiev (Afanasyev), O. N. Krokhin, and G. V. Sklizkov, IEEE J. Quantum. Electron., QE-2:483 (1966).
9. N. G. Basov, O. N. Krokhin, and G. V. Sklizkov, Appl. Opt., 6:1814 (1967).
10. E. W. Sucov, J. L. Pack, A. V. Phelps, and A. G. Engelhardt, Phys. Fluids, 10:2035 (1967).
11. G. V. Sklizkov, Thesis for Candidate's Degree [in Russian], Lebedev Physics Institute, Academy of Sciences of the USSR, Moscow (1967).
12. N. G. Basov, V. A. Gribkov, O. N. Krokhin, and G. V. Sklizkov, Zh. Eksp. Teor. Fiz., 54:1073 (1968).
13. N. G. Basov, O. N. Krokhin, and G. V. Sklizkov, IEEE J.Quantum. Electron., QE-4:988 (1968).
14. N. G. Basov, O. N. Krokhin, and G. V. Sklizkov, Tr. Fiz. Inst. Akad. Nauk SSSR, 52:171 (1970).
15. Laser Applications in Plasma Physics 1962-1968 (Bibliographical Series No. 35), International Atomic Energy Agency, Vienna (1969).
16. C. DeMichelis, IEEE J. Quantum. Electron., QE-6:630 (1970).
17. N. G. Basov, P. G. Kryukov (Kriukov), S. D. Zakharov, Yu. V. Senatskii, and S. V. Chekalin (Tchekalin), J. Quantum. Electron., QE-4:864 (1968).
18. G. W. Gobeli, J. C. Bushnell, P. S. Peercy, and E. D. Jones, Phys. Rev., 188:300 (1969).
19. F. Floux, D. Cognard, L. G. Denoeud, G. Pair, D. Parisot, J. L. Bobin, F. Delobeau, and C. Fauguignon, Phys. Rev. A, 1:821 (1970).
20. N. G. Basov, V. A. Boiko, S. M. Zakharov, O. N. Krokhin, and G. V. Sklizkov, ZhETF Pis'ma Red., 13:691 (1971).
21. S. W. Mead, R. E. Kidder, J. E. Swain, F. Ranier, and J. Petruzzi, Appl. Opt., 11:345 (1972).
22. E. D. Jones, G. W. Gobelli, and J. N. Olsen, "Nanosecond and picosecond Laser irradiation of solid target," in: Laser Interaction and Related Phenomena (ed. by H. J. Schwarz and H. Hora), Vol. 2, Plenum Press, New York (1972).
23. C. Yamanaka, T. Yamanaka, T. Sasaki, et al., Plasma Generation and Heating to Thermonuclear Temperature by Lasers (Research Report No. IPPJ-117), Institute of Plasma Physics, Nagoya University, Japan (1972).
24. K. Büchl, K. Eidmann, P. Mulser, et al., Paper CN 28-D-11 presented at Fourth Conf. on Plasma Physics and Controlled Nuclear Fusion Research, Wisconsin, 1971, publ. by International Atomic Energy Agency, Vienna (1971).
25. J. W. Shearer, S. W. Mead, J. Petruzzi et al., Preprint UCRL-73489, Lawrence Radiation Laboratory, University of California (1971).
26. N. G. Basov, Yu. S. Ivanov, O. N. Krokhin, Yu. A. Mikhailov, G. V. Sklizkov, and S. I. Fedotov, ZhETF Pis'ma Red., 15:589 (1972).
27. B. C. Fawcett, A. H. Gabriel, F. E. Irons, N. J. Peacock, and P. A. H. Saunders, Proc. Phys. Soc. Lond., 88:1051 (1966).
28. N. G. Basov, V. A. Boiko, Yu. P. Voinov, É. Ya. Kononov, S. L. Mandel'shtam, and G. V. Sklizkov, ZhETF Pis'ma Red., 5:177 (1967).
29. N. G. Basov, V. A. Boiko, Yu. P. Voinov, É. Ya. Kononov, S. L. Mandel'shtam, and G. V. Sklizkov, ZhETF Pis'ma Red., 6:849 (1967).
30. N. G. Basov, V. A. Boiko, Yu. P. Voinov, É. Ya. Kononov, O. N. Krokhin, S. L. Mandel'-shtam (Mandelstam), and G. V. Sklizkov, Rev. Roum. Phys., 13:97 (1968).
31. E. V. Aglitskii, V. A. Boiko, S. M. Zakharov, et al., Kratk. Soobshch. Fiz., No. 12, 36 (1971).
32. A. J. DeMaria, R. Gagosz, H. A. Heynau, A. W. Penney Jr, and G. Wisner, J. Appl. Phys., 38:2693 (1967).

33. N. G. Basov, P. G. Kryukov, Yu. V. Senatskii, and S. V. Chekalin, Zh. Eksp. Teor. Fiz., 57, 1175 (1969).
34. M. P. Vanyukov, V. A. Venchikov, V. I. Isaenko, V. A. Serebryakov, A. D. Starikov, and I. M. Buzhinskii, No. 6, 67 (1969); M. P. Vanyukov, V. A. Venchikov, V. I. Isaenko, V. A. Serebryakov, and A. D. Starikov, Opt. Spektrosk., 28:1008 (1970).
35. W. F. Hagen, J. Appl. Phys., 40:511 (1969).
36. C. G. Young, Proc. IEEE, 57:1267 (1969).
37. F. Floux, Onde Elec., 50:582 (1970).
38. N. G. Basov, O. N. Krokhin, G. V. Sklizkov, S. I. Fedotov, and A. S. Shikanov, Zh. Eksp. Teor. Fiz., 62:203 (1972).
39. Yu. V. Afanas'ev and O. N. Krokhin, Zh. Eksp. Teor. Fiz., 52:966 (1967).
40. O. N. Krokhin, "High-temperature and plasma phenomena induced by laser radiation," (Intern. School of Physics, Course 48, Physics of High Energy Density), Academic Press, New York (1971), p. 278-305.
41. N. G. Basov, V. A. Boiko, O. N. Krokhin, O. G. Semenov, and G. V. Sklizkov, Zh. Tekh. Fiz., 38:1973 (1968).
42. N. G. Basov and O. N. Krokhin, Vestn. Akad. Nauk SSSR, No. 6, 55 (1970).
43. I. V. Nemchinov, Prikl. Mat. Mekh., 31:300 (1967).
44. D. W. Gregg and S. J. Thomas, J. Appl. Phys., 38:1729 (1967).
45. J. Cooper, Rep. Prog. Phys., 29:35 (1966).
46. J. Dawson and C. Oberman, Phys. Fluids, 5:517 (1962).
47. A. W. Ehler and G. L. Weissler, Appl. Phys. Lett., 8:89 (1966).
48. A. W. Ehler, J. Appl. Phys., 37:4962 (1966).
49. J. L. Schwob, C. Breton, W. Seka, and C. Minier, Plasma Phys., 12:217 (1970).
50. D. D. Burgess, B. C. Fawcett, and N. J. Peacock, Proc. Phys. Soc. Lond., 92:805 (1967).
51. B. C. Boland, F. E. Irons, and R. W. P. McWhirter, J. Phys. B, 1:1180 (1968).
52. V. A. Boiko, Yu. P. Voinov, V. A. Gribkov, and G. V. Sklizkov, Preprint No. 79 [in Russian], Lebedev Physics Institute, Academy of Sciences of the USSR, Moscow (1970); Plasma Diagnostics (Proc. Third All-Union Conf., Sukhumi, 1970).
53. F. C. Jahoda, E. M. Kittle, W. E. Quinn, G. A. Sawyer, and T. F. Stratton, Phys. Rev., 119:843 (1960).
54. N. G. Basov, V. A. Boiko, V. A. Gribkov, et. al., Preprint No. 111 [in Russian], Lebedev Physics Institute, Academy of Sciences of the USSR, Moscow (1968).
55. N. G. Basov, V. A. Boiko, V. A. Gribkov, S. M. Zakharov, O. N. Krokhin, and G. V. Sklizkov, ZhETF Pis'ma Red., 9:520 (1969).
56. V. A. Boiko, Yu. A. Drozhbin, S. M. Zakharov, et al., Preprint No. 77 [in Russian], Lebedev Physics Institute, Academy of Sciences of the USSR, Moscow (1973).
57. J. L. Bobin, F. Floux, P. Langer, and H. Pignerol, Phys. Lett. A, 28:398 (1968).
58. J. L. Bobin, F. Delobeau, G. DeGiovanni, C. Fauquignon, and F. Floux, Nucl. Fusion, 9:115 (1969).
59. V. A. Boiko, Thesis for Candidate's Degree [in Russian], Lebedev Physics Institute, Academy of Sciences of the USSR, Moscow (1970).
60. W. Seka, C. Breton, J. L. Schwob, and C. Minier, Plasma Phys., 12:73 (1970).
61. H. Puell, H. J. Neusser, and W. Kaiser, Z. Naturforsch. a, 25:1815 (1970).
62. R. Sigel, S. Witkowski, H. Baumhacker, K. Büchl, K. Eidmann, H. Hora, H. Mennicke, P. Mulser, D. Pfirsch, and H. Salzmann, Kvant. Elektron. (Mosc.), No. 2(8), 37 (1972).
63. S. W. Mead, R. E. Kidder, and S. E. Swain, Preprint OCRL-73356, Lawrence Radiation Laboratory, University of California (1971).
64. G. Tonon, D. Schirmann, and M. Rabeau, Proc. Tenth Intern. Conf. on Phenomena in Ionized Gases, Oxford, 1971, publ. by Donald Parsons, Oxford (1971), p. 225.
65. H. Puell, W. Spengler, and W. Kaiser, Phys. Lett. A, 37:35 (1971).

66. A. J. Alcock, P. P. Pashinin, and S. A. Ramsden, Phys. Rev. Lett., 17:528 (1966).
67. M. P. Vanyukov, V. A. Venchikov, V. I. Isaenko, P. P. Pashinin, and A. M. Prokhorov, ZhETF Pis'ma Red., 7:321 (1968).
68. N. J. Neusser, H. Puell, and W. Kaiser, Appl. Phys. Lett., 19:300 (1971).
69. T. F. Stratton, "X-ray spectroscopy," Chap. 8 in: Plasma Diagnostic Techniques (ed. by R. Huddlestone and S. L. Leonard), Academic Press, New York (1965).
70. H. Griem, Plasma Spectroscopy, McGraw-Hill, New York (1964).
71. R. C. Elton and N. V. Roth, Appl. Opt. 6:207 (1967).
72. R. C. Elton and A. D. Anderson, NRL Report 6541, Washington (1967).
73. R. C. Elton, NRL Report 6738, Washington (1968).
74. I. I. Sobelman, Introduction to the Theory of Atomic Spectra, Pergamon Press, Oxford (1973).
75. J. B. Birks, Scintillation Counters, Pergamon Press, London (1953).
76. V. V. Matveev and A. D. Sokolov, Photomultipliers in Scintillation Counters [in Russian], Atomizdat, Moscow (1962).
77. A. J. Meyerott, P. C. Fisher, and D. T. Roething, Rev. Sci. Instrum., 35:669 (1964).
78. A. M. Tyutikov, Usp. Fiz. Nauk, 100:467 (1970).
79. L. I. Andreeva, N. G. Basov, V. A. Boiko, et al., Preprint No. 157 [in Russian], Lebedev Physics Institute Academy of Sciences of the USSR, Moscow (1968); L. I. Andreeva, N. G. Basov, V. A. Boiko, V. A. Boiko, M. I. Vanina, S. M. Zakharov, O. N. Krokhin, G. V. Sklizkov, B. M. Stepanov, V. N. Filinov, and V. P. Churakov, Prib. Tekh. Eksp., No. 6, 217 (1969).
80. N. G. Basov, V. A. Boiko, O. N. Krokhin, and G. V. Sklizkov, Dokl. Akad. Nauk SSSR, 173:538 (1967).
81. F. Floux, D. Cognard, and D. de Giovanni, Proc. Ninth Intern. Conf. on Phenomena in Ionized Gases, Bucharest, 1969, publ. by Editura Akad. RSR, Bucharest (1969), p. 335.
82. C. Fauquignon and F. Floux, Phys. Fluids, 13:386 (1970).
83. N. G. Basov, V. A. Boiko, V. A. Gribkov, S. M. Zakharov, O. N. Krokhin, and G. V. Sklizkov, Zh. Eksp. Teor. Fiz., 61:154 (1971).
84. V. A. Boiko and A. V. Vinogradov, Preprint No. 31 [in Russian], Lebedev Physics Institute, Academy of Sciences of the USSR, Moscow (1971).
85. Yu. V. Afanas'ev, É. M. Belenov, and O. N. Krokhin. Zh. Eksp. Teor. Fiz., 56:256 (1969).
86. Yu. V. Afanas'ev, É. M. Belenov, O. N. Krokhin, and I. A. Poluéktov, Zh. Eksp. Teor. Fiz., 57:580 (1969).
87. Yu. V. Afanas'ev, É. M. Belenov, O. N. Krokhin, and I. A. Poluéktov, ZhETF Pis'ma Red., 10:553 (1969).
88. Yu. V. Afanas'ev, É. M. Belenov, O. N. Krokhin, and I. A. Poluéktov, Zh. Eksp. Teor. Fiz., 60:73 (1971).
89. L. Spitzer, Jr., Physics of Fully Ionized Gases, 2nd ed., Wiley-Interscience, New York (1962).
90. P. K. Kaw and J. M. Dawson, Phys. Fluids, 12:2586 (1969).
91. A. Caruso and R. Gratton, Plasma Phys., 11:839 (1970).
92. V. V. Pustovalov and V. P. Silin, Zh. Eksp. Teor. Fiz., 59:2215 (1970).
93. Ya. B. Zel'dovich and E. V. Levich, ZhETF Pis'ma Red., 11:497 (1970).
94. A. V. Vinogradov and V. V. Pustovalov, ZhETF Pis'ma Red., 13:317 (1971).
95. V. V. Pustovalov and V. P. Silin, ZhETF Pis'ma Red., 14:439 (1971).
96. A. V. Vinogradov and V. V. Pustovalov, Zh. Eksp. Teor. Fiz., 62:980 (1972).
97. H. Puell, Z. Naturforsch, a, 25:1807 (1970).
98. A. Caruso, A. De Angelis, G. Gatti, R. Gratton, and S. Martelluci, Phys. Lett. A, 35:279 (1971).
99. V. L. Ginzburg, Propagation of Electromagnetic Waves in Plasma, Pergamon Press, Oxford (1964).

100. T. S. Green, Phys. Lett. A, 32:530 (1970).
101. N. G. Basov, S. D. Zakharov, O. N. Krokhin, P. G. Kryukov, Yu. V. Senatskii, E. L. Tyurin, A. I. Fedosimov, S. V. Chekalin, and M. Ya. Shchelev, Kvant. Elektron. (Mosc.), No. 1, 4 (1971).
102. N. G. Basov, V. A. Boiko, V. A. Gribkov, S. M. Zakharov, O. N. Krokhin, and G. V. Sklizkov, Proc. Ninth Inter. Conf. on Phenomena in Ionized Gases, Bucharest, 1969, publ. by Editura Akad. RSR, Bucharest (1969), p. 333.
103. N. G. Basov, V. A. Boiko, Yu. A. Drozhbin, S. M. Zakharov, O. N. Krokhin, G. V. Sklizkov, and V. A. Yakovlev, Dokl. Akad. Nauk SSSR, 192:1248 (1970).
104. D. R. Bates, A. E. Kingston, and R. W. P. McWhirter, Proc. R. Soc. A, 267:297 (1962).
105. W. K. Wiese, M. W. Smith, and B. M. Glennon, Atomic Transition Probabilities (Report NSRDS-NBS4), Vol. 1, Washington, DC (1966).
106. F. E. Irons, N. J. Peacock, and R. S. Pease, Kvant. Elektron. (Mosc.), No. 1(7), 20 (1972).
107. F. E. Irons, R. W. P. McWhirter, and N. J. Peacock, The Ion and Velocity Structure in a Laser-Produced Plasma (Preprint CLM-P268), Culham Laboratory (1971).
108. Yu. A. Bykovskii, M. F. Gryukanov, V. G. Degtyarev, N. N. Degtyarenko, V. F. Elesin, I. D. Laptev, and V. N. Nevolin, ZhETF Pis'ma Red., 14:238 (1971).
109. Yu. P. Raizer, Zh. Eksp. Teor. Fiz., 37:580 (1959).
110. Ya. B. Zel'dovich and Yu. P. Raizer, Physics of Shock Waves and High-Temperature Hydrodynamic Phenomena, 2 vols., Academic Press, New York (1966-7).
111. R. Courant and K. O. Friedrichs, Supersonic Flow and Shock Waves, Wiley, New York (1948).
112. W. G. Griffin and J. Schluter, Phys. Lett. A, 26:241 (1968).
113. R. Sigel, Z. Naturforsch. a, 25:488 (1970).
114. N. G. Basov, O. N. Krokhin, and G. V. Sklizkov, ZhETF Pis'ma Red., 6:683 (1967).
115. B. V. Aglitskii (Aglitsky), N. G. Basov, V. A. Boiko, V. A. Gribkov, S. M. Zakharov, O. H. Krokhin, and G. V. Sklizkov, Proc. Tenth Inter. Conf. on Phenomena in Ionized Gases, Oxford, 1971, publ. by Donald Parsons, Oxford (1971), p. 229.
116. E. V. Aglitskii, V. A. Boiko, S. M. Zakharov, and G. V. Sklizkov, Preprint No. 143 [in Russian], Lebedev Physics Institute, Academy of Sciences of the USSR, Moscow (1970).
117. B. C. Boland and F. E. Irons, Proc. Eighth Intern. Conf. on Phenomena in Ionized Gases, Vienna, 1967, publ. by International Atomic Energy Agency, Vienna (1968), p. 462.
118. S. L. Mandel'shtam, P. P. Pashinin, A. V. Prokhindeev, A. M. Prokhorov, and N. K. Sukhodrev, Zh. Eksp. Teor. Fiz., 47:2003 (1964).
119. L. A. Minaeva and I. I. Sobelman, Astron. Tsirk., No. 383 (1966).
120. Yu. V. Afanas'ev, É. M. Belenov, O. N. Krokhin, and I. A. Poluéktov, ZhETF Pis'ma Red., 13:257 (1971).
121. R. W. P. McWhirter, "Spectral Intensities," Chap. 5 in: Plasma Diagnostic Techniques (ed. by R. Huddlestone and S. L. Leonard), Academic Press, New York (1965).
122. I. L. Beigman and L. A. Vainshtein, Preprint No. 104 [in Russian], Lebedev Physics Institute, Academy of Sciences of the USSR, Moscow (1967).
123. I. L. Beigman, L. A. Vainshtein, and R. A. Syunyaev, Usp. Fiz. Nauk, 95:267 (1968).
124. I. L. Beigman and L. A. Vainshtein, Preprint No. 94 [in Russian], Lebedev Physics Institute, Academy of Sciences of the USSR, Moscow (1969).
125. I. L. Beigman, L. A. Vainshtein, and A. V. Vinogradov, Astron. Zh., 46:985 (1969).
126. I. L. Beigman and L. A. Vainshtein, Tr. Fiz. Inst. Akad. Nauk SSSR, 51:8 (1970).
127. L. House, Astrophys. J. Suppl. 8, 307 (1964).
128. C. Jordan, Mon. Not. R. Astron. Soc., 148:17 (1970).
129. I. L. Beigman, L. A. Vainshtein, and A. M. Urnov, Preprint No. 28 [in Russian], Lebedev Physics Institute, Academy of Sciences of the USSR, Moscow (1971).
130. V. A. Boiko, Yu. P. Voinov, and V. A. Gribkov, Preprint No. 207 [in Russian], Lebedev Physics Institute, Academy of Sciences of the USSR, Moscow (1969).

131. C. W. Allen, Astrophysical Quantities, Athlone Press, London, 1st ed. (1956), 3rd ed. (1973).
132. L. A. Vainshtein, Opt. Spektrosk., 3:313 (1957).
133. L. A. Vainshtein and V. P. Shevel'ko, Preprint No. 87 [in Russian], Lebedev Physics Institute, Academy of Sciences of the USSR, Moscow (1970).
134. I. B. Levinson and A. A. Nikitin, Handbook on Theoretical Calculation of Line Intensities in Atomic Spectra [in Russian], Leningrad State University (1962).
135. L. N. Ivanov, E. P. Ivanova, and V. V. Tolmachev, Izv. Vyssh. Uchebn. Zaved. Fiz., No. 12, 84 (1969).
136. B. Edlen, Rep. Prog. Phys., 26:181 (1963).
137. B. Edlen, "Atomic spectra," in: Handbuch der Physik Vol. 27, Springer Verlag, Berlin (1964), p. 80.
138. S. L. Mandelshtam, Appl. Opt., 6:1834 (1967).
139. R. Tousy, Astron. J., 149:239 (1967).
140. A. Chiorso, IEEE Trans. Nucl. Sci., NS-14(3):5 (1967).
141. N. K. Sukhodrev, Tr. Fiz. Inst. Akad. Nauk SSSR, 15:123 (1961).
142. S. V. Lebedev, S. L. Mandel'shtam, and G. M. Rodin, Zh. Eksp. Teor. Fiz., 37:349 (1959).
143. W. M. Neupert and M. Swartz, Astron. J. Lett., 160:189 (1970).
144. L. Cohen, V. Feldman, M. Swartz, and J. H. Underwood, J. Opt. Soc. Am., 58:843 (1968).
145. T. N. Lie and R. C. Elton, Phys. Rev. A, 3:865 (1971).
146. A. H. Gabriel, Nucl. Instrum. Methods, 90:157 (1970).
147. B. C. Fawcett, A. H. Gabriel, and P. A. H. Saunders, Proc. Phys. Soc., 90:863 (1967).
148. B. C. Fawcett, D. D. Burgess, and N. J. Peacock, Proc. Phys. Soc., 91:970 (1967).
149. É. Ya. Kononov, Astron. Zh., 46:340 (1949).
150. V. A. Boiko, Yu. P. Voinov, V. A. Gribkov, and G. V. Sklizkov, Opt. Spektrosk., 29:1023 (1970).
151. B. C. Fawcett, J. Phys. B., 3:1152 (1970).
152. B. C. Fawcett, J. Phys. B., 3:1732 (1970).
153. B. C. Fawcett, J. Phys. B., 4:981 (1971).
154. B. C. Fawcett, J. Phys. B., 4:1115 (1971).
155. V. A. Boiko, Yu. P. Voinov, P. I. Ivashkin, and G. V. Sklizkov, Preprint No. 4 [in Russian], Lebedev Physics Institute, Academy of Sciences of the USSR, Moscow (1972).
156. B. C. Fawcett, Wavelengths and Classification of Emission Lines due to $2s^2 2p^n - 2s2p^{n+1}$ Transitions (Preprint ARU-R2), Culham Laboratory (1971).
157. E. V. Aglitskii, V. A. Boiko, L. A. Vainshtein, S. M. Zakharov, O. N. Krokhin, and G. V. Sklizkov, Opt. Spektrosk., 35:963 (1973).
158. E. V. Aglitskii, V. A. Boiko, L. A. Vainshtein, et al., Preprint No. 146 [in Russian], Lebedev Physics Institute, Academy of Sciences of the USSR, Moscow (1973).
159. A. H. Gabriel and C. Jordan, Nature (Lond.), 221:947 (1969).
160. H. Krupp, Nucl. Instrum. Methods, 90:167 (1970).
161. N. J. Peacock and R. S. Pease, J. Phys. D., 2:1705 (1969).
162. Yu. A. Bykovskii, N. N. Degtyarenko, V. I. Dymovich, V. F. Elesin, Yu. P. Kozyrev, B. I. Nikolaev, S. V. Ryzhik, and S. M. Sil'nov, Zh. Tekh. Fiz., 39:1694 (1969).
163. Yu. A. Bykovskii, V. I. Dymovich, Yu. P. Kozyrev, V. N. Nevolin, and S. M. Sil'nov, Zh. Tekh. Fiz., 40:2401 (1970).
164. Yu. A. Bykovskii, N. N. Degtyarenko, V. F. Elesin, Yu. P. Kozyrev, and S. M. Sil'nov, Izv. Vyssh. Uchebn. Zaved. Radiofiz., 13:891 (1970).
165. V. V. Apollonov, Yu. A. Bykovskii, N. N. Degtyarenko, V. F. Elesin, Yu. P. Kozyrev, and S. M. Sil'nov, Zh. Eksp. Teor. Fiz., 11:377 (1970).
166. Yu. A. Bykovskii, N. N. Degtyarenko, V. F. Elesin, Yu. P. Kozyrev, and S. M. Sil'nov, Zh. Tekh. Fiz., 40:2578 (1970).

167. Yu. A. Bykovskii, N. N. Degtyarenko, V. F. Elesin, Yu. P. Kozyrev, and S. M. Sil'nov, Zh. Eksp. Teor. Fiz., 60:1306 (1971).
168.*
169. Yu. V. Afanas'ev and V. B. Rozanov, Zh. Eksp. Teor. Fiz., 62:247 (1972).
170. N. G. Basov, O. N. Krokhin, G. V. Sklizkov, and S. I. Fedotov, present issue, p. 145.
171. J. Nuckolls, L. Wood, A. Thiessen, and G. Zimmerman, Nature (Lond.), 239:139 (1972).
172. J. S. Clarke, H. N. Fisher, and R. J. Mason, Phys. Rev. Lett., 30:89 (1973).
173. E. G. Gamalii, ZhETF Pis'ma Red., 19:520 (1974).
174. G. V. Sklizkov, Doctoral Thesis [in Russian], Lebedev Physics Institute, Academy of Sciences of the USSR, Moscow (1973).
175. M. Decroisette, G. Piar, and F. Floux, Phys. Lett. A, 32:249 (1970).
176. I. K. Krasyuk, P.P. Pashinin, and A. M. Prokhorov, ZhETF Pis'ma Red., 12:439 (1970).
177. P. Belland, C. DeMichelis, M. Mattioli, and R. Papoular, Appl. Phys. Lett., 18:542 (1971).
178. A. Caruso, A. de Angelis, G. Gatti, R. Gratton, and S. Martelluci, Phys. Lett. A, 33:29 (1970).
179. G. Piar, B. Meyer, and M. Decroisette, Proc. Tenth Intern. Conf. on Phenomena in Ionized Gases, Oxford (1971), publ. by Donald Parsons, Oxford (1971), p. 331.
180. A. E. Kazakov, I. K. Krasyuk, P. P. Pashinin, and A. M. Prokhorov, ZhETF Pis'ma Red., 14:416 (1971).
181. H. J. Kunze, A. H. Gabriel, and H. R. Griem, Phys. Rev., 165:267 (1968).
182. E. V. Aglitskii, V. A. Boiko, A. V. Vinogradov, and E. A. Yukov, Proc. Eleventh Intern. Conf. on Phenomena in Ionized Gases, Prague, 1973, Vol. 1, Institute of Physics, Czechoslovak Academy of Sciences, Prague (1973), p. 432.
183. M. G. Hobby and N. J. Peacock, Proc. Tenth Intern. Conf. on Phenomena in Ionized Gases, Oxford (1971), publ. by Donald Parsons, Oxford (1971), p. 407.

*Reference 168 omitted in Russian original – Publisher.